# Mathematical Proof Theory

Edited by Paul F. Kisak

# Contents

# Chapter 1

# Proof theory

**Proof theory** is a branch of mathematical logic that represents proofs as formal mathematical objects, facilitating their analysis by mathematical techniques. Proofs are typically presented as inductively-defined data structures such as plain lists, boxed lists, or trees, which are constructed according to the axioms and rules of inference of the logical system. As such, proof theory is syntactic in nature, in contrast to model theory, which is semantic in nature. Together with model theory, axiomatic set theory, and recursion theory, proof theory is one of the so-called *four pillars* of the foundations of mathematics.[1]

Some of the major areas of proof theory include structural proof theory, ordinal analysis, provability logic, reverse mathematics, proof mining, automated theorem proving, and proof complexity. Much research also focuses on applications in computer science, linguistics, and philosophy.

## 1.1 History

Although the formalisation of logic was much advanced by the work of such figures as Gottlob Frege, Giuseppe Peano, Bertrand Russell, and Richard Dedekind, the story of modern proof theory is often seen as being established by David Hilbert, who initiated what is called Hilbert's program in the foundations of mathematics. The central idea of this program was that if we could give finitary proofs of consistency for all the sophisticated formal theories needed by mathematicians, then we could ground these theories by means of a metamathematical argument, which shows that all of their purely universal assertions (more technically their provable $\Pi_1^0$ sentences) are finitarily true; once so grounded we do not care about the non-finitary meaning of their existential theorems, regarding these as pseudo-meaningful stipulations of the existence of ideal entities.

The failure of the program was induced by Kurt Gödel's incompleteness theorems, which showed that any $\omega$-consistent theory that is sufficiently strong to express certain simple arithmetic truths, cannot prove its own consistency, which on Gödel's formulation is a $\Pi_1^0$ sentence. However, modified versions of Hilbert's program emerged and research has been carried out on related topics. This has led, in particular, to:

- Refinement of Gödel's result, particularly J. Barkley Rosser's refinement, weakening the above requirement of $\omega$-consistency to simple consistency;

- Axiomatisation of the core of Gödel's result in terms of a modal language, provability logic;

- Transfinite iteration of theories, due to Alan Turing and Solomon Feferman;

- The recent discovery of self-verifying theories, systems strong enough to talk about themselves, but too weak to carry out the diagonal argument that is the key to Gödel's unprovability argument.

In parallel to the rise and fall of Hilbert's program, the foundations of structural proof theory were being founded. Jan Łukasiewicz suggested in 1926 that one could improve on Hilbert systems as a basis for the axiomatic presentation of logic if one allowed the drawing of conclusions from assumptions in the inference rules of the logic. In response to this Stanisław Jaśkowski (1929) and Gerhard Gentzen (1934) independently provided such systems, called calculi of natural deduction, with Gentzen's approach introducing the idea of symmetry between the grounds for asserting propositions, expressed in introduction rules, and the consequences of accepting propositions in the elimination rules, an idea that has proved very important in proof theory.[2] Gentzen (1934) further introduced the idea of the sequent calculus, a calculus advanced in a similar spirit that better expressed the duality of the logical connectives,[3] and went on to make fundamental advances in the formalisation of intuitionistic logic, and provide the first combinatorial proof of the consistency of Peano arithmetic. Together, the presentation of natural deduction and the sequent calculus introduced the fundamental idea of analytic proof to proof theory.

## 1.2   Structural proof theory

Main article: Structural proof theory

Structural proof theory is the subdiscipline of proof theory that studies the specifics of proof calculi. The three most well-known styles of proof calculi are:

- The Hilbert calculi

- The natural deduction calculi

- The sequent calculi

Each of these can give a complete and axiomatic formalization of propositional or predicate logic of either the classical or intuitionistic flavour, almost any modal logic, and many substructural logics, such as relevance logic or linear logic. Indeed it is unusual to find a logic that resists being represented in one of these calculi.

Proof theorists are typically interested in proof calculi that support a notion of analytic proof. The notion of analytic proof was introduced by Gentzen for the sequent calculus; there the analytic proofs are those that are cut-free. Much of the interest in cut-free proofs comes from the subformula property: every formula in the end sequent of a cut-free proof is a subformula of one of the premises. This allows one to show consistency of the sequent calculus easily; if the empty sequent were derivable it would have to be a subformula of some premise, which it is not. Gentzen's midsequent theorem, the Craig interpolation theorem, and Herbrand's theorem also follow as corollaries of the cut-elimination theorem.

Gentzen's natural deduction calculus also supports a notion of analytic proof, as shown by Dag Prawitz. The definition is slightly more complex: we say the analytic proofs are the normal forms, which are related to the notion of normal form in term rewriting. More exotic proof calculi such as Jean-Yves Girard's proof nets also support a notion of analytic proof.

Structural proof theory is connected to type theory by means of the Curry-Howard correspondence, which observes a structural analogy between the process of normalisation in the natural deduction calculus and beta reduction in the typed lambda calculus. This provides the foundation for the intuitionistic type theory developed by Per Martin-Löf, and is often extended to a three way correspondence, the third leg of which are the cartesian closed categories.

Other research topics in structural theory include analytic tableau, which apply the central idea of analytic proof from structural proof theory to provide decision procedures and semi-decision procedures for a wide range of logics, and the proof theory of substructural logics.

## 1.3   Ordinal analysis

Main article: Ordinal analysis

Ordinal analysis is a powerful technique for providing combinatorial consistency proofs for subsystems of arithmetic, analysis, and set theory. Gödel's second incompleteness theorem is often interpreted as demonstrating that finitistic consistency proofs are impossible for theories of sufficient strength. Ordinal analysis allows one to measure precisely the infinitary content of the consistency of theories. For a consistent recursively axiomatized theory T, one can prove in finitistic arithmetic that the well-foundedness of a certain transfinite ordinal implies the consistency of T. Gödel's second incompleteness theorem implies that the well-foundedness of such an ordinal cannot be proved in the theory T.

Consequences of ordinal analysis include (1) consistency of subsystems of classical second order arithmetic and set theory relative to constructive theories, (2) combinatorial independence results, and (3) classifications of provably total recursive functions and provably well-founded ordinals.

Ordinal analysis was originated by Genzten, who proved the consistency of Peano Arithmetic using transfinite induction up to ordinal $\varepsilon_0$. Ordinal analysis has been extended to many fragments of first and second order arithmetic and set theory. One major challenge has been the ordinal analysis of impredicative theories. The first breakthrough in this direction was Takeuti's proof of the consistency of $\Pi^1_1$-$CA_0$ using the method of ordinal diagrams

## 1.4 Provability logic

Main article: Provability logic

*Provability logic* is a modal logic, in which the box operator is interpreted as 'it is provable that'. The point is to capture the notion of a proof predicate of a reasonably rich formal theory. As basic axioms of the provability logic GL, which captures provable in Peano Arithmetic, one takes modal analogues of the Hilbert-Bernays derivability conditions and Löb's theorem (if it is provable that the provability of A implies A, then A is provable).

Some of the basic results concerning the incompleteness of Peano Arithmetic and related theories have analogues in provability logic. For example, it is a theorem in GL that if a contradiction is not provable then it is not provable that a contradiction is not provable (Gödel's second incompleteness theorem). There are also modal analogues of the fixed-point theorem. Robert Solovay proved that the modal logic GL is complete with respect to Peano Arithmetic. That is, the propositional theory of provability in Peano Arithmetic is completely represented by the modal logic GL. This straightforwardly implies that propositional reasoning about provability in Peano Arithmetic is complete and decidable.

Other research in provability logic has focused on first-order provability logic, polymodal provability logic (with one modality representing provability in the object theory and another representing provability in the meta-theory), and interpretability logics intended to capture the interaction between provability and interpretability. Some very recent research has involved applications of graded provability algebras to the ordinal analysis of arithmetical theories.

## 1.5 Reverse mathematics

Main article: Reverse mathematics

**Reverse mathematics** is a program in mathematical logic that seeks to determine which axioms are required to prove theorems of mathematics. It was founded by Harvey Friedman. Its defining method can be described as "going backwards from the theorems to the axioms", in contrast to the ordinary mathematical practice of deriving theorems from axioms. The reverse mathematics program was foreshadowed by results in set theory such as the classical theorem that the axiom of choice and Zorn's lemma are equivalent over ZF set theory. The goal of reverse mathematics, however, is to study possible axioms of ordinary theorems of mathematics rather than possible axioms for set theory.

In reverse mathematics, one starts with a framework language and a base theory—a core axiom system—that is too weak to prove most of the theorems one might be interested in, but still powerful enough to develop the definitions necessary to state these theorems. For example, to study the theorem "Every bounded sequence of real numbers has a supremum" it is necessary to use a base system which can speak of real numbers and sequences of real numbers.

For each theorem that can be stated in the base system but is not provable in the base system, the goal is to determine the particular axiom system (stronger than the base system) that is necessary to prove that theorem. To show that a system *S* is required to prove a theorem $T$, two proofs are required. The first proof shows $T$ is provable from $S$; this is an ordinary mathematical proof along with a justification that it can be carried out in the system $S$. The second proof, known as a **reversal**, shows that $T$ itself implies $S$; this proof is carried out in the base system. The reversal establishes that no axiom system $S'$ that extends the base system can be weaker than $S$ while still proving $T$.

One striking phenomenon in reverse mathematics is the robustness of the *Big Five* axiom systems. In order of increasing strength, these systems are named by the initialisms $RCA_0$, $WKL_0$, $ACA_0$, $ATR_0$, and $\Pi^1_1$-$CA_0$. Nearly every theorem of ordinary mathematics that has been reverse mathematically analyzed has been proven equivalent to one of these five systems. Much recent research has focused on combinatorial principles that do not fit neatly into this framework, like $RT^2_2$ (Ramsey's theorem for pairs).

Research in reverse mathematics often incorporates methods and techniques from recursion theory as well as proof theory.

## 1.6   Functional interpretations

Functional interpretations are interpretations of non-constructive theories in functional ones. Functional interpretations usually proceed in two stages. First, one "reduces" a classical theory C to an intuitionistic one I. That is, one provides a constructive mapping that translates the theorems of C to the theorems of I. Second, one reduces the intuitionistic theory I to a quantifier free theory of functionals F. These interpretations contribute to a form of Hilbert's program, since they prove the consistency of classical theories relative to constructive ones. Successful functional interpretations have yielded reductions of infinitary theories to finitary theories and impredicative theories to predicative ones.

Functional interpretations also provide a way to extract constructive information from proofs in the reduced theory. As a direct consequence of the interpretation one usually obtains the result that any recursive function whose totality can be proven either in I or in C is represented by a term of F. If one can provide an additional interpretation of F in I, which is sometimes possible, this characterization is in fact usually shown to be exact. It often turns out that the terms of F coincide with a natural class of functions, such as the primitive recursive or polynomial-time computable functions. Functional interpretations have also been used to provide ordinal analyses of theories and classify their provably recursive functions.

The study of functional interpretations began with Kurt Gödel's interpretation of intuitionistic arithmetic in a quantifier-free theory of functionals finite type. This interpretation is commonly known as the Dialectica interpretation. Together with the double-negation interpretation of classical logic in intuitionistic logic, it provides a reduction of classical arithmetic to intuitionistic arithmetic.

## 1.7   Formal and informal proof

Main article: Formal proof

The *informal* proofs of everyday mathematical practice are unlike the *formal* proofs of proof theory. They are rather like high-level sketches that would allow an expert to reconstruct a formal proof at least in principle, given enough time and patience. For most mathematicians, writing a fully formal proof is too pedantic and long-winded to be in common use.

Formal proofs are constructed with the help of computers in interactive theorem proving. Significantly, these proofs can be checked automatically, also by computer. (Checking formal proofs is usually simple, whereas *finding* proofs (automated theorem proving) is generally hard.) An informal proof in the mathematics literature, by contrast, requires weeks of peer review to be checked, and may still contain errors.

## 1.8   Proof-theoretic semantics

Main articles: proof-theoretic semantics and logical harmony

In linguistics, type-logical grammar, categorial grammar and Montague grammar apply formalisms based on structural proof theory to give a formal natural language semantics.

## 1.9 See also

- Intermediate logic

- Model theory

- Proof (truth)

- Proof techniques

## 1.10 Notes

[1] E.g., Wang (1981), pp. 3–4, and Barwise (1978).

[2] Prawitz (2006, p. 98).

[3] Girard, Lafont, and Taylor (1988).

## 1.11 References

- J. Avigad, E.H. Reck (2001). "Clarifying the nature of the infinite": the development of metamathematics and proof theory. Carnegie-Mellon Technical Report CMU-PHIL-120.

- J. Barwise (ed., 1978). Handbook of Mathematical Logic. North-Holland.

- S. Buss. Handbook of Proof Theory. Elsevier.

- A. S. Troelstra, H. Schwichtenberg (1996). *Basic Proof Theory*. In series *Cambridge Tracts in Theoretical Computer Science*, Cambridge University Press, ISBN 0-521-77911-1.

- G. Gentzen (1935/1969). Investigations into logical deduction. In M. E. Szabo, editor, *Collected Papers of Gerhard Gentzen*. North-Holland. Translated by Szabo from "Untersuchungen über das logische Schliessen", Mathematisches Zeitschrift 39: 176-210, 405-431.

- Hazewinkel, Michiel, ed. (2001), "Proof theory", *Encyclopedia of Mathematics*, Springer, ISBN 978-1-55608-010-4

- Luis Moreno & Bharath Sriraman (2005).*Structural Stability and Dynamic Geometry: Some Ideas on Situated Proof. International Reviews on Mathematical Education. Vol. 37, no.3, pp. 130–139*

- Prawitz, Dag (2006) [1965]. *Natural deduction: A proof-theoretical study*. Mineola, New York: Dover Publications. ISBN 978-0-486-44655-4.

- J. von Plato (2008). The Development of Proof Theory. Stanford Encyclopedia of Philosophy.

- Wang, Hao (1981). *Popular Lectures on Mathematical Logic*. Van Nostrand Reinhold Company. ISBN 0-442-23109-1.

# Chapter 2

# Epsilon calculus

Hilbert's **epsilon calculus** is an extension of a formal language by the epsilon operator, where the epsilon operator substitutes for quantifiers in that language as a method leading to a proof of consistency for the extended formal language. The *epsilon operator* and *epsilon substitution method* are typically applied to a first-order predicate calculus, followed by a showing of consistency. The epsilon-extended calculus is further extended and generalized to cover those mathematical objects, classes, and categories for which there is a desire to show consistency, building on previously-shown consistency at earlier levels.[1]

## 2.1 Epsilon operator

### 2.1.1 Hilbert notation

For any formal language $L$, extend $L$ by adding the epsilon operator to redefine quantification:

- $(\exists x)A(x) \equiv A(\epsilon x\ A)$

- $(\forall x)A(x) \equiv A(\epsilon x\ (\neg A))$

The intended interpretation of $\epsilon x\ A$ is *some x* that satisfies $A$, if it exists. In other words, $\epsilon x\ A$ returns some term $t$ such that $A(t)$ is true, otherwise it returns some default or arbitrary term. If more than one term can satisfy $A$, then any one of these terms (which make $A$ true) can be chosen, non-deterministically. Equality is required to be defined under $L$, and the only rules required for $L$ extended by the epsilon operator are modus ponens and the substitution of $A(t)$ to replace $A(x)$ for any term $t$.[2]

### 2.1.2 Bourbaki notation

In tau-square notation from N. Bourbaki's *Theory of Sets*, the quantifiers are defined as follows:

- $(\exists x)A(x) \equiv (\tau_x(A)|x)A$

- $(\forall x)A(x) \equiv \neg(\tau_x(\neg A)|x)\neg A \equiv (\tau_x(\neg A)|x)A$

where $A$ is a relation in $L$, $x$ is a variable, and $\tau_x(A)$ juxtaposes a $\tau$ at the front of $A$, replaces all instances of $x$ with $\square$, and links them back to $\tau$. Then let $Y$ be an assembly, $(Y|x)A$ denotes the replacement of all variables $x$ in $A$ with $Y$.

This notation is equivalent to the Hilbert notation and is read the same. It is used by Bourbaki to define cardinal assignment since he does not use the axiom of replacement.

## 2.2 Modern approaches

Hilbert's Program for mathematics was to justify those formal systems as consistent in relation to constructive or semi-constructive systems. While Gödel's results on incompleteness mooted Hilbert's Program to a great extent, modern researchers find the epsilon calculus to provide alternatives for approaching proofs of systemic consistency as described in the epsilon substitution method.

### 2.2.1 Epsilon substitution method

A theory to be checked for consistency is first embedded in an appropriate epsilon calculus. Second, a process is developed for re-writing quantified theorems to be expressed in terms of epsilon operations via the epsilon substitution method. Finally, the process must be shown to normalize the re-writing process, so that the re-written theorems satisfy the axioms of the theory.[3]

## 2.3 References

- Epsilon Calculi entry in the *Internet Encyclopedia of Philosophy*

- Moser, Georg; Richard Zach. *The Epsilon Calculus (Tutorial)*. Berlin: Springer-Verlag. OCLC 108629234.

- The epsilon calculus entry by Jeremy Avigad and Richard Zach in the *Stanford Encyclopedia of Philosophy*, November 27, 2013

- Bourbaki, N. *Theory of Sets*. Berlin: Springer-Verlag. ISBN 3-540-22525-0.

## 2.4 Notes

[1] Stanford, overview paragraphs

[2] Stanford, the epsilon calculus paragraphs

[3] Stanford, more recent developments paragraphs

# Chapter 3

# Analytic proof

In mathematics, an **analytic proof** is a proof of a theorem in analysis that only makes use of methods from analysis, and which does not predominantly make use of algebraic or geometrical methods. The term was first used by Bernard Bolzano, who first provided a non-analytic proof of his intermediate value theorem and then, several years later provided proof of the theorem which was free from intuitions concerning lines crossing each other at a point and so he felt happy calling analytic (Bolzano 1817).

Bolzano's philosophical work encouraged a more abstract reading of when a demonstration could be regarded as analytic, where a proof is analytic if it does not go beyond its subject matter (Sebastik 2007). In proof theory, an analytic proof has come to mean a proof whose structure is simple in a special way, due to conditions on the kind of inferences that ensure none of them go beyond what is contained in the assumptions and what is demonstrated.

## 3.1   Structural proof theory

In proof theory, the notion of analytic proof provides the fundamental concept that brings out the similarities between a number of essentially distinct proof calculi, so defining the subfield of structural proof theory. There is no uncontroversial general definition of analytic proof, but for several proof calculi there is an accepted notion. For example:

- In Gerhard Gentzen's natural deduction calculus the analytic proofs are those in normal form; that is, no formula occurrence is both the principal premise of an elimination rule and the conclusion of an introduction rule;

- In Gentzen's sequent calculus the analytic proofs are those that do not use the cut rule.

However, it is possible to extend the inference rules of both calculi so that there are proofs that satisfy the condition but are not analytic. For example, a particularly tricky example of this is the *analytic cut rule*, used widely in the tableau method, which is a special case of the cut rule where the cut formula is a subformula of side formulae of the cut rule: a proof that contains an analytic cut is by virtue of that rule not analytic.

Furthermore, structural proof theories that are not analogous to Gentzen's theories have other notions of analytic proof. For example, the calculus of structures organises its inference rules into pairs, called the up fragment and the down fragment, and an analytic proof is one that only contains the down fragment.

## 3.2   See also

- Proof-theoretic semantics

## 3.3 References

- Bernard Bolzano (1817). Purely analytic proof of the theorem that between any two values which give results of opposite sign, there lies at least one real root of the equation. In *Abhandlungen der koniglichen bohmischen Gesellschaft der Wissenschaften* Vol. V, pp.225-48.

- Pfenning (1984). Analytic and Non-analytic Proofs. In *Proc. 7th International Conference on Automated Deduction.*

- Sebastik (2007). Bolzano's Logic. Entry in the *Stanford Encyclopedia of Philosophy.*

# Chapter 4

# Bachmann–Howard ordinal

In mathematics, the **Bachmann–Howard ordinal** (or **Howard ordinal**) is a large countable ordinal. It is the proof theoretic ordinal of several mathematical theories, such as Kripke–Platek set theory (with the axiom of infinity) and the system CZF of constructive set theory. It was introduced by Heinz Bachmann (1950) and William Alvin Howard (1972).

## 4.1 Definition

The Bachmann–Howard ordinal is defined using an ordinal collapsing function:

- $\varepsilon\alpha$ enumerates the epsilon numbers, the ordinals $\varepsilon$ such that $\omega^\varepsilon = \varepsilon$.

- $\Omega = \omega_1$ is the first uncountable ordinal.

- $\varepsilon\Omega_{+1}$ is the first epsilon number after $\Omega = \varepsilon\Omega$.

- $\psi(\alpha)$ is defined to be the smallest ordinal that cannot be constructed by starting with 0, 1, $\omega$ and $\Omega$, and repeatedly applying ordinal addition, multiplication and exponentiation, and $\psi$ to previously constructed ordinals (except that $\psi$ can only be applied to arguments less than $\alpha$, to ensure that it is well defined).

- The **Bachmann–Howard ordinal** is $\psi(\varepsilon\Omega_{+1})$.

The Bachmann–Howard ordinal can also be defined as $\phi_{\varepsilon\Omega+1}(0)$ for an extension of the Veblen functions $\varphi\alpha$ to uncountable $\alpha$; this extension is not completely straightforward.

## 4.2 References

- Bachmann, Heinz (1950), "Die Normalfunktionen und das Problem der ausgezeichneten Folgen von Ordnungszahlen", *Vierteljschr. Naturforsch. Ges. Zürich* **95**: 115–147, MR 0036806

- Howard, W. A. (1972), "A system of abstract constructive ordinals.", *J. Symbolic Logic* (Association for Symbolic Logic) **37** (2): 355–374, doi:10.2307/2272979, JSTOR 2272979, MR 0329869

- Pohlers, Wolfram (1989), *Proof theory*, Lecture Notes in Mathematics **1407**, Berlin: Springer-Verlag, ISBN 3-540-51842-8, MR 1026933

- Rathjen, Michael (August 2005). "Proof Theory: Part III, Kripke-Platek Set Theory" (PDF). Retrieved 2008-04-17. (slides of a talk given at Fischbachau)

# Chapter 5

# Bounded quantifier

This article is about bounded quantification in mathematical logic. For bounded quantification in type theory, see Bounded quantification.

In the study of formal theories in mathematical logic, **bounded quantifiers** are often added to a language in addition to the standard quantifiers "∀" and "∃". Bounded quantifiers differ from "∀" and "∃" in that bounded quantifiers restrict the range of the quantified variable. The study of bounded quantifiers is motivated by the fact that determining whether a sentence with only bounded quantifiers is true is often not as difficult as determining whether an arbitrary sentence is true.

Examples of bounded quantifiers in the context of real analysis include "∀$x$>0", "∃$y$<0", and "∀$x \in \mathbb{R}$". Informally "∀$x$>0" says "for all $x$ where $x$ is larger than 0", "∃$y$<0" says "there exists a $y$ where $y$ is less than 0" and "∀$x \in \mathbb{R}$" says "for all $x$ where $x$ is a real number". For example, "∀$x$>0 ∃$y$<0 $(x = y^2)$" says "every positive number is the square of a negative number".

## 5.1 Bounded quantifiers in arithmetic

Suppose that $L$ is the language of Peano arithmetic (the language of second-order arithmetic or arithmetic in all finite types would work as well). There are two types of bounded quantifiers: $\forall n < t$ and $\exists n < t$. These quantifiers bind the number variable $n$ and contain a numeric term $t$ which may not mention $n$ but which may have other free variables. (By "numeric terms" here we mean terms such as "1 + 1", "2", "2 × 3", "$m + 3$", etc.)

These quantifiers are defined by the following rules ($\phi$ denotes formulas):

$$\exists n < t\, \phi \Leftrightarrow \exists n(n < t \wedge \phi)$$

$$\forall n < t\, \phi \Leftrightarrow \forall n(n < t \to \phi)$$

There are several motivations for these quantifiers.

- In applications of the language to recursion theory, such as the arithmetical hierarchy, bounded quantifiers add no complexity. If $\phi$ is a decidable predicate then $\exists n < t\, \phi$ and $\forall n < t\, \phi$ are decidable as well.

- In applications to the study of Peano Arithmetic, formulas are sometimes provable with bounded quantifiers but unprovable with unbounded quantifiers.

For example, there is a definition of primality using only bounded quantifiers. A number $n$ is prime if and only if there are not two numbers strictly less than $n$ whose product is $n$. There is no quantifier-free definition of primality in the language $\langle 0, 1, +, \times, <, = \rangle$ , however. The fact that there is a bounded quantifier formula defining primality shows that the primality of each number can be computably decided.

In general, a relation on natural numbers is definable by a bounded formula if and only if it is computable in the linear-time hierarchy, which is defined similarly to the polynomial hierarchy, but with linear time bounds instead of polynomial. Consequently, all predicates definable by a bounded formula are Kalmár elementary, context-sensitive, and primitive recursive.

In the arithmetical hierarchy, an arithmetical formula which contains only bounded quantifiers is called $\Sigma_0^0$ , $\Delta_0^0$ , and $\Pi_0^0$ . The superscript 0 is sometimes omitted.

## 5.2  Bounded quantifiers in set theory

Suppose that $L$ is the language $\langle \in, \ldots, = \rangle$ of the Zermelo–Fraenkel set theory, where the ellipsis may be replaced by term-forming operations such as a symbol for the powerset operation. There are two bounded quantifiers: $\forall x \in t$ and $\exists x \in t$ . These quantifiers bind the set variable $x$ and contain a term $t$ which may not mention $x$ but which may have other free variables.

The semantics of these quantifiers is determined by the following rules:

$$\exists x \in t\,(\phi) \Leftrightarrow \exists x (x \in t \wedge \phi)$$

$$\forall x \in t\,(\phi) \Leftrightarrow \forall x (x \in t \rightarrow \phi)$$

A ZF formula which contains only bounded quantifiers is called $\Sigma_0$ , $\Delta_0$ , and $\Pi_0$ . This forms the basis of the Levy hierarchy, which is defined analogously with the arithmetical hierarchy.

Bounded quantifiers are important in Kripke-Platek set theory and constructive set theory, where only $\Delta_0$ separation is included. That is, it includes separation for formulas with only bounded quantifiers, but not separation for other formulas. In KP the motivation is the fact that whether a set $x$ satisfies a bounded quantifier formula only depends on the collection of sets that are close in rank to $x$ (as the powerset operation can only be applied finitely many times to form a term). In constructive set theory, it is motivated on predicative grounds.

## 5.3  See also

- Subtyping — bounded quantification in type theory
- System F<: — a polymorphic typed lambda calculus with bounded quantification

## 5.4  References

- Hinman, P. (2005). *Fundamentals of Mathematical Logic*. A K Peters. ISBN 1-56881-262-0.

- Kunen, K. (1980). *Set theory: An introduction to independence proofs*. Elsevier. ISBN 0-444-86839-9.

# Chapter 6

# Büchi arithmetic

**Büchi arithmetic** of base $k$ is the first-order theory of the natural numbers with addition and the function $V_k(x)$ which is defined as the largest power of $k$ dividing $x$, named in honor of the Swiss mathematician Julius Richard Büchi. The signature of Presburger arithmetic contains only the addition operation, $V_k$ and equality, omitting the multiplication operation entirely.

Unlike Peano arithmetic, Büchi arithmetic is a decidable theory. This means it is possible to effectively determine, for any sentence in the language of Büchi arithmetic, whether that sentence is provable from the axioms of Presburger arithmetic.

## 6.1 Büchi arithmetic and automaton

A subset $X \subseteq \mathbb{N}^n$ is definable in Bûchi arithmetic of base $k$ if and only if it is $k$-recognisable.

If $n = 1$ this means that the set of integers of $X$ in base $k$ is accepted by an automaton. Similarly if $n > 1$ there exists an automata that read firsts digit, then second digits, and so on, of $n$ integers in base $k$, and accept the words if the $n$ integers are in the relation $X$.

## 6.2 Properties of Büchi arithmetic

Two numbers $k$ and $l$ are multiplicatively dependent if there exists integers $n$ and $m$ such that $k^n = l^m$ .

If $k$ and $l$ are multiplicatively dependent then Büchi arithmetic of base $k$ and $l$ have the same expressivity. Indeed $V_l$ can be defined in $\mathrm{FO}(V_k, +)$ .

Else the theory with both $V_k$ and $V_l$ function is equivalent to Peano arithmetic the logic with addition and multiplication since multiplication is definable in $\mathrm{FO}(V_k, V_l, +)$ .

On the other hand by the Cobham–Semenov theorem, if a relation is definable in both $k$ and $l$ Büchi arithmetics it is definable in Presburger arithmetic.[1][2]

## 6.3 References

[1] Cobham, Alan (1969). "On the base-dependence of sets of numbers recognizable by finite automata". *Math. Systems Theory* **3**: 186–192. doi:10.1007/BF01746527.

[2] Semenov, A. L. (1977). "Presburgerness of predicates regular in two number systems". *Sibirsk. Mat. Zh.* (in Russian) **18**: 403–418.

- Bès, Alexis. "A survey of Arithmetical Definability". Retrieved 27 June 2012.

## 6.4   Further reading

- Bès, Alexis (1997). "Undecidable extensions of Büchi arithmetic and Cobham-Semënov theorem". *J. Symb. Log.* **62** (4): 1280–1296. doi:10.2307/2275643. Zbl 0896.03011.

# Chapter 7

# Church–Kleene ordinal

In mathematics, the **Church–Kleene ordinal**, $\omega_1^{CK}$ , named after Alonzo Church and S. C. Kleene, is a large countable ordinal. It is the set of all recursive ordinals and the smallest non-recursive ordinal. It is also the first ordinal which is not hyperarithmetical, and the first admissible ordinal after $\omega$.

## 7.1 References

- Church, Alonzo; Kleene, S. C. (1937), "Formal definitions in the theory of ordinal numbers.", *Fundamenta mathematicae, Warszawa,* **28**: 11–21, JFM 63.0029.02

- Church, Alonzo (1938), "The constructive second number class", *Bull. Amer. Math. Soc.* **44** (4): 224–232, doi:10.1090/S0002-9904-1938-06720-1

- Kleene, S. C. (1938), "On Notation for Ordinal Numbers", *The Journal of Symbolic Logic* (The Journal of Symbolic Logic, Vol. 3, No. 4) **3** (4): 150–155, doi:10.2307/2267778, JSTOR 2267778

- Rogers, Hartley (1987) [1967], *The Theory of Recursive Functions and Effective Computability*, First MIT press paperback edition, ISBN 978-0-262-68052-3

# Chapter 8

# Cirquent calculus

**Cirquent calculus** is a proof calculus which manipulates graph-style constructs termed *cirquents*, as opposed to the traditional tree-style objects such as formulas or sequents. Cirquents come in a variety of forms, but they all share one main characteristic feature, making them different from the more traditional objects of syntactic manipulation. This feature is the ability to explicitly account for possible sharing of subcomponents between different components. For instance, it is possible to write an expression where two subexpressions $F$ and $E$, while neither one is a subexpression of the other, still have a common occurrence of a subexpression $G$ (as opposed to having two different occurrences of $G$, one in $F$ and one in $E$).

The approach was introduced by G. Japaridze in[1] as an alternative proof theory capable of "taming" various nontrivial fragments his computability logic, which had otherwise resisted all axiomatization attempts within the traditional proof-theoretic frameworks.[2] [3]

The basic version of cirquent calculus in[4] was accompanied with an "*abstract resource semantics*" and the claim that the latter was an adequate formalization of the resource philosophy traditionally associated with linear logic. Based on that claim and the fact that the semantics induced a logic properly stronger than (affine) linear logic, Japaridze argued that linear logic was incomplete as a logic of resources. Furthermore, he argued that not only the deductive power but also the expressive power of linear logic was weak, for it, unlike cirquent calculus, failed to capture the ubiquitous phenomenon of resource sharing.[5]

Among the later-found applications of cirquent calculus was the use of it to define a semantics for purely propositional independence-friendly logic.[6] The corresponding logic was axiomatized by W. Xu.[7]

## 8.1   Literature

- M.Bauer, "The computational complexity of propositional cirquent calculus". Logical Methods is Computer Science 11 (2015),

Issue 1, Paper 12, pp. 1–16.

- G.Japaridze, "Introduction to cirquent calculus and abstract resource semantics". Journal of Logic and Computation 16 (2006), pp. 489–532.

- G.Japaridze, "Cirquent calculus deepened." Journal of Logic and Computation 18 (2008), pp. 983–1028.

- G.Japaridze, "From formulas to cirquents in computability logic". Logical Methods is Computer Science 7 (2011), Issue 2 , Paper 1, pp. 1–55.

- G.Japaridze, "The taming of recurrences in computability logic through cirquent calculus, Part I".Archive for Mathematical Logic 52 (2013), pages 173–212.

- G.Japaridze, "The taming of recurrences in computability logic through cirquent calculus, Part II" Archive for Mathematical Logic 52 (2013), pages 213–259.

- W.Xu and S.Liu, "Soundness and completeness of the cirquent calculus system CL6 for computability logic". Logic Journal of the IGPL 20 (2012), pp. 317–330.

- W.Xu and S.Liu, "Cirquent calculus system CL8S versus calculus of structures system SKSg for propositional logic". In: Quantitative Logic and Soft Computing. Guojun Wang, Bin Zhao and Yongming Li, eds. Singapore, World Scientific, 2012, pp. 144–149.

- W.Xu, "A propositional system induced by Japaridze's approach to IF logic". Logic Journal of the IGPL 22 (2014), pages 982–991.

## 8.2 References

[1] G.Japaridze, "Introduction to cirquent calculus and abstract resource semantics". Journal of Logic and Computation 16 (2006), pp. 489–532.

[2] G.Japaridze, "The taming of recurrences in computability logic through cirquent calculus, Part I".Archive for Mathematical Logic 52 (2013), pages 173-212.

[3] G.Japaridze, "The taming of recurrences in computability logic through cirquent calculus, Part II" Archive for Mathematical Logic 52 (2013), pages 213–259.

[4] G.Japaridze, "Introduction to cirquent calculus and abstract resource semantics". Journal of Logic and Computation 16 (2006), pp. 489–532.

[5] G.Japaridze, "Cirquent calculus deepened." Journal of Logic and Computation 18 (2008), pp. 983–1028.

[6] G.Japaridze, "From formulas to cirquents in computability logic". Logical Methods is Computer Science 7 (2011), Issue 2 , Paper 1, pp. 1–55.

[7] W.Xu, "A propositional system induced by Japaridze's approach to IF logic". Logic Journal of the IGPL 22 (2014), pages 982–991.

# Chapter 9

# Completeness (logic)

Not to be confused with Complete (complexity).

In mathematical logic and metalogic, a formal system is called **complete** with respect to a particular property if every formula having the property can be derived using that system, i.e. is one of its theorems; otherwise the system is said to be **incomplete**. The term "complete" is also used without qualification, with differing meanings depending on the context, mostly referring to the property of semantical validity. Intuitively, a system is called complete in this particular sense, if it can derive every formula that is true. Kurt Gödel, Leon Henkin, and Emil Leon Post all published proofs of completeness. (See History of the Church–Turing thesis.)

## 9.1 Other properties related to completeness

Main articles: Soundness and Consistency

The property converse to completeness is called soundness, or consistency: a system is sound with respect to a property (mostly semantical validity) if each of its theorems has that property.

## 9.2 Forms of completeness

### 9.2.1 Expressive completeness

A formal language is **expressively complete** if it can express the subject matter for which it is intended.

### 9.2.2 Functional completeness

Main article: Functional completeness

A set of logical connectives associated with a formal system is **functionally complete** if it can express all propositional functions.

### 9.2.3 Semantic completeness

**Semantic completeness** is the converse of soundness for formal systems. A formal system is complete with respect to tautologousness or "semantically complete" when all its tautologies are theorems, whereas a formal system is "sound" when all theorems are tautologies (that is, they are semantically valid formulas: formulas that are true under every interpretation of the language of the system that is consistent with the rules of the system). That is,

$$\models_{\mathcal{S}} \varphi \ \to\ \vdash_{\mathcal{S}} \varphi. \text{[1]}$$

### 9.2.4 Strong completeness

A formal system S is **strongly complete** or **complete in the strong sense** if for every set of premises Γ, any formula that semantically follows from Γ is derivable from Γ. That is:

$$\Gamma \models_{\mathcal{S}} \varphi \ \to\ \Gamma \vdash_{\mathcal{S}} \varphi.$$

### 9.2.5 Refutation completeness

A formal system S is **refutation-complete** if it is able to derive *false* from every unsatisfiable set of formulas. That is,

$$\Gamma \models_{\mathcal{S}} \bot \ \to\ \Gamma \vdash_{\mathcal{S}} \bot. \text{[2]}$$

Every strongly complete system is also refutation-complete. Intuitively, strong completeness means that, given a formula set Γ, it is possible to *compute* every semantical consequence $\varphi$ of Γ, while refutation-completeness means that, given a formula set Γ and a formula $\varphi$, it is possible to *check* whether $\varphi$ is a semantical consequence of Γ.

Examples of refutation-complete systems include: SLD resolution on Horn clauses, superposition on equational clausal first-order logic, Robinson's resolution on clause sets.[3] The latter is not strongly complete: e.g. $\{a\} \models a \vee b$ holds even in the propositional subset of first-order logic, but $a \vee b$ cannot be derived from $\{a\}$ by resolution. However, $\{a, \neg(a \vee b)\} \vdash \bot$ can be derived.

### 9.2.6 Syntactical completeness

A formal system S is **syntactically complete** or **deductively complete** or **maximally complete** if for each sentence (closed formula) φ of the language of the system either φ or ¬φ is a theorem of S. This is also called **negation completeness**. In another sense, a formal system is **syntactically complete** if and only if no unprovable sentence can be added to it without introducing an inconsistency. Truth-functional propositional logic and first-order predicate logic are semantically complete, but not syntactically complete (for example, the propositional logic statement consisting of a single propositional variable **A** is not a theorem, and neither is its negation, but these are not tautologies). Gödel's incompleteness theorem shows that any recursive system that is sufficiently powerful, such as Peano arithmetic, cannot be both consistent and syntactically complete.

## 9.3 References

[1] Hunter, Geoffrey, Metalogic: An Introduction to the Metatheory of Standard First-Order Logic, University of California Pres, 1971

[2] David A. Duffy (1991). *Principles of Automated Theorem Proving*. Wiley. Here: sect. 2.2.3.1, p.33

[3] Stuart J. Russell, Peter Norvig (1995). *Artificial Intelligence: A Modern Approach*. Prentice Hall. Here: sect. 9.7, p.286

# Chapter 10

# Completeness of atomic initial sequents

In sequent calculus, the **completeness of atomic initial sequents** states that initial sequents $A \vdash A$ (where $A$ is an arbitrary formula) can be derived from only atomic initial sequents $p \vdash p$ (where $p$ is an atomic formula). This theorem plays a role analogous to eta expansion in lambda calculus, and dual to cut-elimination and beta reduction. Typically it can be established by induction on the structure of $A$, much more easily than cut-elimination.

## 10.1 References

- Gaisi Takeuti. *Proof theory*. Volume 81 of *Studies in Logic and the Foundation of Mathematics*. North-Holland, Amsterdam, 1975.

- Anne Sjerp Troelstra and Helmut Schwichtenberg. *Basic Proof Theory*. Edition: 2, illustrated, revised. Published by Cambridge University Press, 2000.

# Chapter 11

# Conservative extension

In mathematical logic, a logical theory $T_2$ is a (proof theoretic) **conservative extension** of a theory $T_1$ if the language of $T_2$ extends the language of $T_1$ ; every theorem of $T_1$ is a theorem of $T_2$ ; and any theorem of $T_2$ that is in the language of $T_1$ is already a theorem of $T_1$ .

More generally, if $\Gamma$ is a set of formulas in the common language of $T_1$ and $T_2$ , then $T_2$ is **$\Gamma$-conservative** over $T_1$ if every formula from $\Gamma$ provable in $T_2$ is also provable in $T_1$ .

To put it informally, the new theory may possibly be more convenient for proving theorems, but it proves no new theorems about the language of the old theory.

Note that a conservative extension of a consistent theory is consistent. [If it were not, then by the principle of explosion ("everything follows from a contradiction"), every theorem in the original theory *as well as its negation* would belong to the new theory, which then would not be a conservative extension.] Hence, conservative extensions do not bear the risk of introducing new inconsistencies. This can also be seen as a methodology for writing and structuring large theories: start with a theory, $T_0$ , that is known (or assumed) to be consistent, and successively build conservative extensions $T_1$ , $T_2$ , ... of it.

The theorem provers Isabelle and ACL2 adopt this methodology by providing a language for conservative extensions by definition.

Recently, conservative extensions have been used for defining a notion of module for ontologies: if an ontology is formalized as a logical theory, a subtheory is a module if the whole ontology is a conservative extension of the subtheory.

An extension which is not conservative may be called a **proper extension**.

## 11.1   Examples

- $ACA_0$ (a subsystem of second-order arithmetic) is a conservative extension of first-order Peano arithmetic.

- Von Neumann–Bernays–Gödel set theory is a conservative extension of Zermelo–Fraenkel set theory with the axiom of choice (ZFC).

- Internal set theory is a conservative extension of Zermelo–Fraenkel set theory with the axiom of choice (ZFC).

- Extensions by definitions are conservative.

- Extensions by unconstrained predicate or function symbols are conservative.

- $I\Sigma_1$ (a subsystem of Peano arithmetic with induction only for $\Sigma^0_1$-formulas) is a $\Pi^0_2$-conservative extension of the primitive recursive arithmetic (PRA).[1]

- ZFC is a $\Pi^1_3$-conservative extension of ZF by Shoenfield's absoluteness theorem.

- ZFC with the continuum hypothesis is a $\Pi^2_1$-conservative extension of ZFC.

## 11.2 Model-theoretic conservative extension

With model-theoretic means, a stronger notion is obtained: an extension $T_2$ of a theory $T_1$ is **model-theoretically conservative** if every model of $T_1$ can be expanded to a model of $T_2$ . It is straightforward to see that each model-theoretic conservative extension also is a (proof-theoretic) conservative extension in the above sense. The model theoretic notion has the advantage over the proof theoretic one that it does not depend so much on the language at hand; on the other hand, it is usually harder to establish model theoretic conservativity.

Further information: Conservativity theorem

## 11.3 References

[1]  Notre Dame Journal of Formal Logic, Fernando Ferreira, A simple proof of Parsons' theorem

## 11.4 External links

- The importance of conservative extensions for the foundations of mathematics

# Chapter 12

# Conservativity theorem

In mathematical logic, the **conservativity theorem** states the following: Suppose that a *closed* formula

$$\exists x_1 \ldots \exists x_m \, \varphi(x_1, \ldots, x_m)$$

is a theorem of a first-order theory $T$. Let $T_1$ be a theory obtained from $T$ by extending its language with new constants

$$a_1, \ldots, a_m$$

and adding a new axiom

$$\varphi(a_1, \ldots, a_m)$$

Then $T_1$ is a conservative extension of $T$, which means that the theory $T_1$ has the same set of theorems in the original language (i.e., without constants $a_i$) as the theory $T$.

In a more general setting, the **conservativity theorem** is formulated for extensions of a first-order theory by introducing a new functional symbol:

> Suppose that a *closed* formula $\forall \vec{y} \exists x \, \varphi(x, \vec{y})$ is a theorem of a first-order theory $T$, where we denote $\vec{y} := (y_1, \ldots, y_n)$. Let $T_1$ be a theory obtained from $T$ by extending its language with new functional symbol $f$ (of arity $n$) and adding a new axiom $\forall \vec{y} \varphi(f(\vec{y}), \vec{y})$. Then $T_1$ is a conservative extension of $T$, i.e. the theories $T$ and $T_1$ prove the same theorems not involving the functional symbol $f$).

## 12.1 References

- Elliott Mendelson (1997). *Introduction to Mathematical Logic* (4th ed.) Chapman & Hall.

- J.R. Shoenfield (1967). *Mathematical Logic*. Addison-Wesley Publishing Company.

# Chapter 13

# Consistency

For other uses, see Consistency (disambiguation).

In classical deductive logic, a **consistent** theory is one that does not contain a contradiction.[1][2] The lack of contradiction can be defined in either semantic or syntactic terms. The semantic definition states that a theory is consistent if and only if it has a model, i.e. there exists an interpretation under which all formulas in the theory are true. This is the sense used in traditional Aristotelian logic, although in contemporary mathematical logic the term **satisfiable** is used instead. The syntactic definition states that a theory is consistent if and only if there is no formula $P$ such that both $P$ and its negation are provable from the axioms of the theory under its associated deductive system.

If these semantic and syntactic definitions are equivalent for any theory formulated using a particular deductive logic, the logic is called **complete**. The completeness of the sentential calculus was proved by Paul Bernays in 1918[3] and Emil Post in 1921,[4] while the completeness of predicate calculus was proved by Kurt Gödel in 1930,[5] and consistency proofs for arithmetics restricted with respect to the induction axiom schema were proved by Ackermann (1924), von Neumann (1927) and Herbrand (1931).[6] Stronger logics, such as second-order logic, are not complete.

A **consistency proof** is a mathematical proof that a particular theory is consistent. The early development of mathematical proof theory was driven by the desire to provide finitary consistency proofs for all of mathematics as part of Hilbert's program. Hilbert's program was strongly impacted by incompleteness theorems, which showed that sufficiently strong proof theories cannot prove their own consistency (provided that they are in fact consistent).

Although consistency can be proved by means of model theory, it is often done in a purely syntactical way, without any need to reference some model of the logic. The cut-elimination (or equivalently the normalization of the underlying calculus if there is one) implies the consistency of the calculus: since there is obviously no cut-free proof of falsity, there is no contradiction in general.

## 13.1 Consistency and completeness in arithmetic and set theory

In theories of arithmetic, such as Peano arithmetic, there is an intricate relationship between the consistency of the theory and its completeness. A theory is complete if, for every formula $\varphi$ in its language, at least one of $\varphi$ or $\neg\,\varphi$ is a logical consequence of the theory.

Presburger arithmetic is an axiom system for the natural numbers under addition. It is both consistent and complete.

Gödel's incompleteness theorems show that any sufficiently strong effective theory of arithmetic cannot be both complete and consistent. Gödel's theorem applies to the theories of Peano arithmetic (PA) and Primitive recursive arithmetic (PRA), but not to Presburger arithmetic.

Moreover, Gödel's second incompleteness theorem shows that the consistency of sufficiently strong effective theories of arithmetic can be tested in a particular way. Such a theory is consistent if and only if it does *not* prove a particular sentence, called the Gödel sentence of the theory, which is a formalized statement of the claim that the theory is indeed consistent. Thus the consistency of a sufficiently strong, effective, consistent theory of arithmetic can never be proven in that system itself. The same result is true for effective theories that can describe a strong enough fragment of arithmetic – including set theories such as Zermelo–Fraenkel set theory. These set theories cannot prove their own Gödel sentences – provided that they are consistent, which is generally believed.

Because consistency of ZF is not provable in ZF, the weaker notion **relative consistency** is interesting in set theory (and in other sufficiently expressive axiomatic systems). If $T$ is a theory and $A$ is an additional axiom, $T + A$ is said to be consistent relative to $T$ (or simply that $A$ is consistent with $T$) if it can be proved that if $T$ is consistent then $T + A$ is consistent. If both $A$ and $\neg A$ are consistent with $T$, then $A$ is said to be independent of $T$.

## 13.2 First-order logic

### 13.2.1 Notation

$\vdash$ (Turnstile symbol) in the following context of Mathematical logic, means "provable from". That is, a $\vdash$ b reads: b is provable from a (in some specified formal system) -- see List of logic symbols) . In other cases, the turnstile symbol may stand to mean infers; derived from. See: List of mathematical symbols.

### 13.2.2 Definition

A set of formulas $\Phi$ in first-order logic is **consistent** (written Con $\Phi$ ) if and only if there is no formula $\phi$ such that $\Phi \vdash \phi$ and $\Phi \vdash \neg\phi$ . Otherwise $\Phi$ is **inconsistent** and is written Inc $\Phi$ .

$\Phi$ is said to be **simply consistent** if and only if for no formula $\phi$ of $\Phi$ , both $\phi$ and the negation of $\phi$ are theorems of $\Phi$ .

$\Phi$ is said to be **absolutely consistent** or **Post consistent** if and only if at least one formula of $\Phi$ is not a theorem of $\Phi$ .

$\Phi$ is said to be **maximally consistent** if and only if for every formula $\phi$ , if Con ( $\Phi \cup \phi$ ) then $\phi \in \Phi$ .

$\Phi$ is said to **contain witnesses** if and only if for every formula of the form $\exists x\phi$ there exists a term $t$ such that $(\exists x\phi \to \phi\frac{t}{x}) \in \Phi$ . See First-order logic.

### 13.2.3 Basic results

1. The following are equivalent:

    (a) Inc $\Phi$
    (b) For all $\phi$, $\Phi \vdash \phi$.

2. Every satisfiable set of formulas is consistent, where a set of formulas $\Phi$ is satisfiable if and only if there exists a model $\mathfrak{I}$ such that $\mathfrak{I} \models \Phi$ .

3. For all $\Phi$ and $\phi$ :

    (a) if not $\Phi \vdash \phi$ , then Con ( $\Phi \cup \{\neg\phi\}$ ) ;
    (b) if Con $\Phi$ and $\Phi \vdash \phi$ , then Con ( $\Phi \cup \{\phi\}$ ) ;
    (c) if Con $\Phi$ , then Con ( $\Phi \cup \{\phi\}$ ) or Con ( $\Phi \cup \{\neg\phi\}$ ) .

4. Let $\Phi$ be a maximally consistent set of formulas and contain witnesses. For all $\phi$ and $\psi$ :

    (a) if $\Phi \vdash \phi$ , then $\phi \in \Phi$ ,
    (b) either $\phi \in \Phi$ or $\neg\phi \in \Phi$ ,
    (c) $(\phi \vee \psi) \in \Phi$ if and only if $\phi \in \Phi$ or $\psi \in \Phi$ ,
    (d) if $(\phi \to \psi) \in \Phi$ and $\phi \in \Phi$ , then $\psi \in \Phi$ ,
    (e) $\exists x\phi \in \Phi$ if and only if there is a term $t$ such that $\phi\frac{t}{x} \in \Phi$ .

### 13.2.4  Henkin's theorem

Let $\Phi$ be a maximally consistent set of $S$-formulas containing witnesses.

Define a binary relation $\sim$ on the set of $S$-terms such that $t_0 \sim t_1$ if and only if $t_0 \equiv t_1 \in \Phi$; and let $\bar{t}$ denote the equivalence class of terms containing $t$; and let $T_\Phi := \{ \bar{t} \mid t \in T^S \}$ where $T^S$ is the set of terms based on the symbol set $S$.

Define the $S$-structure $\mathfrak{T}_\Phi$ over $T_\Phi$ the **term-structure** corresponding to $\Phi$ by:

1. for $n$-ary $R \in S$, $R^{\mathfrak{T}_\Phi}\overline{t_0}\ldots\overline{t_{n-1}}$ if and only if $Rt_0 \ldots t_{n-1} \in \Phi$;

2. for $n$-ary $f \in S$, $f^{\mathfrak{T}_\Phi}(\overline{t_0}\ldots\overline{t_{n-1}}) := \overline{ft_0 \ldots t_{n-1}}$;

3. for $c \in S$, $c^{\mathfrak{T}_\Phi} := \bar{c}$.

Let $\mathfrak{I}_\Phi := (\mathfrak{T}_\Phi, \beta_\Phi)$ be the **term interpretation** associated with $\Phi$, where $\beta_\Phi(x) := \bar{x}$.

$$\text{For all } \phi, \ \mathfrak{I}_\Phi \vDash \phi \text{ if and only if } \phi \in \Phi.$$

### 13.2.5  Sketch of proof

There are several things to verify. First, that $\sim$ is an equivalence relation. Then, it needs to be verified that (1), (2), and (3) are well defined. This falls out of the fact that $\sim$ is an equivalence relation and also requires a proof that (1) and (2) are independent of the choice of $t_0, \ldots, t_{n-1}$ class representatives. Finally, $\mathfrak{I}_\Phi \vDash \Phi$ can be verified by induction on formulas.

## 13.3  See also

- Equiconsistency

- Hilbert's problems

- Hilbert's second problem

- Jan Łukasiewicz

- Paraconsistent logic

- ω-consistency

- Gentzen's consistency proof

## 13.4  Footnotes

[1] Tarski 1946 states it this way: "A deductive theory is called CONSISTENT or NON-CONTRADICTORY if no two asserted statements of this theory contradict each other, or in other words, if of any two contradictory sentences . . . at least one cannot be proved," (p. 135) where Tarski defines *contradictory* as follows: "With the help of the word *not* one forms the NEGATION of any sentence; two sentences, of which the first is a negation of the second, are called CONTRADICTORY SENTENCES" (p. 20). This definition requires a notion of "proof". Gödel in his 1931 defines the notion this way: "The class of *provable formulas* is defined to be the smallest class of formulas that contains the axioms and is closed under the relation "immediate consequence", i.e. formula $c$ of $a$ and $b$ is defined as an *immediate consequence* in terms of *modus ponens* or substitution; cf Gödel 1931 van Heijenoort 1967:601. Tarski defines "proof" informally as "statements follow one another in a definite order according to certain principles . . . and accompanied by considerations intended to establish their validity[true conclusion for all true premises -- Reichenbach 1947:68]" cf Tarski 1946:3. Kleene 1952 defines the notion with respect to either an induction

or as to paraphrase) a finite sequence of formulas such that each formula in the sequence is either an axiom or an "immediate consequence" of the preceding formulas; "A *proof is said to be a proof* of *its last formula, and this formula is said to be* (formally) provable *or be a* (formal) theorem" cf Kleene 1952:83.

[2] Paraconsistent logic *tolerates* contradictions, but toleration of contradiction does not entail consistency.

[3] van Heijenoort 1967:265 states that Bernays determined the *independence* of the axioms of *Principia Mathematica*, a result not published until 1926, but he says nothing about Bernays proving their *consistency*.

[4] Post proves both consistency and completeness of the propositional calculus of PM, cf van Heijenoort's commentary and Post's 1931 *Introduction to a general theory of elementary propositons* in van Heijenoort 1967:264ff. Also Tarski 1946:134ff.

[5] cf van Heijenoort's commentary and Gödel's 1930 *The completeness of the axioms of the functional calculus of logic* in van Heijenoort 1967:582ff

[6] cf van Heijenoort's commentary and Herbrand's 1930 *On the consistency of arithmetic* in van Heijenoort 1967:618ff.

## 13.5 References

- Stephen Kleene, 1952 10th impression 1991, *Introduction to Metamathematics*, North-Holland Publishing Company, Amsterday, New York, ISBN 0-7204-2103-9.

- Hans Reichenbach, 1947, *Elements of Symbolic Logic*, Dover Publications, Inc. New York, ISBN 0-486-24004-5,

- Alfred Tarski, 1946, *Introduction to Logic and to the Methodology of Deductive Sciences, Second Edition*, Dover Publications, Inc., New York, ISBN 0-486-28462-X.

- Jean van Heijenoort, 1967, *From Frege to Gödel: A Source Book in Mathematical Logic*, Harvard University Press, Cambridge, MA, ISBN 0-674-32449-8 (pbk.)

- The Cambridge Dictionary of Philosophy, *consistency*

- H.D. Ebbinghaus, J. Flum, W. Thomas, **Mathematical Logic**

- Jevons, W.S., 1870, *Elementary Lessons in Logic*

## 13.6 External links

- Chris Mortensen, Inconsistent Mathematics, Stanford Encyclopedia of Philosophy

# Chapter 14

# Curry–Howard correspondence

```
plus_comm =
fun n m : nat =>
nat_ind (fun n0 : nat => n0 + m = m + n0)
   (plus_n_0 m)
   (fun (y : nat) (H : y + m = m + y) =>
     eq_ind (S (m + y))
       (fun n0 : nat => S (y + m) = n0)
       (f_equal S H)
       (m + S y)
       (plus_n_Sm m y)) n
       : forall n m : nat, n + m = m + n
```

*A proof written as a functional program: the proof of commutativity of addition on natural numbers in the proof assistant Coq. nat_ind stands for mathematical induction, eq_ind for substitution of equals and f_equal for taking the same function on both sides of the equality. Earlier theorems are referenced showing m = m + 0 and S (m + y) = m + S y.*

In programming language theory and proof theory, the **Curry–Howard correspondence** (also known as the **Curry–Howard isomorphism** or **equivalence**, or the **proofs-as-programs** and **propositions-** or **formulae-as-types interpretation**) is the direct relationship between computer programs and mathematical proofs. It is a generalization of a syntactic analogy between systems of formal logic and computational calculi that was first discovered by the American mathematician Haskell Curry and logician William Alvin Howard.[1] It is the link between logic and computation that is usually attributed to Curry and Howard, although the idea is related to the operational interpretation of intuitionistic logic given in various formulations by L. E. J. Brouwer, Arend Heyting and Andrey Kolmogorov (see Brouwer–Heyting–Kolmogorov interpretation)[2] and Stephen Kleene (see Realizability). The relationship has been extended to include category theory as the three-way **Curry–Howard–Lambek correspondence**.

# 14.1   Origin, scope, and consequences

At the very beginning, the **Curry–Howard correspondence** is

1. the observation in 1934 by Curry that the types of the combinators could be seen as axiom-schemes for intuitionistic implicational logic,[3]

2. the observation in 1958 by Curry that a certain kind of proof system, referred to as Hilbert-style deduction systems, coincides on some fragment to the typed fragment of a standard model of computation known as combinatory logic,[4]

3. the observation in 1969 by Howard that another, more "high-level" proof system, referred to as natural deduction, can be directly interpreted in its intuitionistic version as a typed variant of the model of computation known as lambda calculus.[5]

In other words, the Curry–Howard correspondence is the observation that two families of formalisms that had seemed unrelated—namely, the proof systems on one hand, and the models of computation on the other—were, in the two examples considered by Curry and Howard, in fact structurally the same kind of objects.

If one now abstracts on the peculiarities of this or that formalism, the immediate generalization is the following claim: *a proof is a program, the formula it proves is a type for the program*. More informally, this can be seen as an analogy that states that the return type of a function (i.e., the type of values returned by a function) is analogous to a logical theorem, subject to hypotheses corresponding to the types of the argument values passed to the function; and that the program to compute that function is analogous to a proof of that theorem. This sets a form of logic programming on a rigorous foundation: *proofs can be represented as programs, and especially as lambda terms*, or *proofs can be **run***.

The correspondence has been the starting point of a large spectrum of new research after its discovery, leading in particular to a new class of formal systems designed to act both as a proof system and as a typed functional programming language. This includes Martin-Löf's intuitionistic type theory and Coquand's Calculus of Constructions, two calculi in which proofs are regular objects of the discourse and in which one can state properties of proofs the same way as of any program. This field of research is usually referred to as modern type theory.

Such typed lambda calculi derived from the Curry–Howard paradigm led to software like Coq in which proofs seen as programs can be formalized, checked, and run.

A converse direction is to *use a program to extract a proof*, given its correctness— an area of research closely related to proof-carrying code. This is only feasible if the programming language the program is written for is very richly typed: the development of such type systems has been partly motivated by the wish to make the Curry–Howard correspondence practically relevant.

The Curry–Howard correspondence also raised new questions regarding the computational content of proof concepts that were not covered by the original works of Curry and Howard. In particular, classical logic has been shown to correspond to the ability to manipulate the continuation of programs and the symmetry of sequent calculus to express the duality between the two evaluation strategies known as call-by-name and call-by-value.

Speculatively, the Curry–Howard correspondence might be expected to lead to a substantial unification between mathematical logic and foundational computer science:

Hilbert-style logic and natural deduction are but two kinds of proof systems among a large family of formalisms. Alternative syntaxes include sequent calculus, proof nets, calculus of structures, etc. If one admits the Curry–Howard correspondence as the general principle that any proof system hides a model of computation, a theory of the underlying untyped computational structure of these kinds of proof system should be possible. Then, a natural question is whether something mathematically interesting can be said about these underlying computational calculi.

Conversely, combinatory logic and simply typed lambda calculus are not the only models of computation, either. Girard's linear logic was developed from the fine analysis of the use of resources in some models of lambda calculus; can we imagine a typed version of Turing's machine that would behave as a proof system? Typed assembly languages are such an instance of "low-level" models of computation that carry types.

Because of the possibility of writing non-terminating programs, Turing-complete models of computation (such as languages with arbitrary recursive functions) must be interpreted with care, as naive application of the correspondence leads to an inconsistent logic. The best way of dealing with arbitrary computation from a logical point of view is still an actively debated research question, but one popular approach is based on using monads to segregate provably terminating from potentially non-terminating code (an approach that also generalizes to much richer models of computation,[6] and is itself related to modal logic by a natural extension of the Curry–Howard isomorphism[ext 1]). A more radical approach, advocated by total functional programming, is to eliminate unrestricted recursion (and forgo Turing completeness, although still retaining high computational complexity), using more controlled corecursion wherever non-terminating behavior is actually desired.

## 14.2   General formulation

In its more general formulation, the Curry–Howard correspondence is a correspondence between formal proof calculi and type systems for models of computation. In particular, it splits into two correspondences. One at the level of formulas and types that is independent of which particular proof system or model of computation is considered, and one at the level of proofs and programs which, this time, is specific to the particular choice of proof system and model of computation considered.

At the level of formulas and types, the correspondence says that implication behaves the same as a function type, conjunction as a "product" type (this may be called a tuple, a struct, a list, or some other term depending on the language), disjunction as a sum type (this type may be called a union), the false formula as the empty type and the true formula as the singleton type (whose sole member is the null object). Quantifiers correspond to dependent function space or products (as appropriate). This is summarized in the following table:

At the level of proof systems and models of computations, the correspondence mainly shows the identity of structure, first, between some particular formulations of systems known as Hilbert-style deduction system and combinatory logic, and, secondly, between some particular formulations of systems known as natural deduction and lambda calculus.

Between the natural deduction system and the lambda calculus there are the following correspondences:

## 14.3   Corresponding systems

### 14.3.1   Hilbert-style deduction systems and combinatory logic

It was at the beginning a simple remark in Curry and Feys's 1958 book on combinatory logic: the simplest types for the basic combinators K and S of combinatory logic surprisingly corresponded to the respective axiom schemes $\alpha \to (\beta \to \alpha)$ and $(\alpha \to (\beta \to \gamma)) \to ((\alpha \to \beta) \to (\alpha \to \gamma))$ used in Hilbert-style deduction systems. For this reason, these schemes are now often called axioms K and S. Examples of programs seen as proofs in a Hilbert-style logic are given below.

If one restricts to the implicational intuitionistic fragment, a simple way to formalize logic in Hilbert's style is as follows. Let $\Gamma$ be a finite collection of formulas, considered as hypotheses. We say that $\delta$ is derivable from $\Gamma$, and we write $\Gamma \vdash \delta$, in the following cases:

- $\delta$ is an hypothesis, i.e. it is a formula of $\Gamma$,

- $\delta$ is an instance of an axiom scheme; i.e., under the most common axiom system:

- δ has the form $\alpha \to (\beta \to \alpha)$, or

- δ has the form $(\alpha \to (\beta \to \gamma)) \to ((\alpha \to \beta) \to (\alpha \to \gamma))$,

- δ follows by deduction, i.e., for some α, both $\alpha \to \delta$ and α are already derivable from Γ (this is the rule of modus ponens)

This can be formalized using inference rules, what we do in the left column of the following table.

We can formulate typed combinatory logic using a similar syntax: let Γ be a finite collection of variables, annotated with their types. A term T (also annotated with its type) will depend on these variables $[\Gamma \vdash T:\delta]$ when:

- T is one of the variables in Γ,

- T is a basic combinator; i.e., under the most common combinator basis:

  - T is K:$\alpha \to (\beta \to \alpha)$ [where α and β denote the types of its arguments], or

  - T is S:$(\alpha \to (\beta \to \gamma)) \to ((\alpha \to \beta) \to (\alpha \to \gamma))$,

- T is the composition of two subterms which depend on the variables in Γ.

The generation rules defined here are given in the right-column below. Curry's remark simply states that both columns are in one-to-one correspondence. The restriction of the correspondence to intuitionistic logic means that some classical tautologies, such as Peirce's law $((\alpha \to \beta) \to \alpha) \to \alpha$, are excluded from the correspondence.

Seen at a more abstract level, the correspondence can be restated as shown in the following table. Especially, the deduction theorem specific to Hilbert-style logic matches the process of abstraction elimination of combinatory logic.

Thanks to the correspondence, results from combinatory logic can be transferred to Hilbert-style logic and vice versa. For instance, the notion of reduction of terms in combinatory logic can be transferred to Hilbert-style logic and it provides a way to canonically transform proofs into other proofs of the same statement. One can also transfer the notion of normal terms to a notion of normal proofs, expressing that the hypotheses of the axioms never need to be all detached (since otherwise a simplification can happen).

Conversely, the non provability in intuitionistic logic of Peirce's law can be transferred back to combinatory logic: there is no typed term of combinatory logic that is typable with type $((\alpha \to \beta) \to \alpha) \to \alpha$.

Results on the completeness of some sets of combinators or axioms can also be transferred. For instance, the fact that the combinator **X** constitutes a one-point basis of (extensional) combinatory logic implies that the single axiom scheme

$$(((\alpha \to (\beta \to \gamma)) \to ((\alpha \to \beta) \to (\alpha \to \gamma))) \to ((\delta \to (\varepsilon \to \delta)) \to \zeta)) \to \zeta,$$

which is the principal type of **X**, is an adequate replacement to the combination of the axiom schemes

$\alpha \to (\beta \to \alpha)$ and

$(\alpha \to (\beta \to \gamma)) \to ((\alpha \to \beta) \to (\alpha \to \gamma))$.

## 14.3.2 Natural deduction and lambda calculus

After Curry emphasized the syntactic correspondence between Hilbert-style deduction and combinatory logic, Howard made explicit in 1969 a syntactic analogy between the programs of simply typed lambda calculus and the proofs of natural deduction. Below, the left-hand side formalizes intuitionistic implicational natural deduction as a calculus of sequents (the use of sequents is standard in discussions of the Curry–Howard isomorphism as it allows the deduction rules to be stated more cleanly) with implicit weakening and the right-hand side shows the typing rules of lambda calculus. In the left-hand side, Γ, $\Gamma_1$ and $\Gamma_2$ denote ordered sequences of formulas while in the right-hand side, they denote sequences of named (i.e., typed) formulas with all names different.

To paraphrase the correspondence, proving $\Gamma \vdash \alpha$ means having a program that, given values with the types listed in $\Gamma$, manufactures an object of type $\alpha$. An axiom corresponds to the introduction of a new variable with a new, unconstrained type, the $\to$ I rule corresponds to function abstraction and the $\to$ E rule corresponds to function application. Observe that the correspondence is not exact if the context $\Gamma$ is taken to be a set of formulas as, e.g., the $\lambda$-terms $\lambda x.\lambda y.x$ and $\lambda x.\lambda y.y$ of type $\alpha \to \alpha \to \alpha$ would not be distinguished in the correspondence. Examples are given below.

Howard showed that the correspondence extends to other connectives of the logic and other constructions of simply typed lambda calculus. Seen at an abstract level, the correspondence can then be summarized as shown in the following table. Especially, it also shows that the notion of normal forms in lambda calculus matches Prawitz's notion of normal deduction in natural deduction, from what we deduce, among others, that the algorithms for the type inhabitation problem can be turned into algorithms for deciding intuitionistic provability.

Howard's correspondence naturally extends to other extensions of natural deduction and simply typed lambda calculus. Here is a non exhaustive list:

- Girard-Reynolds System F as a common language for both second-order propositional logic and polymorphic lambda calculus,

- higher-order logic and Girard's System F$\omega$

- inductive types as algebraic data type

- necessity $\Box$ in modal logic and staged computation[ext 2]

- possibility $\Diamond$ in modal logic and monadic types for effects[ext 1]

- The $\lambda$I calculus corresponds to relevant logic.[7]

- The local truth ($\nabla$) modality in Grothendieck topology or the equivalent "lax" modality ($\circ$) of Benton, Bierman, and de Paiva (1998) correspond to CL-logic describing "computation types".[8]

### 14.3.3   Classical logic and control operators

At the time of Curry, and also at the time of Howard, the proofs-as-programs correspondence concerned only intuitionistic logic, i.e. a logic in which, in particular, Peirce's law was *not* deducible. The extension of the correspondence to Peirce's law and hence to classical logic became clear from the work of Griffin on typing operators that capture the evaluation context of a given program execution so that this evaluation context can be later on reinstalled. The basic Curry–Howard-style correspondence for classical logic is given below. Note the correspondence between the double-negation translation used to map classical proofs to intuitionistic logic and the continuation-passing-style translation used to map lambda terms involving control to pure lambda terms. More particularly, call-by-name continuation-passing-style translations relates to Kolmogorov's double negation translation and call-by-value continuation-passing-style translations relates to a kind of double-negation translation due to Kuroda.

A finer Curry–Howard correspondence exists for classical logic if one defines classical logic not by adding an axiom such as Peirce's law, but by allowing several conclusions in sequents. In the case of classical natural deduction, there exists a proofs-as-programs correspondence with the typed programs of Parigot's $\lambda\mu$-calculus.

### 14.3.4   Sequent calculus

A proofs-as-programs correspondence can be settled for the formalism known as Gentzen's sequent calculus but it is not a correspondence with a well-defined pre-existing model of computation as it was for Hilbert-style and natural deductions.

Sequent calculus is characterized by the presence of left introduction rules, right introduction rule and a cut rule that can be eliminated. The structure of sequent calculus relates to a calculus whose structure is close to the one of some abstract machines. The informal correspondence is as follows:

# 14.4 Related proofs-as-programs correspondences

## 14.4.1 The role of de Bruijn

N. G. de Bruijn used the lambda notation for representing proofs of the theorem checker Automath, and represented propositions as "categories" of their proofs. It was in the late 1960s at the same period of time Howard wrote his manuscript; de Bruijn was likely unaware of Howard's work, and stated the correspondence independently (Sørensen & Urzyczyn [1998] 2006, pp 98–99). Some researchers tend to use the term Curry–Howard–de Bruijn correspondence in place of Curry–Howard correspondence.

## 14.4.2 BHK interpretation

The BHK interpretation interprets intuitionistic proofs as functions but it does not specify the class of functions relevant for the interpretation. If one takes lambda calculus for this class of function, then the BHK interpretation tells the same as Howard's correspondence between natural deduction and lambda calculus.

## 14.4.3 Realizability

Kleene's recursive realizability splits proofs of intuitionistic arithmetic into the pair of a recursive function and of a proof of a formula expressing that the recursive function "realizes", i.e. correctly instantiates the disjunctions and existential quantifiers of the initial formula so that the formula gets true.

Kreisel's modified realizability applies to intuitionistic higher-order predicate logic and shows that the simply typed lambda term inductively extracted from the proof realizes the initial formula. In the case of propositional logic, it coincides with Howard's statement: the extracted lambda term is the proof itself (seen as an untyped lambda term) and the realizability statement is a paraphrase of the fact that the extracted lambda term has the type that the formula means (seen as a type).

Gödel's dialectica interpretation realizes (an extension of) intuitionistic arithmetic with computable functions. The connection with lambda calculus is unclear, even in the case of natural deduction.

## 14.4.4 Curry–Howard–Lambek correspondence

Joachim Lambek showed in the early 1970s that the proofs of intuitionistic propositional logic and the combinators of typed combinatory logic share a common equational theory which is the one of cartesian closed categories. The expression Curry–Howard–Lambek correspondence is now used by some people to refer to the three way isomorphism between intuitionistic logic, typed lambda calculus and cartesian closed categories, with objects being interpreted as types or propositions and morphisms as terms or proofs. The correspondence works at the equational level and is not the expression of a syntactic identity of structures as it is the case for each of Curry's and Howard's correspondences: i.e. the structure of a well-defined morphism in a cartesian-closed category is not comparable to the structure of a proof of the corresponding judgment in either Hilbert-style logic or natural deduction. To clarify this distinction, the underlying syntactic structure of cartesian closed categories is rephrased below.

Objects (types) are defined by

- $\top$ is an object

- if $\alpha$ and $\beta$ are objects then $\alpha \times \beta$ and $\alpha \rightarrow \beta$ are objects.

Morphisms (terms) are defined by

- $id$ , $\star$ , eval , $\pi_1$ and $\pi_2$ are morphisms

- if $t$ is a morphism, $\lambda t$ is a morphism

- if $t$ and $u$ are morphisms, $(t, u)$ and $u \circ t$ are morphisms.

Well-defined morphisms (typed terms) are defined by the following typing rules (in which the usual categorical morphism notation $f : \alpha \to \beta$ is replaced with sequent calculus notation $f :- \alpha \vdash \beta$ ).

Identity:

$$\overline{id :- \alpha \vdash \alpha}$$

Composition:

$$\frac{t :- \alpha \vdash \beta \qquad u :- \beta \vdash \gamma}{u \circ t :- \alpha \vdash \gamma}$$

Unit type (terminal object):

$$\overline{\star :- \alpha \vdash \top}$$

Cartesian product:

$$\frac{t :- \alpha \vdash \beta \qquad u :- \alpha \vdash \gamma}{(t, u) :- \alpha \vdash \beta \times \gamma}$$

Left and right projection:

$$\overline{\pi_1 :- \alpha \times \beta \vdash \alpha} \qquad \overline{\pi_2 :- \alpha \times \beta \vdash \beta}$$

Currying:

$$\frac{t :- \alpha \times \beta \vdash \gamma}{\lambda t :- \alpha \vdash \beta \to \gamma}$$

Application:

$$\overline{eval :- (\alpha \to \beta) \times \alpha \vdash \beta}$$

Finally, the equations of the category are

- $id \circ t = t$ , $t \circ id = t$ , $(v \circ u) \circ t = v \circ (u \circ t)$

- $\star \circ t = \star$

- $\pi_1 \circ (t, u) = t, \pi_2 \circ (t, u) = u, (\pi_1 \circ t, \pi_2 \circ t) = t$

- $eval \circ (\lambda t \circ \pi_1, \pi_2) = t, \lambda eval = id$

Now, there exists $t$ such that $t :- \alpha_1 \times \ldots \times \alpha_n \vdash \beta$ iff $\alpha_1, \ldots, \alpha_n \vdash \beta$ is provable in implicational intuitionistic logic,.

## 14.5 Examples

Thanks to the Curry–Howard correspondence, a typed expression whose type corresponds to a logical formula is analogous to a proof of that formula. Here are examples.

**The identity combinator seen as a proof of $\alpha \to \alpha$ in Hilbert-style logic**

As a simple example, we construct a proof of the theorem $\alpha \to \alpha$. In lambda calculus, this is the type of the identity function $\mathbf{I} = \lambda x.x$ and in combinatory logic, the identity function is obtained by applying $S$ twice to $\mathbf{K}$. That is, we have $\mathbf{I} = ((\mathbf{S}\ \mathbf{K})\ \mathbf{K})$. As a description of a proof, this says that to prove $\alpha \to \alpha$, we can proceed as follows:

- instantiate the second axiom scheme with the formulas $\alpha$, $\beta \to \alpha$ and $\alpha$, so that to obtain a proof of $(\alpha \to ((\beta \to \alpha) \to \alpha)) \to ((\alpha \to (\beta \to \alpha)) \to (\alpha \to \alpha))$,

- instantiate the first axiom scheme once with $\alpha$ and $\beta \to \alpha$, so that to obtain a proof of $\alpha \to ((\beta \to \alpha) \to \alpha)$,

- instantiate the first axiom scheme a second time with $\alpha$ and $\beta$, so that to obtain a proof of $\alpha \to (\beta \to \alpha)$,

- apply modus ponens twice so that to obtain a proof of $\alpha \to \alpha$

In general, the procedure is that whenever the program contains an application of the form $(P\ Q)$, we should first prove theorems corresponding to the types of $P$ and $Q$. Since $P$ is being applied to $Q$, the type of $P$ must have the form $\alpha \to \beta$ and the type of $Q$ must have the form $\alpha$ for some $\alpha$ and $\beta$. We can then detach the conclusion, $\beta$, via the modus ponens rule.

**The composition combinator seen as a proof of $(\beta \to \alpha) \to (\gamma \to \beta) \to \gamma \to \alpha$ in Hilbert-style logic**

As a more complicated example, let's look at the theorem that corresponds to the $\mathbf{B}$ function. The type of $\mathbf{B}$ is $(\beta \to \alpha) \to (\gamma \to \beta) \to \gamma \to \alpha$. $\mathbf{B}$ is equivalent to $(\mathbf{S}\ (\mathbf{K}\ \mathbf{S})\ \mathbf{K})$. This is our roadmap for the proof of the theorem $(\beta \to \alpha) \to (\gamma \to \beta) \to \gamma \to \alpha$.

First we need to construct $(\mathbf{K}\ \mathbf{S})$. We make the antecedent of the $\mathbf{K}$ axiom look like the $\mathbf{S}$ axiom by setting $\alpha$ equal to $(\alpha \to \beta \to \gamma) \to (\alpha \to \beta) \to \alpha \to \gamma$, and $\beta$ equal to $\delta$ (to avoid variable collisions):

$\mathbf{K} : \alpha \to \beta \to \alpha$

$\mathbf{K}[\alpha = (\alpha \to \beta \to \gamma) \to (\alpha \to \beta) \to \alpha \to \gamma, \beta=\delta] : ((\alpha \to \beta \to \gamma) \to (\alpha \to \beta) \to \alpha \to \gamma) \to \delta \to (\alpha \to \beta \to \gamma) \to (\alpha \to \beta) \to \alpha \to \gamma$

Since the antecedent here is just $\mathbf{S}$, we can detach the consequent using Modus Ponens:

$\mathbf{K}\ \mathbf{S} : \delta \to (\alpha \to \beta \to \gamma) \to (\alpha \to \beta) \to \alpha \to \gamma$

This is the theorem that corresponds to the type of $(\mathbf{K}\ \mathbf{S})$. We now apply $\mathbf{S}$ to this expression. Taking $\mathbf{S}$

$\mathbf{S} : (\alpha \to \beta \to \gamma) \to (\alpha \to \beta) \to \alpha \to \gamma$

we put $\alpha = \delta$, $\beta = \alpha \to \beta \to \gamma$, and $\gamma = (\alpha \to \beta) \to \alpha \to \gamma$, yielding

$\mathbf{S}[\alpha = \delta, \beta = \alpha \to \beta \to \gamma, \gamma = (\alpha \to \beta) \to \alpha \to \gamma] : (\delta \to (\alpha \to \beta \to \gamma) \to (\alpha \to \beta) \to \alpha \to \gamma) \to (\delta \to (\alpha \to \beta \to \gamma)) \to \delta \to (\alpha \to \beta) \to \alpha \to \gamma$

and we then detach the consequent:

$$\mathbf{S}\,(\mathbf{K}\,\mathbf{S}) : (\delta \to \alpha \to \beta \to \gamma) \to \delta \to (\alpha \to \beta) \to \alpha \to \gamma$$

This is the formula for the type of $(\mathbf{S}\,(\mathbf{K}\,\mathbf{S}))$. A special case of this theorem has $\delta = (\beta \to \gamma)$:

$$\mathbf{S}\,(\mathbf{K}\,\mathbf{S})[\delta = \beta \to \gamma] : ((\beta \to \gamma) \to \alpha \to \beta \to \gamma) \to (\beta \to \gamma) \to (\alpha \to \beta) \to \alpha \to \gamma$$

We need to apply this last formula to $\mathbf{K}$. Again, we specialize $\mathbf{K}$, this time by replacing $\alpha$ with $(\beta \to \gamma)$ and $\beta$ with $\alpha$:

$$\mathbf{K} : \alpha \to \beta \to \alpha$$
$$\mathbf{K}[\alpha = \beta \to \gamma, \beta = \alpha] : (\beta \to \gamma) \to \alpha \to (\beta \to \gamma)$$

This is the same as the antecedent of the prior formula, so we detach the consequent:

$$\mathbf{S}\,(\mathbf{K}\,\mathbf{S})\,\mathbf{K} : (\beta \to \gamma) \to (\alpha \to \beta) \to \alpha \to \gamma$$

Switching the names of the variables $\alpha$ and $\gamma$ gives us

$$(\beta \to \alpha) \to (\gamma \to \beta) \to \gamma \to \alpha$$

which was what we had to prove.

**The normal proof of $(\beta \to \alpha) \to (\gamma \to \beta) \to \gamma \to \alpha$ in natural deduction seen as a $\lambda$-term**

We give below a proof of $(\beta \to \alpha) \to (\gamma \to \beta) \to \gamma \to \alpha$ in natural deduction and show how it can be interpreted as the $\lambda$-expression $\lambda\,a.\,\lambda b.\,\lambda\,g.(a\,(b\,g))$ of type $(\beta \to \alpha) \to (\gamma \to \beta) \to \gamma \to \alpha$.

$$
\cfrac{
  \cfrac{
    \cfrac{
      a{:}\beta \to \alpha, b{:}\gamma \to \beta, g{:}\gamma \vdash b : \gamma \to \beta \quad a{:}\beta \to \alpha, b{:}\gamma \to \beta, g{:}\gamma \vdash g : \gamma
    }{
      a{:}\beta \to \alpha, b{:}\gamma \to \beta, g{:}\gamma \vdash b\,g : \beta
    } \quad
    \cfrac{
      \cfrac{
        a{:}\beta \to \alpha, b{:}\gamma \to \beta, g{:}\gamma \vdash a : \beta \to \alpha
      }{
        a{:}\beta \to \alpha, b{:}\gamma
      }
    }{
      a{:}\beta \to \alpha, b{:}\gamma
    }
  }{
    \vdash a\,(b\,g) : \alpha
  }
}{}
$$

$$
\cfrac{
\cfrac{
\cfrac{
\to \beta \vdash \lambda\,g.\,a\,(b\,g) : \gamma \to \alpha
}{
a{:}\beta \to \alpha \vdash \lambda\,b.\,\lambda\,g.\,a\,(b\,g) : (\gamma \to \beta) \to \gamma \to \alpha
}
}{
\vdash \lambda\,a.\,\lambda\,b.\,\lambda\,g.\,a\,(b\,g) : (\beta \to \alpha) \to (\gamma \to \beta) \to \gamma \to \alpha
}
}{}
$$

## 14.6  Other applications

Recently, the isomorphism has been proposed as a way to define search space partition in Genetic programming.[9] The method indexes sets of genotypes (the program trees evolved by the GP system) by their Curry–Howard isomorphic proof (referred to as a species).

## 14.7  Generalizations

The correspondences listed here go much farther and deeper. For example, cartesian closed categories are generalized by closed monoidal categories. The internal language of these categories is the linear type system (corresponding to linear logic), which generalizes simply-typed lambda calculus as the internal language of cartesian closed categories. What's more, these can be shown to correspond to cobordisms,[10] which play a vital role in string theory.

An extended set of equivalences is also explored in homotopy type theory, which is a very active area of research at this time (2013). Here, type theory is extended by the univalence axiom, ('equivalence is equivalent to equality') which permits homotopy type theory to be used as a foundation for all of mathematics (including set theory and classical logic, providing new ways to discuss the axiom of choice and many other things). That is, the Curry–Howard correspondence that proofs are elements of inhabited types is generalized to the notion homotopic equivalence of proofs (as paths in space, the identity type or equality type of type theory being interpreted as a path).[11]

# 14.8 References

[1] The correspondence was first made explicit in Howard 1980. See, for example section 4.6, p.53 Gert Smolka and Jan Schwinghammer (2007-8), Lecture Notes in Semantics

[2] The Brouwer–Heyting–Kolmogorov interpretation is also called the 'proof interpretation': p. 161 of Juliette Kennedy, Roman Kossak, eds. 2011. *Set Theory, Arithmetic, and Foundations of Mathematics: Theorems, Philosophies* ISBN 978-1-107-00804-5

[3] Curry 1934.

[4] Curry & Feys 1958.

[5] Howard 1980.

[6] Moggi, Eugenio (1991), "Notions of Computation and Monads" (PDF), *Information and Computation* **93** (1): 55–92, doi:10.1016/0890-5401(91)90052-4

[7] Sørenson, Morten; Urzyczyn, Paweł, *Lectures on the Curry-Howard Isomorphism*, CiteSeerX: 10.1.1.17.7385

[8] Goldblatt, "7.6 Grothendieck Topology as Intuitionistic Modality", *Mathematical Modal Logic: A Model of its Evolution* (PDF), pp. 76–81. The "lax" modality referred to is from Benton; Bierman; de Paiva (1998), "Computational types from a logical perspective", *Journal of Functional Programming* **8**: 177–193, doi:10.1017/s0956796898002998

[9] F. Binard and A. Felty, "Genetic programming with polymorphic types and higher-order functions." In *Proceedings of the 10th annual conference on Genetic and evolutionary computation,* pages 1187 1194, 2008.

[10] John c. Baez and Mike Stay, "Physics, Topology, Logic and Computation: A Rosetta Stone", (2009) ArXiv 0903.0340 in *New Structures for Physics*, ed. Bob Coecke, *Lecture Notes in Physics* vol. **813**, Springer, Berlin, 2011, pp. 95-174.

[11] *Homotopy Type Theory: Univalent Foundations of Mathematics.* (2013) The Univalent Foundations Program. Institute for Advanced Study.

## 14.8.1 Seminal references

- Curry, Haskell (1934), "Functionality in Combinatory Logic", *Proceedings of the National Academy of Sciences* **20**, pp. 584–590.

- Curry, Haskell B.; Feys, Robert (1958), Craig, William, ed., *Combinatory Logic Vol. I*, Amsterdam: North-Holland, with 2 sections by William Craig, see paragraph 9E.

- De Bruijn, Nicolaas (1968), *Automath, a language for mathematics*, Department of Mathematics, Eindhoven University of Technology, TH-report 68-WSK-05. Reprinted in revised form, with two pages commentary, in: *Automation and Reasoning, vol 2, Classical papers on computational logic 1967–1970*, Springer Verlag, 1983, pp. 159–200.

- Howard, William A. (September 1980) [original paper manuscript from 1969], "The formulae-as-types notion of construction", in Seldin, Jonathan P.; Hindley, J. Roger, *To H.B. Curry: Essays on Combinatory Logic, Lambda Calculus and Formalism*, Boston, MA: Academic Press, pp. 479–490, ISBN 978-0-12-349050-6.

## 14.8.2    Extensions of the correspondence

[1]  Pfenning, Frank; Davies, Rowan (2001), "A Judgmental Reconstruction of Modal Logic" (PDF), *Mathematical Structures in Computer Science* **11**: 511–540, doi:10.1017/S0960129501003322

[2]  Davies, Rowan; Pfenning, Frank (2001), "A Modal Analysis of Staged Computation" (PDF), *Journal of the ACM* **48** (3): 555–604, doi:10.1145/382780.382785

- Griffin, Timothy G. (1990), "The Formulae-as-Types Notion of Control", *Conf. Record 17th Annual ACM Symp. on Principles of Programming Languages, POPL '90, San Francisco, CA, USA, 17–19 Jan 1990*, pp. 47–57.

- Parigot, Michel (1992), "Lambda-mu-calculus: An algorithmic interpretation of classical natural deduction", *InternationalConference on Logic Programming and Automated Reasoning: LPAR '92 Proceedings, St. Petersburg, Russia*, Lec-ture Notes in Computer Science**624**, Berlin; New York: Springer-Verlag, pp. 190–201,ISBN978-3-540 -55727-2.

- Herbelin, Hugo (1995), "A Lambda-Calculus Structure Isomorphic to Gentzen-Style Sequent Calculus Structure", in Pacholski, Leszek; Tiuryn, Jerzy, *Computer Science Logic, 8th International Workshop, CSL '94, Kazimierz, Poland, September 25–30, 1994, Selected Papers*, Lecture Notes in Computer Science **933**, Berlin; New York: Springer-Verlag, pp. 61–75, ISBN 978-3-540-60017-6.

- Gabbay, Dov; de Queiroz, Ruy (1992), "Extending the Curry–Howard interpretation to linear, relevant and other resource logics", *Journal of Symbolic Logic* **57**, pp. 1319–1365. (Full version of the paper presented at *Logic Colloquium '90*, Helsinki. Abstract in *JSL* 56(3):1139–1140, 1991.)

- de Queiroz, Ruy; Gabbay, Dov (1994), "Equality in Labelled Deductive Systems and the Functional Interpretation of Propositional Equality", in Dekker, Paul; Stokhof, Martin, *Proceedings of the Ninth Amsterdam Colloquium*, ILLC/Department of Philosophy, University of Amsterdam, pp. 547–565, ISBN 90-74795-07-2.

- de Queiroz, Ruy; Gabbay, Dov (1995), "The Functional Interpretation of the Existential Quantifier", *Bulletin of the Interest Group in Pure and Applied Logics*, 3(2–3), pp. 243–290. (Full version of a paper presented at *Logic Colloquium '91*, Uppsala. Abstract in *JSL* 58(2):753–754, 1993.)

- de Queiroz, Ruy; Gabbay, Dov (1997), "The Functional Interpretation of Modal Necessity", in de Rijke, Maarten, *Advances in Intensional Logic*, Applied Logic Series **7**, Springer-Verlag, pp. 61–91, ISBN 978-0-7923-4711-8.

- de Queiroz, Ruy; Gabbay, Dov (1999), "Labelled Natural Deduction", in Ohlbach, Hans-Juergen; Reyle, Uwe, *Logic, Language and Reasoning. Essays in Honor of Dov Gabbay*, Trends in Logic **7**, Kluwer Acad. Pub., pp. 173–250, ISBN 978-0-7923-5687-5.

- de Oliveira, Anjolina; de Queiroz, Ruy (1999), "A Normalization Procedure for the Equational Fragment of La- belled Natural Deduction", *Logic Journal of the Interest Group in Pure and Applied Logics* **7** (2), Oxford University Press, pp. 173–215. (Full version of a paper presented at *2nd WoLLIC'95*, Recife. Abstract in *Journal of the Interest Group in Pure and Applied Logics* 4(2):330–332, 1996.)

- Poernomo, Iman; Crossley, John; Wirsing; Martin (2005) [2005], *Adapting Proofs-as-Programs: The Curry– Howard Protocol*, Monographs in Computer Science, Springer, ISBN 978-0-387-23759-6, concerns the adaptation of proofs-as-programs program synthesis to coarse-grain and imperative program development problems, via a method the authors call the Curry–Howard protocol. Includes a discussion of the Curry–Howard correspondence from a Computer Science perspective.

- de Queiroz, Ruy J.G.B.; de Oliveira, Anjolina (2011), "The Functional Interpretation of Direct Computations", *Electronic Notes in Theoretical Computer Science* **269**, Elsevier, pp. 19–40, doi:10.1016/j.entcs.2011.03.003. (Full version of a paper presented at *LSFA 2010*, Natal, Brazil.)

### 14.8.3 Philosophical interpretations

- de Queiroz, Ruy J.G.B. (1994), "Normalisation and language-games", *Dialectica* **48** (2), pp. 83–123. (Early version presented at *Logic Colloquium '88*, Padova. Abstract in *JSL* 55:425, 1990.)

- de Queiroz, Ruy J.G.B. (2001), "Meaning, function, purpose, usefulness, consequences – interconnected concepts", *Logic Journal of the Interest Group in Pure and Applied Logics* **9** (5), pp. 693–734. (Early version presented at *Fourteenth International Wittgenstein Symposium (Centenary Celebration)* held in Kirchberg/Wechsel, August 13– 20, 1989.)

- de Queiroz, Ruy J.G.B. (2008), "On reduction rules, meaning-as-use and proof-theoretic semantics", *Studia Logica* **90** (2), pp. 211–247.

### 14.8.4 Synthetic papers

- De Bruijn, Nicolaas Govert (1995), "On the roles of types in mathematics" (PDF), in Groote, Philippe de, *The Curry–Howard isomorphism*, Cahiers du Centre de logique (Université catholique de Louvain) **8**, Academia-Bruyland, pp. 27–54, ISBN 978-2-87209-363-2, the contribution of de Bruijn by himself.

- Geuvers, Herman (1995), "The Calculus of Constructions and Higher Order Logic", in Groote, Philippe de, *The Curry–Howard isomorphism*, Cahiers du Centre de logique (Université catholique de Louvain) **8**, Academia-Bruyland, pp. 139–191, ISBN 978-2-87209-363-2, contains a synthetic introduction to the Curry–Howard correspondence.

- Gallier, Jean H. (1995), "On the Correspondence between Proofs and Lambda-Terms" (PDF), in Groote, Philippe de, *The Curry–Howard isomorphism*, Cahiers du Centre de logique (Université catholique de Louvain) **8**, Academia-Bruyland, pp. 55–138, ISBN 978-2-87209-363-2, contains a synthetic introduction to the Curry–Howard correspondence.

- Wadler, Philip (2014), "Propositions as Types" (PDF), *Communications of the ACM*

### 14.8.5 Books

- edited by Ph. de Groote. (1995), De Groote, Philippe, ed., *The Curry–Howard Isomorphism*, Cahiers du Centre de Logique (Université catholique de Louvain) **8**, Academia-Bruylant, ISBN 978-2-87209-363-2, reproduces the seminal papers of Curry-Feys and Howard, a paper by de Bruijn and a few other papers.

- Sørensen, Morten Heine; Urzyczyn, Paweł (2006) [1998], *Lectures on the Curry–Howard isomorphism*, Studies in Logic and the Foundations of Mathematics **149**, Elsevier Science, ISBN 978-0-444-52077-7, CiteSeerX: 10.1.1.17.7385, notes on proof theory and type theory, that includes a presentation of the Curry–Howard correspondence, with a focus on the formulae-as-types correspondence

- Girard, Jean-Yves (1987–90). *Proof and Types*. Translated by and with appendices by Lafont, Yves and Taylor, Paul. Cambridge University Press (Cambridge Tracts in Theoretical Computer Science, 7), ISBN 0-521-37181-3, notes on proof theory with a presentation of the Curry–Howard correspondence.

- Thompson, Simon (1991). *Type Theory and Functional Programming* Addison–Wesley. ISBN 0-201-41667-0.

- Poernomo, Iman; Crossley, John; Wirsing; Martin (2005) [2005], *Adapting Proofs-as-Programs: The Curry–Howard Protocol*, Monographs in Computer Science, Springer, ISBN 978-0-387-23759-6, concerns the adaptation of proofs-as-programs program synthesis to coarse-grain and imperative program development problems, via a method the authors call the Curry–Howard protocol. Includes a discussion of the Curry–Howard correspondence from a Computer Science perspective.

- F. Binard and A. Felty, "Genetic programming with polymorphic types and higher-order functions." In *Proceedings of the 10th annual conference on Genetic and evolutionary computation*, pages 1187 1194, 2008.

- de Queiroz, Ruy J.G.B.; de Oliveira, Anjolina G.; Gabbay, Dov M. (2011), *The Functional Interpretation of Logical Deduction*, Advances in Logic **5**, Imperial College Press/World Scientific, ISBN 978-981-4360-95-1.

## 14.9   Further reading

- P.T. Johnstone, 2002, *Sketches of an Elephant*, section D4.2 (vol 2) gives a categorical view of "what happens" in the Curry–Howard correspondence.

## 14.10   External links

- Howard on Curry-Howard

- The Curry–Howard Correspondence in Haskell

- The Monad Reader 6: Adventures in Classical-Land: Curry–Howard in Haskell, Pierce's law.

# Chapter 15

# Cut-elimination theorem

The **cut-elimination theorem** (or **Gentzen's** *Hauptsatz*) is the central result establishing the significance of the sequent calculus. It was originally proved by Gerhard Gentzen 1934 in his landmark paper "Investigations in Logical Deduction" for the systems LJ and LK formalising intuitionistic and classical logic respectively. The cut-elimination theorem states that any judgement that possesses a proof in the sequent calculus that makes use of the **cut rule** also possesses a **cut-free proof**, that is, a proof that does not make use of the cut rule.[1][2]

A sequent is a logical expression relating multiple sentences, in the form " $A_1, A_2, A_3, \ldots \vdash B_1, B_2, B_3, \ldots$ ", which is to be read as " $A_1, A_2, A_3, \ldots$ proves $B_1, B_2, B_3, \ldots$ ", and (as glossed by Gentzen) should be understood as equivalent to the truth-function "If ( $A_1$ and $A_2$ and $A_3$ ...) then ( $B_1$ or $B_2$ or $B_3$ ...)."[3] Note that the left-hand side (LHS) is a conjunction (and) and the RHS is a disjunction (or).

The LHS may have arbitrarily many or few formulae; when the LHS is empty, the RHS is a tautology. In LK, the RHS may also have any number of formulae—if it has none, the LHS is a contradiction, whereas in LJ the RHS may only have one formula or none: here we see that allowing more than one formula in the RHS is equivalent, in the presence of the right contraction rule, to the admissibility of the law of the excluded middle. However, the sequent calculus is a fairly expressive framework, and there have been sequent calculi for intuitionistic logic proposed that allow many formulae in the RHS. From Jean-Yves Girard's logic LC it is easy to obtain a rather natural formalisation of classical logic where the RHS contains at most one formula; it is the interplay of the logical and structural rules that is the key here.

"Cut" is a rule in the normal statement of the sequent calculus, and equivalent to a variety of rules in other proof theories, which, given

1. $\Gamma \vdash A, \Delta$

and

1. $\Pi, A \vdash \Lambda$

allows one to infer

1. $\Gamma, \Pi \vdash \Delta, \Lambda$

That is, it "cuts" the occurrences of the formula $A$ out of the inferential relation. The cut-elimination theorem states that (for a given system) any sequent provable using the rule Cut can be proved without use of this rule.

For sequent calculi that have only one formula in the RHS, the "Cut" rule reads, given

1. $\Gamma \vdash A$

and

1. $\Pi, A \vdash B$

allows one to infer

1. $\Gamma, \Pi \vdash B$

If we think of $B$ as a theorem, then cut-elimination in this case simply says that a lemma $A$ used to prove this theorem can be inlined. Whenever the theorem's proof mentions lemma $A$, we can substitute the occurrences for the proof of $A$. Consequently, the cut rule is admissible.

For systems formulated in the sequent calculus, analytic proofs are those proofs that do not use Cut. Typically such a proof will be longer, of course, and not necessarily trivially so. In his essay "Don't Eliminate Cut!" George Boolos demonstrated that there was a derivation that could be completed in a page using cut, but whose analytic proof could not be completed in the lifespan of the universe.

The theorem has many, rich consequences:

- A system is inconsistent if it admits a proof of the absurd. If the system has a cut elimination theorem, then if it has a proof of the absurd, or of the empty sequent, it should also have a proof of the absurd (or the empty sequent), without cuts. It is typically very easy to check that there are no such proofs. Thus, once a system is shown to have a cut elimination theorem, it is normally immediate that the system is consistent.

- Normally also the system has, at least in first order logic, the subformula property, an important property in several approaches to proof-theoretic semantics.

Cut elimination is one of the most powerful tools for proving interpolation theorems. The possibility of carrying out proof search based on resolution, the essential insight leading to the Prolog programming language, depends upon the admissibility of Cut in the appropriate system.

For proof systems based on higher-order typed lambda calculus through a Curry–Howard isomorphism, cut elimination algorithms correspond to the strong normalization property (every proof term reduces in a finite number of steps into a normal form).

## 15.1   See also

- Gentzen's consistency proof for Peano's axioms

## 15.2   Notes

[1] Curry 1977, pp. 208–213, gives a 5-page proof of the elimination theorem. See also pages 188, 250.

[2] Kleene 2009, pp. 453, gives a very brief proof of the cut-elimination theorem.

[3] Wilfried Buchholz, Beweistheorie (university lecture notes about cut-elimination, German, 2002-2003)

## 15.3   References

- Gentzen, Gerhard (1934–1935). "Untersuchungen über das logische Schließen". *Mathematische Zeitschrift* **39**: 405–431. doi:10.1007/BF01201363.

- Gentzen, Gerhard (1964). "Investigations into logical deduction". *American Philosophical Quarterly* **1** (4): 249–287.

- Gentzen, Gerhard (1965). "Investigations into logical deduction". *American Philosophical Quarterly* **2** (3): 204–218.

- Untersuchungen über das logische Schließen I

- Untersuchungen über das logische Schließen II

- Curry, Haskell Brooks (1977) [1963]. *Foundations of mathematical logic*. New York: Dover Publications Inc. ISBN 978-0-486-63462-3.

- Kleene, Stephen Cole (2009) [1952]. *Introduction to metamathematics*. Ishi Press International. ISBN 978-0-923891-57-2.

## 15.4 External links

- Alex Sakharov, "Cut Elimination Theorem", *MathWorld*.

# Chapter 16

# Decidability (logic)

In logic, the term **decidable** refers to the decision problem, the question of the existence of an effective method for determining membership in a set of formulas, or, more precisely, an algorithm that can and will return a Boolean true or false value (instead of looping indefinitely). Logical systems such as propositional logic are decidable if membership in their set of logically valid formulas (or theorems) can be effectively determined. A theory (set of sentences closed under logical consequence) in a fixed logical system is decidable if there is an effective method for determining whether arbitrary formulas are included in the theory. Many important problems are undecidable, that is, it has been proven that no effective method can exist for them.

## 16.1    Relationship to computability

As with the concept of a decidable set, the definition of a decidable theory or logical system can be given either in terms of *effective methods* or in terms of *computable functions*. These are generally considered equivalent per Church's thesis. Indeed, the proof that a logical system or theory is undecidable will use the formal definition of computability to show that an appropriate set is not a decidable set, and then invoke Church's thesis to show that the theory or logical system is not decidable by any effective method (Enderton 2001, pp. 206*ff*.).

## 16.2    Decidability of a logical system

Each logical system comes with both a syntactic component, which among other things determines the notion of provability, and a semantic component, which determines the notion of logical validity. The logically valid formulas of a system are sometimes called the **theorems** of the system, especially in the context of first-order logic where Gödel's completeness theorem establishes the equivalence of semantic and syntactic consequence. In other settings, such as linear logic, the syntactic consequence (provability) relation may be used to define the theorems of a system.

A logical system is decidable if there is an effective method for determining whether arbitrary formulas are theorems of the logical system. For example, propositional logic is decidable, because the truth-table method can be used to determine whether an arbitrary propositional formula is logically valid.

First-order logic is not decidable in general; in particular, the set of logical validities in any signature that includes equality and at least one other predicate with two or more arguments is not decidable.[1] Logical systems extending first-order logic, such as second-order logic and type theory, are also undecidable.

The validities of monadic predicate calculus with identity are decidable, however. This system is first-order logic restricted to signatures that have no function symbols and whose relation symbols other than equality never take more than one argument.

Some logical systems are not adequately represented by the set of theorems alone. (For example, Kleene's logic has no theorems at all.) In such cases, alternative definitions of decidability of a logical system are often used, which ask for an effective method for determining something more general than just validity of formulas; for instance, validity of sequents, or the consequence relation $\{(\Gamma, A) \mid \Gamma \vDash A\}$ of the logic.

## 16.3 Decidability of a theory

A theory is a set of formulas, which here is assumed to be closed under logical consequence. The question of decidability for a theory is whether there is an effective procedure that, given an arbitrary formula in the signature of the theory, decides whether the formula is a member of the theory or not. This problem arises naturally when a theory is defined as the set of logical consequences of a fixed set of axioms. Examples of decidable first-order theories include the theory of real closed fields, and Presburger arithmetic, while the theory of groups and Robinson arithmetic are examples of undecidable theories.

There are several basic results about decidability of theories. Every inconsistent theory is decidable, as every formula in the signature of the theory will be a logical consequence of, and thus member of, the theory. Every complete recursively enumerable first-order theory is decidable. An extension of a decidable theory may not be decidable. For example, there are undecidable theories in propositional logic, although the set of validities (the smallest theory) is decidable.

A consistent theory that has the property that every consistent extension is undecidable is said to be **essentially undecidable**. In fact, every consistent extension will be essentially undecidable. The theory of fields is undecidable but not essentially undecidable. Robinson arithmetic is known to be essentially undecidable, and thus every consistent theory that includes or interprets Robinson arithmetic is also (essentially) undecidable.

## 16.4 Some decidable theories

Some decidable theories include (Monk 1976, p. 234):[2]

- The set of first-order logical validities in the signature with only equality, established by Leopold Löwenheim in 1915.

- The set of first-order logical validities in a signature with equality and one unary function, established by Ehrenfeucht in 1959.

- The first-order theory of the integers in the signature with equality and addition, also called Presburger arithmetic. The completeness was established by Mojżesz Presburger in 1929.

- The first-order theory of Boolean algebras, established by Alfred Tarski in 1949.

- The first-order theory of algebraically closed fields of a given characteristic, established by Tarski in 1949.

- The first-order theory of real-closed ordered fields, established by Tarski in 1949 (see also Tarski's exponential function problem).

- The first-order theory of Euclidean geometry, established by Tarski in 1949.

- The first-order theory of hyperbolic geometry, established by Schwabhäuser in 1959.

- Specific decidable sublanguages of set theory investigated in the 1980s through today.(Cantone *et al.*, 2001)

Methods used to establish decidability include quantifier elimination, model completeness, and Vaught's test.

## 16.5   Some undecidable theories

Some undecidable theories include (Monk 1976, p. 279):[2]

- The set of logical validities in any first-order signature with equality and either: a relation symbol of arity no less than 2, or two unary function symbols, or one function symbol of arity no less than 2, established by Trakhtenbrot in 1953.

- The first-order theory of the natural numbers with addition, multiplication, and equality, established by Tarski and Andrzej Mostowski in 1949.

- The first-order theory of the rational numbers with addition, multiplication, and equality, established by Julia Robinson in 1949.

- The first-order theory of groups, established by Alfred Tarski in 1953.[3] Remarkably, not only the general theory of groups is undecidable, but also several more specific theories, for example (as established by Mal'cev 1961) the theory of finite groups. Mal'cev also established that the theory of semigroups and the theory of rings are undecidable. Robinson established in 1949 that the theory of fields is undecidable.

- Robinson arithmetic (and therefore any consistent extension, such as Peano arithmetic) is essentially undecidable, as established by Raphael Robinson in 1950.

- The first-order theory with equality and two function symbols[4]

The interpretability method is often used to establish undecidability of theories. If an essentially undecidable theory $T$ is interpretable in a consistent theory $S$, then $S$ is also essentially undecidable. This is closely related to the concept of a many-one reduction in computability theory.

## 16.6   Semidecidability

A property of a theory or logical system weaker than decidability is **semidecidability**. A theory is semidecidable if there is an effective method which, given an arbitrary formula, will always tell correctly when the formula is in the theory, but may give either a negative answer or no answer at all when the formula is not in the theory. A logical system is semidecidable if there is an effective method for generating theorems (and only theorems) such that every theorem will eventually be generated. This is different from decidability because in a semidecidable system there may be no effective procedure for checking that a formula is *not* a theorem.

Every decidable theory or logical system is semidecidable, but in general the converse is not true; a theory is decidable if and only if both it and its complement are semi-decidable. For example, the set of logical validities $V$ of first-order logic is semi-decidable, but not decidable. In this case, it is because there is no effective method for determining for an arbitrary formula $A$ whether $A$ is not in $V$. Similarly, the set of logical consequences of any recursively enumerable set of first-order axioms is semidecidable. Many of the examples of undecidable first-order theories given above are of this form.

## 16.7   Relationship with completeness

Decidability should not be confused with completeness. For example, the theory of algebraically closed fields is decidable but incomplete, whereas the set of all true first-order statements about nonnegative integers in the language with $+$ and $\times$ is complete but undecidable. Unfortunately, as a terminological ambiguity, the term "undecidable statement" is sometimes used as a synonym for independent statement.

## 16.8   See also

- László Kalmár (1936)

- Alonzo Church (1956)

- W.V.O. Quine (1953)

- Meyer and Lambert (1967)

## 16.9   References

### 16.9.1   Notes

[1] Trakhtenbrot, 1953

[2] Donald Monk (1976). *Mathematical Logic*. Springer-Verlag. ISBN 9780387901701.

[3] Tarski, A.; Mostovski, A.; Robinson, R. (1953), *Undecidable Theories*, Studies in Logic and the Foundation of Mathematics, North-Holland, Amsterdam

[4] Gurevich, Yuri (1976). "The Decision Problem for Standard Classes". *J. Symb. Log.* **41** (2): 460—464. Retrieved 5 August 2014.

### 16.9.2   Bibliography

- Barwise, Jon (1982), "Introduction to first-order logic", in Barwise, Jon, *Handbook of Mathematical Logic*, Studies in Logic and the Foundations of Mathematics, Amsterdam: North-Holland, ISBN 978-0-444-86388-1

- Cantone, D., E. G. Omodeo and A. Policriti, "Set Theory for Computing. From Decision Procedures to Logic Programming with Sets," Monographs in Computer Science, Springer, 2001.

- Chagrov, Alexander; Zakharyaschev, Michael (1997), *Modal logic*, Oxford Logic Guides **35**, The Clarendon Press Oxford University Press, ISBN 978-0-19-853779-3, MR 1464942

- Davis, Martin (1958), *Computability and Unsolvability*, McGraw-Hill Book Company, Inc, New York

- Enderton, Herbert (2001), *A mathematical introduction to logic* (2nd ed.), Boston, MA: Academic Press, ISBN 978-0-12-238452-3

- Keisler, H. J. (1982), "Fundamentals of model theory", in Barwise, Jon, *Handbook of Mathematical Logic*, Studies in Logic and the Foundations of Mathematics, Amsterdam: North-Holland, ISBN 978-0-444-86388-1

- Monk, J. Donald (1976), *Mathematical Logic*, Berlin, New York: Springer-Verlag

# Chapter 17

# Decidable sublanguages of set theory

In mathematical logic, various sublanguages of set theory are decidable.[1][2] These include:

- Sets with Monotone, Additive, and Multiplicative Functions.[3]
- Sets with restricted quantifiers.[4]

## 17.1   References

[1] Cantone, D., E. G. Omodeo and A. Policriti, "Set Theory for Computing. From Decision Procedures to Logic Programming with Sets," Monographs in Computer Science, Springer, 2001.

[2] "Decision procedures for elementary sublanguages of set theory: XIII. Model graphs, reflection and decidability", by Franco Parlamento and Alberto Policriti Journal of Automated Reasoning, Volume 7 , Issue 2 (June 1991), Pages: 271 - 284

[3] "A Decision Procedure for a Sublanguage of Set Theory Involving Monotone, Additive, and Multiplicative Functions", by Domenico Cantone and et al.

[4] "A tableau-based decision procedure for a fragment of set theory involving a restricted form of quantification", by Domenico Cantone, Calogero G. Zarba, Viale A. Doria, 1997

# Chapter 18

# Deduction theorem

In mathematical logic, the **deduction theorem** is a metatheorem of first-order logic.[1] It is a formalization of the common proof technique in which an implication $A \to B$ is proved by assuming $A$ and then deriving $B$ from this assumption conjoined with known results. The deduction theorem explains why proofs of conditional sentences in mathematics are logically correct. Though it has seemed "obvious" to mathematicians literally for centuries that proving B from A conjoined with a set of theorems is sufficient to proving the implication $A \to B$ based on those theorems alone, it was left to Herbrand and Tarski to show (independently) this was logically correct in the general case—another instance, perhaps, of modern logic "cleaning up" mathematical practice.

The deduction theorem states that if a formula $B$ is deducible from a set of assumptions $\Delta \cup \{A\}$, where $A$ is a closed formula, then the implication $A \to B$ is deducible from $\Delta$. In symbols, $\Delta \cup \{A\} \vdash B$ implies $\Delta \vdash A \to B$.. In the special case where $\Delta$ is the empty set, the deduction theorem shows that $\{A\} \vdash B$ implies $\vdash A \to B$.

The deduction theorem holds for all first-order theories with the usual deductive systems for first-order logic. However, there are first-order systems in which new inference rules are added for which the deduction theorem fails.[2]

The deduction rule is an important property of Hilbert-style systems because the use of this metatheorem leads to much shorter proofs than would be possible without it. Although the deduction theorem could be taken as primitive rule of inference in such systems, this approach is not generally followed; instead, the deduction theorem is obtained as an admissible rule using the other logical axioms and modus ponens. In other formal proof systems, the deduction theorem is sometimes taken as a primitive rule of inference. For example, in natural deduction, the deduction theorem is recast as an introduction rule for "→".

## 18.1 Examples of deduction

"Prove" axiom 1:

- - *P* 1. hypothesis
    - *Q* 2. hypothesis
    - *P* 3. reiteration of 1
  - *Q*→*P* 4. deduction from 2 to 3
- *P*→(*Q*→*P*) 5. deduction from 1 to 4 QED

"Prove" axiom 2:

- - *P*→(*Q*→*R*) 1. hypothesis
    - *P*→*Q* 2. hypothesis

- $P$ 3. hypothesis
- $Q$ 4. modus ponens 3,2
- $Q{\rightarrow}R$ 5. modus ponens 3,1
- $R$ 6. modus ponens 4,5
- $P{\rightarrow}R$ 7. deduction from 3 to 6
- $(P{\rightarrow}Q){\rightarrow}(P{\rightarrow}R)$ 8. deduction from 2 to 7
- $(P{\rightarrow}(Q{\rightarrow}R)){\rightarrow}((P{\rightarrow}Q){\rightarrow}(P{\rightarrow}R))$ 9. deduction from 1 to 8 QED

Using axiom 1 to show $((P{\rightarrow}(Q{\rightarrow}P)){\rightarrow}R){\rightarrow}R$:

- - $(P{\rightarrow}(Q{\rightarrow}P)){\rightarrow}R$ 1. hypothesis
  - $P{\rightarrow}(Q{\rightarrow}P)$ 2. axiom 1
  - $R$ 3. modus ponens 2,1
- $((P{\rightarrow}(Q{\rightarrow}P)){\rightarrow}R){\rightarrow}R$ 4. deduction from 1 to 3 QED

## 18.2  Virtual rules of inference

From the examples, you can see that we have added three virtual (or extra and temporary) rules of inference to our normal axiomatic logic. These are "hypothesis", "reiteration", and "deduction". The normal rules of inference (i.e. "modus ponens" and the various axioms) remain available.

1. **Hypothesis** is a step where one adds an additional premise to those already available. So, if your previous step $S$ was deduced as:

$$E_1, E_2, ..., E_{n-1}, E_n \vdash S,$$

then one adds another premise $H$ and gets:

$$E_1, E_2, ..., E_{n-1}, E_n, H \vdash H.$$

This is symbolized by moving from the n-th level of indentation to the n+1-th level and saying

- - - - - $S$ previous step
          - $H$ hypothesis

2. **Reiteration** is a step where one re-uses a previous step. In practice, this is only necessary when one wants to take a hypothesis which is not the most recent hypothesis and use it as the final step before a deduction step.

3. **Deduction** is a step where one removes the most recent hypothesis (still available) and prefixes it to the previous step. This is shown by unindenting one level as follows:

- - - - - - $H$ hypothesis
          - ......... (other steps)
          - $C$ (conclusion drawn from $H$)
        - $H{\rightarrow}C$ deduction

## 18.3 Conversion from proof using the deduction meta-theorem to axiomatic proof

In axiomatic versions of propositional logic, one usually has among the axiom schemas (where $P$, $Q$, and $R$ are replaced by any propositions):

- Axiom 1 is: $P \to (Q \to P)$

- Axiom 2 is: $(P \to (Q \to R)) \to ((P \to Q) \to (P \to R))$

- Modus ponens is: from $P$ and $P \to Q$ infer $Q$

These axiom schemas are chosen to enable one to derive the deduction theorem from them easily. So it might seem that we are begging the question. However, they can be justified by checking that they are tautologies using truth tables and that modus ponens preserves truth.

From these axiom schemas one can quickly deduce the theorem schema $P \to P$ (reflexivity of implication) which is used below:

1. $(P \to ((Q \to P) \to P)) \to ((P \to (Q \to P)) \to (P \to P))$ from axiom schema 2 with $P$, $(Q \to P)$, $P$

2. $P \to ((Q \to P) \to P)$ from axiom schema 1 with $P$, $(Q \to P)$

3. $(P \to (Q \to P)) \to (P \to P)$ from modus ponens applied to step 2 and step 1

4. $P \to (Q \to P)$ from axiom schema 1 with $P$, $Q$

5. $P \to P$ from modus ponens applied to step 4 and step 3

Suppose that we have that $\Gamma$ and $H$ prove $C$, and we wish to show that $\Gamma$ proves $H \to C$. For each step $S$ in the deduction which is a premise in $\Gamma$ (a reiteration step) or an axiom, we can apply modus ponens to the axiom 1, $S \to (H \to S)$, to get $H \to S$. If the step is $H$ itself (a hypothesis step), we apply the theorem schema to get $H \to H$. If the step is the result of applying modus ponens to $A$ and $A \to S$, we first make sure that these have been converted to $H \to A$ and $H \to (A \to S)$ and then we take the axiom 2, $(H \to (A \to S)) \to ((H \to A) \to (H \to S))$, and apply modus ponens to get $(H \to A) \to (H \to S)$ and then again to get $H \to S$. At the end of the proof we will have $H \to C$ as required, except that now it only depends on $\Gamma$, not on $H$. So the deduction step will disappear, consolidated into the previous step which was the conclusion derived from $H$.

To minimize the complexity of the resulting proof, some preprocessing should be done before the conversion. Any steps (other than the conclusion) which do not actually depend on $H$ should be moved up before the hypothesis step and unindented one level. And any other unnecessary steps (which are not used to get the conclusion or can be bypassed), such as reiterations which are not the conclusion, should be eliminated.

During the conversion, it may be useful to put all the applications of modus ponens to axiom 1 at the beginning of the deduction (right after the $H \to H$ step).

When converting a modus ponens, if $A$ is outside the scope of $H$, then it will be necessary to apply axiom 1, $A \to (H \to A)$, and modus ponens to get $H \to A$. Similarly, if $A \to S$ is outside the scope of $H$, apply axiom 1, $(A \to S) \to (H \to (A \to S))$, and modus ponens to get $H \to (A \to S)$. It should not be necessary to do both of these, unless the modus ponens step is the conclusion, because if both are outside the scope, then the modus ponens should have been moved up before $H$ and thus be outside the scope also.

Under the Curry–Howard correspondence, the above conversion process for the deduction meta-theorem is analogous to the conversion process from lambda calculus terms to terms of combinatory logic, where axiom 1 corresponds to the K combinator, and axiom 2 corresponds to the S combinator. Note that the I combinator corresponds to the theorem schema $P \to P$.

## 18.4   The deduction theorem in predicate logic

The deduction theorem is also valid in first-order logic in the following form:

- If $T$ is a theory and $F$, $G$ are formulas with $F$ closed, and $T \cup \{F\} \vdash G$, then $T \vdash F \rightarrow G$.

Here, the symbol $\vdash$ means "is a syntactical consequence of." We indicate below how the proof of this deduction theorem differs from that of the deduction theorem in propositional calculus.

In the most common versions of the notion of formal proof, there are, in addition to the axiom schemes of propositional calculus (or the understanding that all tautologies of propositional calculus are to be taken as axiom schemes in their own right), quantifier axioms, and in addition to modus ponens, one additional rule of inference, known as the rule of *generalization*: "From $K$, infer $\forall v K$."

In order to convert a proof of $G$ from $T \cup \{F\}$ to one of $F \rightarrow G$ from $T$, one deals with steps of the proof of $G$ which are axioms or result from application of modus ponens in the same way as for proofs in propositional logic. Steps which result from application of the rule of generalization are dealt with via the following quantifier axiom (valid whenever the variable $v$ is not free in formula $H$):

- $(H \rightarrow K) \rightarrow (H \rightarrow \forall v K)$.

Since in our case $F$ is assumed to be closed, we can take $H$ to be $F$. This axiom allows one to deduce $F \rightarrow \forall v K$ from $F \rightarrow K$, which is just what is needed whenever the rule of generalization is applied to some $K$ in the proof of $G$.

## 18.5   Example of conversion

To illustrate how one can convert a natural deduction to the axiomatic form of proof, we apply it to the tautology $Q \rightarrow ((Q \rightarrow R) \rightarrow R)$. In practice, it is usually enough to know that we could do this. We normally use the natural-deductive form in place of the much longer axiomatic proof.

First, we write a proof using a natural-deduction like method:

- - $Q$ 1. hypothesis
    - - $Q \rightarrow R$ 2. hypothesis
      - $R$ 3. modus ponens 1,2
    - $(Q \rightarrow R) \rightarrow R$ 4. deduction from 2 to 3
  - $Q \rightarrow ((Q \rightarrow R) \rightarrow R)$ 5. deduction from 1 to 4 QED

Second, we convert the inner deduction to an axiomatic proof:

- $(Q \rightarrow R) \rightarrow (Q \rightarrow R)$ 1. theorem schema $(A \rightarrow A)$

- $((Q \rightarrow R) \rightarrow (Q \rightarrow R)) \rightarrow (((Q \rightarrow R) \rightarrow Q) \rightarrow ((Q \rightarrow R) \rightarrow R))$ 2. axiom 2

- $((Q \rightarrow R) \rightarrow Q) \rightarrow ((Q \rightarrow R) \rightarrow R)$ 3. modus ponens 1,2

- $Q \rightarrow ((Q \rightarrow R) \rightarrow Q)$ 4. axiom 1

  - $Q$ 5. hypothesis
  - $(Q \rightarrow R) \rightarrow Q$ 6. modus ponens 5,4
  - $(Q \rightarrow R) \rightarrow R$ 7. modus ponens 6,3

- $Q{\rightarrow}((Q{\rightarrow}R){\rightarrow}R)$ 8. deduction from 5 to 7 QED

Third, we convert the outer deduction to an axiomatic proof:

- $(Q{\rightarrow}R){\rightarrow}(Q{\rightarrow}R)$ 1. theorem schema $(A{\rightarrow}A)$

- $((Q{\rightarrow}R){\rightarrow}(Q{\rightarrow}R)){\rightarrow}(((Q{\rightarrow}R){\rightarrow}Q){\rightarrow}((Q{\rightarrow}R){\rightarrow}R))$ 2. axiom 2

- $((Q{\rightarrow}R){\rightarrow}Q){\rightarrow}((Q{\rightarrow}R){\rightarrow}R)$ 3. modus ponens 1,2

- $Q{\rightarrow}((Q{\rightarrow}R){\rightarrow}Q)$ 4. axiom 1

- $[((Q{\rightarrow}R){\rightarrow}Q){\rightarrow}((Q{\rightarrow}R){\rightarrow}R)]{\rightarrow}[Q{\rightarrow}(((Q{\rightarrow}R){\rightarrow}Q){\rightarrow}((Q{\rightarrow}R){\rightarrow}R))]$ 5. axiom 1

- $Q{\rightarrow}(((Q{\rightarrow}R){\rightarrow}Q){\rightarrow}((Q{\rightarrow}R){\rightarrow}R))$ 6. modus ponens 3,5

- $[Q{\rightarrow}(((Q{\rightarrow}R){\rightarrow}Q){\rightarrow}((Q{\rightarrow}R){\rightarrow}R))]{\rightarrow}([Q{\rightarrow}((Q{\rightarrow}R){\rightarrow}Q)]{\rightarrow}[Q{\rightarrow}((Q{\rightarrow}R){\rightarrow}R))])$ 7. axiom 2

- $[Q{\rightarrow}((Q{\rightarrow}R){\rightarrow}Q)]{\rightarrow}[Q{\rightarrow}((Q{\rightarrow}R){\rightarrow}R))]$ 8. modus ponens 6,7

- $Q{\rightarrow}((Q{\rightarrow}R){\rightarrow}R))$ 9. modus ponens 4,8 QED

These three steps can be stated succinctly using the Curry–Howard correspondence:

- first, in lambda calculus, the function f = λa. λb. b a has type $q \rightarrow (q \rightarrow r) \rightarrow r$

- second, by lambda elimination on b, f = λa. s i (k a)

- third, by lambda elimination on a, f = s (k (s i)) k

## 18.6 Paraconsistent deduction theorem

In general, the classical deduction theorem doesn't hold in paraconsistent logic. However, the following "two-way deduction theorem" does hold in one form of paraconsistent logic:[3]

$$\vdash E \rightarrow F \text{ if and only if } ( E \vdash F \text{ and } \neg F \vdash \neg E )$$

that requires the contrapositive inference to hold in addition to the requirement of the classical deduction theorem.

## 18.7 The resolution theorem

The **resolution theorem** is the converse of the deduction theorem. It follows immediately from modus ponens which is the elimination rule for implication.

## 18.8 See also

- Conditional proof

- Propositional calculus

- Peirce's law

## 18.9   Notes

[1]  Kleene 1967, p. 39, 112; Shoenfield 1967, p. 33

[2]  Kohlenbach 2008, p. 148

[3]  Hewitt 2008

## 18.10   References

- Carl Hewitt (2008), "ORGs for Scalable, Robust, Privacy-Friendly Client Cloud Computing", *IEEE Internet Computing* **12** (5): 96, doi:10.1109/MIC.2008.107. September/October 2008

- Kohlenbach, Ulrich (2008), *Applied proof theory: proof interpretations and their use in mathematics*, Springer Monographs in Mathematics, Berlin, New York: Springer-Verlag, ISBN 978-3-540-77532-4, MR 2445721

- Kleene, Stephen Cole (2002) [1967], *Mathematical logic*, New York: Dover Publications, ISBN 978-0-486-42533-7, MR 1950307

- Rautenberg, Wolfgang (2010), *A Concise Introduction to Mathematical Logic* (3rd ed.), New York: Springer Science+Business Media, doi:10.1007/978-1-4419-1221-3, ISBN 978-1-4419-1220-6.

- Shoenfield, Joseph R. (2001) [1967], *Mathematical Logic* (2nd ed.), A K Peters, ISBN 978-1-56881-135-2

## 18.11   External links

- *Introduction to Mathematical Logic* by Vilnis Detlovs and Karlis Podnieks Podnieks is a comprehensive tutorial. See Section 1.5.

# Chapter 19

# Deep inference

**Deep inference** names a general idea in structural proof theory that breaks with the classical sequent calculus by generalising the notion of structure to permit inference to occur in contexts of high structural complexity. The term *deep inference* is generally reserved for proof calculi where the structural complexity is unbounded; in this article we will use **non-shallow inference** to refer to calculi that have structural complexity greater than the sequent calculus, but not unboundedly so, although this is not at present established terminology.

Deep inference is not important in logic outside of structural proof theory, since the phenomena that lead to the proposal of formal systems with deep inference are all related to the cut-elimination theorem. The first calculus of deep inference was proposed by Kurt Schütte,[1] but the idea did not generate much interest at the time.

Nuel Belnap proposed display logic in an attempt to characterise the essence of structural proof theory. The calculus of structures was proposed in order to give a cut-free characterisation of noncommutative logic.

## 19.1   Notes

[1] Kurt Schütte. Proof Theory. Springer-Verlag, 1977.

## 19.2   Further reading

- Kai Brünnler, "Deep Inference and Symmetry in Classical Proofs" (Ph.D. thesis 2004) , also published in book form by Logos Verlag (ISBN 978-3-8325-0448-9).

- Deep Inference and the Calculus of Structures Intro and reference web page about ongoing research in deep inference.

# Chapter 20

# Dialectica interpretation

In proof theory, the **Dialectica interpretation**[1] is a proof interpretation of intuitionistic arithmetic (Heyting arithmetic) into a finite type extension of primitive recursive arithmetic, the so-called **System T**. It was developed by Kurt Gödel to provide a consistency proof of arithmetic. The name of the interpretation comes from the journal *Dialectica*, where Gödel's paper was published in a special issue dedicated to Paul Bernays on his 70th birthday.

## 20.1  Motivation

Via the Gödel–Gentzen negative translation, the consistency of classical Peano arithmetic had already been reduced to the consistency of intuitionistic Heyting arithmetic. Gödel's motivation for developing the dialectica interpretation was to obtain a relative consistency proof for Heyting arithmetic (and hence for Peano arithmetic).

## 20.2  Dialectica interpretation of intuitionistic logic

The interpretation has two components: a formula translation and a proof translation. The formula translation describes how each formula $A$ of Heyting arithmetic is mapped to a quantifier-free formula $A_D(x; y)$ of the system T, where $x$ and $y$ are tuples of fresh variables (not appearing free in $A$ ). Intuitively, $A$ is interpreted as $\exists x \forall y A_D(x; y)$ . The proof translation shows how a proof of $A$ has enough information to witness the interpretation of $A$ , i.e. the proof of $A$ can be converted into a closed term $t$ and a proof of $A_D(t; y)$ in the system T.

### 20.2.1  Formula translation

The quantifier-free formula $A_D(x; y)$ is defined inductively on the logical structure of $A$ as follows, where $P$ is an atomic formula:

$$
\begin{aligned}
(P)_D & \equiv P \\
(A \wedge B)_D(x, v; y, w) & \equiv A_D(x; y) \wedge B_D(v; w) \\
(A \vee B)_D(x, v, z; y, w) & \equiv (z = 0 \to A_D(x; y)) \wedge (z \neq 0 \to B_D(v; w)) \\
(A \to B)_D(f, g; x, w) & \equiv A_D(x; fxw) \to B_D(gx; w) \\
(\exists z A)_D(x, z; y) & \equiv A_D(x; y) \\
(\forall z A)_D(f; y, z) & \equiv A_D(fz; y)
\end{aligned}
$$

56

### 20.2.2 Proof translation (soundness)

The formula interpretation is such that whenever $A$ is provable in Heyting arithmetic then there exists a sequence of closed terms $t$ such that $A_D(t; y)$ is provable in the system T. The sequence of terms $t$ and the proof of $A_D(t; y)$ are constructed from the given proof of $A$ in Heyting arithmetic. The construction of $t$ is quite straightforward, except for the contraction axiom $A \to A \wedge A$ which requires the assumption that quantifier-free formulas are decidable.

### 20.2.3 Characterisation principles

It has also been shown that Heyting arithmetic extended with the following principles

- Axiom of choice

- Markov's principle

- Independence of premise for universal formulas

is necessary and sufficient for characterising the formulas of HA which are interpretable by the Dialectica interpretation.

## 20.3 Extensions of basic interpretation

The basic dialectica interpretation of intuitionistic logic has been extended to various stronger systems. Intuitively, the dialectica interpretation can be applied to a stronger system, as long as the dialectica interpretation of the extra principle can be witnessed by terms in the system T (or an extension of system T).

### 20.3.1 Induction

Given Gödel's incompleteness theorem (which implies that the consistency of PA cannot be proven by finitistic means) it is reasonable to expect that system T must contain non-finitistic constructions. Indeed this is the case. The non-finitistic constructions show up in the interpretation of mathematical induction. To give a Dialectica interpretation of induction, Gödel makes use of what is nowadays called Gödel's primitive recursive functionals, which are higher order functions with primitive recursive descriptions.

### 20.3.2 Classical logic

Formulas and proofs in classical arithmetic can also be given a dialectica interpretation via an initial embedding into Heyting arithmetic followed the dialectica interpretation of Heyting arithmetic. Shoenfield, in his book, combines the negative translation and the dialectica interpretation into a single interpretation of classical arithmetic.

### 20.3.3 Comprehension

In 1962 Spector [2] extended Gödel's Dialectica interpretation of arithmetic to full mathematical analysis, by showing how the schema of countable choice can be given a Dialectica interpretation by extending system T with bar recursion.

## 20.4 Dialectica interpretation of linear logic

The Dialectica interpretation has been used to build a model of Girard's refinement of intuitionistic logic known as linear logic, via the so-called Dialectica spaces.[3] Since linear logic is a refinement of intuitionistic logic, the dialectica interpretation of linear logic can also be viewed as a refinement of the dialectica interpretation of intuitionistic logic.

Although the linear interpretation in Shirarata's work [4] validates the weakening rule (it is actually an interpretation of affine logic), de Paiva's dialectica spaces interpretation does not validate weakening for arbitrary formulas.

## 20.5   Variants of the Dialectica interpretation

Several variants of the Dialectica interpretation have been proposed since. Most notably the Diller-Nahm variant (to avoid the contraction problem) and Kohlenbach's monotone and Ferreira-Oliva bounded interpretations (to interpret weak König's lemma). Comprehensive treatments of the interpretation can be found at ,[5] [6] and .[7]

## 20.6   References

[1]  Kurt Gödel (1958). *Über eine bisher noch nicht benützte Erweiterung des finiten Standpunktes.* Dialectica. pp. 280–287.

[2]  Clifford Spector (1962). *Provably recursive functionals of analysis: a consistency proof of analysis by an extension of principles in current intuitionistic mathematics.* Recursive Function Theory: Proc. Symposia in Pure Mathematics. pp. 1–27.

[3]  Valeria de Paiva (1991). *The Dialectica Categories* (PDF). University of Cambridge, Computer Laboratory, PhD Thesis, Technical Report 213.

[4]  Masaru Shirahata (2006). *The Dialectica interpretation of first-order classical affine logic.* Theory and Applications of Categories, Vol. 17, No. 4. pp. 49–79.

[5]  Jeremy Avigad and Solomon Feferman (1999). *Gödel's functional ("Dialectica") interpretation* (PDF). in S. Buss ed., The Handbook of Proof Theory, North-Holland. pp. 337–405.

[6]  Ulrich Kohlenbach (2008). *Applied Proof Theory: Proof Interpretations and Their Use in Mathematics.* Springer Verlag, Berlin. pp. 1–536.

[7]  Anne S. Troelstra (with C.A. Smoryński, J.I. Zucker, W.A.Howard) (1973). *Metamathematical Investigation of intuitionistic Arithmetic and Analysis.* Springer Verlag, Berlin. pp. 1–323.

# Chapter 21

# Disjunction and existence properties

In mathematical logic, the **disjunction and existence properties** are the "hallmarks" of constructive theories such as Heyting arithmetic and constructive set theories (Rathjen 2005). The **disjunction property** is satisfied by a theory if, whenever a sentence $A \vee B$ is a theorem, then either $A$ is a theorem, or $B$ is a theorem. The **existence property** or **witness property** is satisfied by a theory if, whenever a sentence $(\exists x)A(x)$ is a theorem, where $A(x)$ has no other free variables, then there is some term $t$ such that the theory proves $A(t)$.

## 21.1 Related properties

Rathjen (2005) lists five properties that a theory may possess. These include the disjunction property (**DP**), the existence property (**EP**), and three additional properties:

- The **numerical existence property (NEP)** states that if the theory proves $(\exists x \in \mathbb{N})\varphi(x)$, where $\varphi$ has no other free variables, then the theory proves $\varphi(\bar{n})$ for some $n \in \mathbb{N}$. Here $\bar{n}$ is a term in $T$ representing the number $n$.

- **Church's rule (CR)** states that if the theory proves $(\forall x \in \mathbb{N})(\exists y \in \mathbb{N})\varphi(x, y)$ then there is a natural number $e$ such that, letting $f_e$ be the computable function with index $e$, the theory proves $(\forall x)\varphi(x, f_e(x))$.

- A variant of Church's rule, $\mathbf{CR}_1$, states that if the theory proves $(\exists f \colon \mathbb{N} \to \mathbb{N})\psi(f)$ then there is a natural number $e$ such that the theory proves $f_e$ is total and proves $\psi(f_e)$.

These properties can only be directly expressed for theories that have the ability to quantify over natural numbers and, for $CR_1$, quantify over functions from $\mathbb{N}$ to $\mathbb{N}$. In practice, one may say that a theory has one of these properties if a definitional extension of the theory has the property stated above (Rathjen 2005).

## 21.2 Background and History

Kurt Gödel (1932) proved that intuitionistic propositional logic (with no additional axioms) has the disjunction property; this result was extended to intuitionistic predicate logic by Gerhard Gentzen (1934,1935). Stephen Cole Kleene (1945) proved that Heyting arithmetic has the disjunction property and the existence property. Kleene's method introduced the technique of realizability, which is now one of the main methods in the study of constructive theories (Kohlenbach 2008; Troelstra 1973).

While the earliest results were for constructive theories of arithmetic, many results are also known for constructive set theories (Rathjen 2005). John Myhill (1973) showed that IZF with the axiom of Replacement eliminated in favor of the axiom of Collection has the disjunction property, the numerical existence property, and the existence property. Michael Rathjen (2005) proved that CZF has the disjunction property and the numerical existence property.

Most classical theories, such as Peano Arithmetic and ZFC do not have the existence or disjunction property. Some classical theories, such as ZFC plus the axiom of constructibility, do have a weaker form of the existence property (Rathjen 2005).

## 21.3   In topoi

Freyd and Scedrov (1990) observed that the disjunction property holds in free Heyting algebras and free topoi. In categorical terms, in the free topos, that corresponds to the fact that the terminal object, **1** , is not the join of two proper subobjects. Together with the existence property it translates to the assertion that **1** is an indecomposable projective object – the functor it represents (the global-section functor) preserves epimorphisms and coproducts.

## 21.4   Relationships

There are several relationship between the five properties discussed above.

The numerical existence property implies the disjunction property. The proof uses the fact that a disjunction can be rewritten as an existential formula quantifying over natural numbers:

$$A \vee B \equiv (\exists n)[(n = 0 \to A) \wedge (n \neq 0 \to B)]$$

Therefore, if $A \vee B$ is a theorem of $T$ , so is $\exists n \colon (n = 0 \to A) \wedge (n \neq 0 \to B)$ . Thus, assuming the numerical existence property, there exists some $s$ such that $(\bar{s} = 0 \to A) \wedge (\bar{s} \neq 0 \to B)$ is a theorem. Since $\bar{s}$ is a numeral, one may concretely check the value of $s$ : if $s = 0$ then $A$ is a theorem and if $s \neq 0$ then $B$ is a theorem.

Harvey Friedman (1974) proved that in any recursively enumerable extension of intuitionistic arithmetic, the disjunction property implies the numerical existence property. The proof uses self-referential sentences in way similar to the proof of Gödel's incompleteness theorems. The key step is to find a bound on the existential quantifier in a formula $(\exists x)A(x)$, producing a because bounded existential formula $(\exists x{<}n)A(x)$. The bounded formula may then be written as a finite disjunction $A(1) \vee A(2) \vee ... \vee A(n)$. Finally, disjunction elimination may be used to show that one of the disjuncts is provable.

## 21.5   References

- Petre J. Freyd and Andre Scedrov, 1990, *Categories, Allegories*. North-Holland.

- Harvey Friedman, 1975, *The disjunction property implies the numerical existence property*, State University of New York at Buffalo.

- Gerhard Gentzen, 1934, "Untersuchungen über das logische Schließen. I", *Mathematische Zeitschrift* v. 39 n. 2, pp. 176–210.

- Gerhard Gentzen, 1935, "Untersuchungen über das logische Schließen. II", *Mathematische Zeitschrift* v. 39 n. 3, pp. 405–431.

- Kurt Gödel, 1932, "Zum intuitionistischen Aussagenkalkül", *Anzeiger der Akademie der Wissenschaftischen in Wien*, v. 69, pp. 65–66.

- Stephen Cole Kleene, 1945, "On the interpretation of intuitionistic number theory," *Journal of Symbolic Logic*, v. 10, pp. 109–124.

- Ulrich Kohlenbach, 2008, *Applied proof theory*, Springer.

- John Myhill, 1973, "Some properties of Intuitionistic Zermelo-Fraenkel set theory", in A. Mathias and H. Rogers, *Cambridge Summer School in Mathematical Logic*, Lectures Notes in Mathematics v. 337, pp. 206–231, Springer.

- Michael Rathjen, 2005, "The Disjunction and Related Properties for Constructive Zermelo-Fraenkel Set Theory", *Journal of Symbolic Logic*, v. 70 n. 4, pp. 1233–1254.

- Anne S. Troelstra, ed. (1973), *Metamathematical investigation of intuitionistic arithmetic and analysis*, Springer.

## 21.6 External links

- Intuitionistic Logic by Joan Moschovakis, Stanford Encyclopedia of Philosophy

# Chapter 22

# Elementary function arithmetic

In proof theory, a branch of mathematical logic, **elementary function arithmetic**, also called **EFA**, **elementary arithmetic** and **exponential function arithmetic**, is the system of arithmetic with the usual elementary properties of 0, 1, +, ×, $x^y$, together with induction for formulas with bounded quantifiers.

EFA is a very weak logical system, whose proof theoretic ordinal is $\omega^3$, but still seems able to prove much of ordinary mathematics that can be stated in the language of first-order arithmetic.

## 22.1   Definition

EFA is a system in first order logic (with equality). Its language contains:

- two constants 0, 1,

- three binary operations +, ×, exp, with exp(x,y) usually written as $x^y$,

- a binary relation symbol < (This is not really necessary as it can be written in terms of the other operations and is sometimes omitted, but is convenient for defining bounded quantifiers).

Bounded quantifiers are those of the form $\forall (x<y)$ and $\exists (x<y)$ which are abbreviations for $\forall x\ (x<y) \rightarrow,,,$ and $\exists x\ (x<y) \wedge ...$ in the usual way.

The axioms of EFA are

- The axioms of Robinson arithmetic for 0, 1, +, ×, <

- The axioms for exponentiation: $x^0 = 1$, $x^{y+1} = x^y \times x$.

- Induction for formulas all of whose quantifiers are bounded (but which may contain free variables).

## 22.2   Friedman's grand conjecture

Harvey Friedman's **grand conjecture** implies that many mathematical theorems, such as Fermat's last theorem, can be proved in very weak systems such as EFA.

The original statement of the conjecture from Friedman (1999) is:

> "Every theorem published in the *Annals of Mathematics* whose statement involves only finitary mathematical objects (i.e., what logicians call an arithmetical statement) can be proved in EFA. EFA is the weak fragment

of Peano Arithmetic based on the usual quantifier-free axioms for 0, 1, +, ×, exp, together with the scheme of induction for all formulas in the language all of whose quantifiers are bounded."

While it is easy to construct artificial arithmetical statements that are true but not provable in EFA, the point of Friedman's conjecture is that natural examples of such statements in mathematics seem to be rare. Some natural examples include consistency statements from logic, several statements related to Ramsey theory such as the Szemerédi regularity lemma and the graph minor theorem, and Tarjan's algorithm for the disjoint-set data structure.

## 22.3 Related systems

One can omit the binary function symbol exp from the language, by taking Robinson arithmetic together with induction for all formulas with bounded quantifiers and an axiom stating roughly that exponentiation is a function defined everywhere. This is similar to EFA and has the same proof theoretic strength, but is more cumbersome to work with.

There are weak fragments of second-order arithmetic called RCA*
0 and WKL*
0 that have the same consistency strength as EFA and are conservative over it for $\Pi$0
2 sentences, which are sometimes studied in reverse mathematics (Simpson 2009).

**Elementary recursive arithmetic** (ERA) is a subsystem of primitive recursive arithmetic (PRA) in which recursion is restricted to bounded sums and products. This also has the same $\Pi$0
2 sentences as EFA, in the sense that whenever EFA proves $\forall x \exists y\ P(x,y)$, with P quantifier-free, ERA proves the open formula $P(x,T(x))$, with T a term definable in ERA.

Like PRA, ERA can be defined in an entirely logic-free manner, with just the rules of substitution and induction, and defining equations for all elementary recursive functions. Unlike PRA, however, the elementary recursive functions can be charactized by the closure under composition and projection of a *finite* number of basis functions, and thus only a finite number of defining equations are needed.

## 22.4 See also

- ELEMENTARY, a related computational complexity class

- Grzegorczyk hierarchy

- Reverse mathematics

- Tarski's high school algebra problem

## 22.5 References

- Avigad, Jeremy (2003), "Number theory and elementary arithmetic", *Philosophia Mathematica. Philosophy of Mathematics, its Learning, and its Application. Series III* **11** (3): 257–284, doi:10.1093/philmat/11.3.257, ISSN 0031-8019, MR 2006194

- Friedman, Harvey (1999), *grand conjectures*

- Simpson, Stephen G. (2009), *Subsystems of second order arithmetic*, Perspectives in Logic (2nd ed.), Cambridge University Press, ISBN 978-0-521-88439-6, MR 1723993

# Chapter 23

# Extension by definitions

In mathematical logic, more specifically in the proof theory of first-order theories, **extensions by definitions** formalize the introduction of new symbols by means of a definition. For example, it is common in naive set theory to introduce a symbol $\emptyset$ for the set which has no member. In the formal setting of first-order theories, this can be done by adding to the theory a new constant $\emptyset$ and the new axiom $\forall x(x \notin \emptyset)$, meaning 'for all $x$, $x$ is not a member of $\emptyset$'. It can then be proved that doing so adds essentially nothing to the old theory, as should be expected from a definition. More precisely, the new theory is a conservative extension of the old one.

## 23.1   Definition of relation symbols

*Let* $T$ be a first-order theory and $\phi(x_1, \ldots, x_n)$ a formula of $T$ such that $x_1$, ..., $x_n$ are distinct and include the variables free in $\phi(x_1, \ldots, x_n)$. Form a new first-order theory $T'$ from $T$ by adding a new $n$-ary relation symbol $R$, the logical axioms featuring the symbol $R$ and the new axiom

$$\forall x_1 \ldots \forall x_n (R(x_1, \ldots, x_n) \leftrightarrow \phi(x_1, \ldots, x_n))$$

called the *defining axiom* of $R$.

If $\psi$ is a formula of $T'$, let $\psi^*$ be the formula of $T$ obtained from $\psi$ by replacing any occurrence of $R(t_1, \ldots, t_n)$ by $\phi(t_1, \ldots, t_n)$ (changing the bound variables in $\phi$ if necessary so that the variables occurring in the $t_i$ are not bound in $\phi(t_1, \ldots, t_n)$). Then the following hold:

1. $\psi \leftrightarrow \psi^*$ is provable in $T'$, and

2. $T'$ is a conservative extension of $T$.

The fact that $T'$ is a conservative extension of $T$ shows that the defining axiom of $R$ cannot be used to prove new theorems. The formula $\psi^*$ is called a *translation* of $\psi$ into $T$. Semantically, the formula $\psi^*$ has the same meaning as $\psi$, but the defined symbol $R$ has been eliminated.

## 23.2   Definition of function symbols

Let $T$ be a first-order theory (with equality) and $\phi(y, x_1, \ldots, x_n)$ a formula of $T$ such that $y$, $x_1$, ..., $x_n$ are distinct and include the variables free in $\phi(y, x_1, \ldots, x_n)$. Assume that we can prove

$$\forall x_1 \ldots \forall x_n \exists! y \phi(y, x_1, \ldots, x_n)$$

in $T$ , i.e. for all $x_1$ , ..., $x_n$ , there exists a unique $y$ such that $\phi(y, x_1, \ldots, x_n)$ . Form a new first-order theory $T'$ from $T$ by adding a new $n$ -ary function symbol $f$ , the logical axioms featuring the symbol $f$ and the new axiom

$$\forall x_1 \ldots \forall x_n \phi(f(x_1, \ldots, x_n), x_1, \ldots, x_n)$$

called the *defining axiom* of $f$ .

If $\psi$ is an atomic formula of $T'$ , define a formula $\psi^*$ of $T$ recursively as follows. If the new symbol $f$ does not occur in $\psi$ , let $\psi^*$ be $\psi$ . Otherwise, choose an occurrence of $f(t_1, \ldots, t_n)$ in $\psi$ such that $f$ does not occur in the terms $t_i$ , and let $\chi$ be obtained from $\psi$ by replacing that occurrence by a new variable $z$ . Then since $f$ occurs in $\chi$ one less time than in $\psi$ , the formula $\chi^*$ has already been defined, and we let $\psi^*$ be

$$\forall z(\phi(z, t_1, \ldots, t_n) \to \chi^*)$$

(changing the bound variables in $\phi$ if necessary so that the variables occurring in the $t_i$ are not bound in $\phi(z, t_1, \ldots, t_n)$ ). For a general formula $\psi$ , the formula $\psi^*$ is formed by replacing every occurrence of an atomic subformula $\chi$ by $\chi^*$ . Then the following hold:

1. $\psi \leftrightarrow \psi^*$ is provable in $T'$ , and

2. $T'$ is a conservative extension of $T$ .

The formula $\psi^*$ is called a *translation* of $\psi$ into $T$ . As in the case of relation symbols, the formula $\psi^*$ has the same meaning as $\psi$ , but the new symbol $f$ has been eliminated.

The construction of this paragraph also works for constants, which can be viewed as 0-ary function symbols.

## 23.3 Extensions by definitions

A first-order theory $T'$ obtained from $T$ by successive introductions of relation symbols and function symbols as above is called an **extension by definitions** of $T$ . Then $T'$ is a conservative extension of $T$ , and for any formula $\psi$ of $T'$ we can form a formula $\psi^*$ of $T$ , called a *translation* of $\psi$ into $T$ , such that $\psi \leftrightarrow \psi^*$ is provable in $T'$ . Such a formula is not unique, but any two of them can be proved to be equivalent in $T$ .

In practice, an extension by definitions $T'$ of $T$ is not distinguished from the original theory $T$ . In fact, the formulas of $T'$ can be thought of as *abbreviating* their translations into $T$ . The manipulation of these abbreviations as actual formulas is then justified by the fact that extensions by definitions are conservative.

## 23.4 Examples

- Traditionally, the first-order set theory ZF has $=$ (equality) and $\in$ (membership) as its only primitive relation symbols, and no function symbols. In everyday mathematics, however, many other symbols are used such as the binary relation symbol $\subseteq$ , the constant $\emptyset$ , the unary function symbol $P$ (the power-set operation), etc. All of these symbols belong in fact to extensions by definitions of ZF.

- Let $T$ be a first-order theory for groups in which the only primitive symbol is the binary product $\cdot$ . In $T$, we can prove that there exists a unique element $y$ such that $x.y=y.x=x$ for every $x$. Therefore we can add to $T$ a new constant $e$ and the axiom

$$\forall x(x \cdot e = x \text{ and } e \cdot x = x)$$

and what we obtain is an extension by definitions $T'$ of $T$. Then in $T'$ we can prove that for every $x$, there exists a unique $y$ such that $x.y=y.x=e$. Consequently, the first-order theory $T''$ obtained from $T'$ by adding a unary function symbol $f$ and the axiom

$$\forall x(x \cdot f(x) = e \text{ and } f(x) \cdot x = e)$$

is an extension by definitions of $T$. Usually, $f(x)$ is denoted $x^{-1}$ .

## 23.5   Bibliography

- S.C. Kleene (1952), *Introduction to Metamathematics*, D. Van Nostrand

- E. Mendelson (1997). *Introduction to Mathematical Logic* (4th ed.), Chapman & Hall.

- J.R. Shoenfield (1967).  *Mathematical Logic*, Addison-Wesley Publishing Company (reprinted in 2001 by AK Peters)

# Chapter 24

# Fast-growing hierarchy

In computability theory, computational complexity theory and proof theory, a **fast-growing hierarchy** (also called an **extended Grzegorczyk hierarchy**) is an ordinal-indexed family of rapidly increasing functions $f\alpha$: $\mathbf{N} \to \mathbf{N}$ (where $\mathbf{N}$ is the set of natural numbers $\{0, 1, ...\}$, and $\alpha$ ranges up to some large countable ordinal). A primary example is the **Wainer hierarchy**, or **Löb–Wainer hierarchy**, which is an extension to all $\alpha < \varepsilon_0$. Such hierarchies provide a natural way to classify computable functions according to rate-of-growth and computational complexity.

## 24.1   Definition

Let $\mu$ be a large countable ordinal such that a fundamental sequence (a strictly increasing sequence of ordinals whose supremum is a limit ordinal) is assigned to every limit ordinal less than $\mu$. A **fast-growing hierarchy** of functions $f\alpha$: $\mathbf{N} \to \mathbf{N}$, for $\alpha < \mu$, is then defined as follows:

- $f_0(n) = n + 1,$

- $f_{\alpha+1}(n) = f_\alpha^n(n),$

- $f_\alpha(n) = f_{\alpha[n]}(n)$ if $\alpha$ is a limit ordinal.

Here $f\alpha^n(n) = f\alpha(f\alpha(...(f\alpha(n))...))$ denotes the $n^{\text{th}}$ iterate of $f\alpha$ applied to $n$, and $\alpha[n]$ denotes the $n^{\text{th}}$ element of the fundamental sequence assigned to the limit ordinal $\alpha$. (An alternative definition takes the number of iterations to be $n+1$, rather than $n$, in the second line above.)

The initial part of this hierarchy, comprising the functions $f\alpha$ with *finite* index (i.e., $\alpha < \omega$), is often called the **Grzegorczyk hierarchy** because of its close relationship to the Grzegorczyk hierarchy; note, however, that the former is here an indexed family of functions $fn$, whereas the latter is an indexed family of *sets* of functions $\mathcal{E}^n$. (See Points of Interest below.)

Generalizing the above definition even further, a **fast iteration hierarchy** is obtained by taking $f_0$ to be any increasing function g: $\mathbf{N} \to \mathbf{N}$.

For limit ordinals not greater than $\varepsilon_0$, there is a straightforward natural definition of the fundamental sequences (see the **Wainer hierarchy** below), but beyond $\varepsilon_0$ the definition is much more complicated. However, this is possible well beyond the Feferman–Schütte ordinal, $\Gamma_0$, up to at least the Bachmann–Howard ordinal. Using Buchholz psi functions one can extend this definition easily to the ordinal of transfinitely iterated $\Pi_1^1$-comprehension (see Analytical hierarchy).

A fully specified extension beyond the recursive ordinals is thought to be unlikely; e.g., Prömel *et al.* [1991](p. 348) note that in such an attempt "there would even arise problems in ordinal notation".

## 24.2   The Wainer hierarchy

The **Wainer hierarchy** is the particular fast-growing hierarchy of functions $f\alpha$ ($\alpha \le \varepsilon_0$) obtained by defining the fundamental sequences as follows [Gallier 1991][Prömel, et al., 1991]:

For limit ordinals $\lambda < \varepsilon_0$, written in Cantor normal form,

- if $\lambda = \omega^{\alpha_1} + ... + \omega^{\alpha_{k-1}} + \omega^{\alpha_k}$ for $\alpha_1 \ge ... \ge \alpha_{k-1} \ge \alpha_k$, then $\lambda[n] = \omega^{\alpha_1} + ... + \omega^{\alpha_{k-1}} + \omega^{\alpha_k}[n]$,

- if $\lambda = \omega^{\alpha+1}$, then $\lambda[n] = \omega^{\alpha}n$,

- if $\lambda = \omega^{\alpha}$ for a limit ordinal $\alpha$, then $\lambda[n] = \omega^{\alpha[n]}$,

and

- if $\lambda = \varepsilon_0$, take $\lambda[0] = 0$ and $\lambda[n + 1] = \omega^{\lambda[n]}$ as in [Gallier 1991]; alternatively, take the same sequence except starting with $\lambda[0] = 1$ as in [Prömel, et al., 1991].

  For $n > 0$, the alternative version has one additional $\omega$ in the resulting exponential tower, i.e. $\lambda[n] = \omega^{\omega^{\cdot^{\cdot^{\omega}}}}$ with $n$ omegas.

Some authors use slightly different definitions (e.g., $\omega^{\alpha+1}[n] = \omega^{\alpha}(n+1)$, instead of $\omega^{\alpha}n$), and some define this hierarchy only for $\alpha < \varepsilon_0$ (thus excluding $f\varepsilon_0$ from the hierarchy).

To continue beyond $\varepsilon_0$, see the Fundamental sequences for the Veblen hierarchy.

## 24.3   Points of interest

Following are some relevant points of interest about fast-growing hierarchies:

- Every $f\alpha$ is a total function. If the fundamental sequences are computable (e.g., as in the Wainer hierarchy), then every $f\alpha$ is a total computable function.

- In the Wainer hierarchy, $f\alpha$ is dominated by $f_\beta$ if $\alpha < \beta$. (For any two functions $f, g$: $\mathbf{N} \to \mathbf{N}$, $f$ is said to **dominate** $g$ if $f(n) > g(n)$ for all sufficiently large $n$.) The same property holds in any fast-growing hierarchy with fundamental sequences satisfying the so-called Bachmann property. (This property holds for most natural well orderings.)

- In the Grzegorczyk hierarchy, every primitive recursive function is dominated by some $f\alpha$ with $\alpha < \omega$. Hence, in the Wainer hierarchy, every primitive recursive function is dominated by $f\omega$, which is a variant of the Ackermann function.

- For $n \ge 3$, the set $\mathcal{E}^n$ in the Grzegorczyk hierarchy is composed of just those total multi-argument functions which, for sufficiently large arguments, are computable within time bounded by some fixed iterate $fn-_1{}^k$ evaluated at the maximum argument.

- In the Wainer hierarchy, every $f\alpha$ with $\alpha < \varepsilon_0$ is computable and provably total in Peano Arithmetic.

- Every computable function that's provably total in Peano Arithmetic is dominated by some $f\alpha$ with $\alpha < \varepsilon_0$ in the Wainer hierarchy. Hence $f\varepsilon_0$ in the Wainer hierarchy is not provably total in Peano Arithmetic.

- The Goodstein function has approximately the same growth rate (*i.e.* each is dominated by some fixed iterate of the other) as $f\varepsilon_0$ in the Wainer hierarchy, dominating every $f\alpha$ for which $\alpha < \varepsilon_0$, and hence is not provably total in Peano Arithmetic.

- In the Wainer hierarchy, if $\alpha < \beta < \varepsilon_0$, then $f_\beta$ dominates every computable function within time and space bounded by some fixed iterate $f\alpha^k$.

- Friedman's TREE function dominates $f\Gamma_0$ in a fast-growing hierarchy described by Gallier (1991).

- The Wainer hierarchy of functions $f\alpha$ and the Hardy hierarchy of functions $h\alpha$ are related by $f\alpha = h\omega^\alpha$ for all $\alpha < \varepsilon_0$. The Hardy hierarchy "catches up" to the Wainer hierarchy at $\alpha = \varepsilon_0$, such that $f\varepsilon_0$ and $h\varepsilon_0$ have the same growth rate, in the sense that $f\varepsilon_0(n-1) \le h\varepsilon_0(n) \le f\varepsilon_0(n+1)$ for all $n \ge 1$. (Gallier 1991)

- Girard (1981) and Cichon & Wainer (1983) showed that the *slow-growing hierarchy* of functions $g\alpha$ attains the same growth rate as the function $f\varepsilon_0$ in the Wainer hierarchy when $\alpha$ is the Bachmann-Howard ordinal. Girard (1981) further showed that the slow-growing hierarchy $g\alpha$ attains the same growth rate as $f\alpha$ (in a particular fast-growing hierarchy) when $\alpha$ is the ordinal of the theory $ID_{<\omega}$ of arbitrary finite iterations of an inductive definition. (Wainer 1989)

## 24.4 Functions in fast-growing hierarchies

The functions at finite levels ($\alpha < \omega$) of any fast-growing hierarchy coincide with those of the Grzegorczyk hierarchy: (using hyperoperation)

- $f_0(n) = n + 1 = 2\,[1]\,n - 1$

- $f_1(n) = f_0{}^n(n) = n + n = 2n = 2\,[2]\,n$

- $f_2(n) = f_1{}^n(n) = 2^n \cdot n > 2^n = 2\,[3]\,n$ for $n \ge 2$

- $fk_{+1}(n) = fk^n(n) > (2\,[k+1])^n\, n \ge 2\,[k+2]\,n$ for $n \ge 2$, $k < \omega$.

Beyond the finite levels are the functions of the Wainer hierarchy ($\omega \le \alpha \le \varepsilon_0$):

- $f\omega(n) = fn(n) > 2\,[n+1]\,n > 2\,[n]\,(n+3) - 3 = A(n, n)$ for $n \ge 4$, where $A$ is the Ackermann function (of which $f\omega$ is a unary version).

- $f\omega_{+1}(n) = f\omega^n(n) \ge fn\,[n+2]\,n(n)$ for all $n > 0$, where $n\,[n+2]\,n$ is the $n^{\text{th}}$ Ackermann number.

- $f\omega_{+1}(64) > f\omega^{64}(6) >$ Graham's number ($= g_{64}$ in the sequence defined by $g_0 = 4$, $gk_{+1} = 3\,[gk+2]\,3$). This follows by noting $f\omega(n) > 2\,[n+1]\,n > 3\,[n]\,3 + 2$, and hence $f\omega(gk+2) > gk_{+1} + 2$.

- $f\varepsilon_0(n)$ is the first function in the Wainer hierarchy that dominates the Goodstein function.

## 24.5 References

- Buchholz, W.; Wainer, S.S (1987). "Provably Computable Functions and the Fast Growing Hierarchy". *Logic and Combinatorics*, edited by S. Simpson, Contemporary Mathematics, Vol. 65, AMS, 179-198.

- Cichon, E. A.; Wainer, S. S. (1983), "The slow-growing and the Grzegorczyk hierarchies", *The Journal of Symbolic Logic* **48** (2): 399–408, doi:10.2307/2273557, ISSN 0022-4812, MR 704094

- Gallier, Jean H. (1991), "What's so special about Kruskal's theorem and the ordinal $\Gamma_0$? A survey of some results in proof theory", *Ann. Pure Appl. Logic* **53** (3): 199–260, doi:10.1016/0168-0072(91)90022-E, MR 1129778 PDF's: part 1 2 3. (In particular part 3, Section 12, pp. 59–64, "A Glimpse at Hierarchies of Fast and Slow Growing Functions".)

- Girard, Jean-Yves (1981), "$\Pi^1_2$-logic. I. Dilators", *Annals of Mathematical Logic* **21** (2): 75–219, doi:10.1016/0003-4843(81)90016-4, ISSN 0003-4843, MR 656793

- Löb, M.H.; Wainer, S.S. (1970), "Hierarchies of number theoretic functions", *Arch. Math. Logik*, 13. Correction, *Arch. Math. Logik*, 14, 1971.  Part I doi:10.1007/BF01967649, Part 2 doi:10.1007/BF01973616, Corrections doi:10.1007/BF01991855.

- Prömel, H. J.; Thumser, W.; Voigt, B. "Fast growing functions based on Ramsey theorems", *Discrete Mathematics*, v.95 n.1-3, p. 341-358, Dec. 1991 doi:10.1016/0012-365X(91)90346-4.

- Wainer, S.S (1989), "Slow Growing Versus Fast Growing". *Journal of Symbolic Logic* **54**(2): 608-614.

# Chapter 25

# Feferman–Schütte ordinal

In mathematics, the **Feferman–Schütte ordinal** $\Gamma_0$ is a large countable ordinal. It is the proof theoretic ordinal of several mathematical theories, such as arithmetical transfinite recursion. It is named after Solomon Feferman and Kurt Schütte.

It is sometimes said to be the first impredicative ordinal, though this is controversial, partly because there is no generally accepted precise definition of "predicative". Sometimes an ordinal is said to be predicative if it is less than $\Gamma_0$.

Unfortunately there is no standard notation for ordinals at and beyond the Feferman–Schütte ordinal, so there are several ways of representing it, some of which use ordinal collapsing functions: $\psi(\Omega^\Omega)$, $\theta(\Omega)$ or $\phi_\Omega(0)$

## 25.1 Definition

The Feferman–Schütte ordinal can be defined as the smallest ordinal that cannot be obtained by starting with 0 and using the operations of ordinal addition and the Veblen functions $\varphi\alpha(\beta)$. That is, it is the smallest $\alpha$ such that $\varphi\alpha(0) = \alpha$.

## 25.2 References

- Pohlers, Wolfram (1989), *Proof theory*, Lecture Notes in Mathematics **1407**, Berlin: Springer-Verlag, ISBN 3-540-51842-8, MR 1026933

- Weaver, Nik (2005), *Predicativity beyond Gamma_0*, arXiv:math/0509244

# Chapter 26

# Formal proof

A **formal proof** or **derivation** is a finite sequence of sentences (called well-formed formulas in the case of a formal language) each of which is an axiom, an assumption, or follows from the preceding sentences in the sequence by a rule of inference. The last sentence in the sequence is a theorem of a formal system. The notion of theorem is not in general effective, therefore there may be no method by which we can always find a proof of a given sentence or determine that none exists. The concept of natural deduction is a generalization of the concept of proof.[1]

The theorem is a syntactic consequence of all the well-formed formulas preceding it in the proof. For a well-formed formula to qualify as part of a proof, it must be the result of applying a rule of the deductive apparatus of some formal system to the previous well-formed formulae in the proof sequence.

Formal proofs often are constructed with the help of computers in interactive theorem proving. Significantly, these proofs can be checked automatically, also by computer. Checking formal proofs is usually simple, while the problem of *finding* proofs (automated theorem proving) is usually computationally intractable and/or only semi-decidable, depending upon the formal system in use.

## 26.1 Background

### 26.1.1 Formal language

Main article: Formal language

A *formal language* is a set of finite sequences of symbols. Such a language can be defined without reference to any meanings of any of its expressions; it can exist before any interpretation is assigned to it – that is, before it has any meaning. Formal proofs are expressed in some formal language.

### 26.1.2 Formal grammar

Main articles: Formal grammar and Formation rule

A *formal grammar* (also called *formation rules*) is a precise description of the well-formed formulas of a formal language. It is synonymous with the set of strings over the alphabet of the formal language which constitute well formed formulas. However, it does not describe their semantics (i.e. what they mean).

### 26.1.3 Formal systems

Main article: Formal system

A *formal system* (also called a *logical calculus*, or a *logical system*) consists of a formal language together with a deductive apparatus (also called a *deductive system*). The deductive apparatus may consist of a set of transformation rules (also called *inference rules*) or a set of axioms, or have both. A formal system is used to derive one expression from one or more other expressions.

### 26.1.4 Interpretations

Main articles: Formal semantics (logic) and Interpretation (logic)

An *interpretation* of a formal system is the assignment of meanings to the symbols, and truth-values to the sentences of a formal system. The study of interpretations is called formal semantics. *Giving an interpretation* is synonymous with *constructing a model.*

## 26.2   See also

- Proof (truth)

- Mathematical proof

- Proof theory

- Axiomatic system

## 26.3   References

[1] The Cambridge Dictionary of Philosophy, *deduction*

## 26.4   External links

- "A Special Issue on Formal Proof". *Notices of the American Mathematical Society*. December 2008.

- 2πix.com: Logic Part of a series of articles covering mathematics and logic.

# Chapter 27

# Friedman translation

In mathematical logic, the **Friedman translation** is a certain transformation of intuitionistic formulas. Among other things it can be used to show that the $\Pi^0_2$-theorems of various first-order theories of classical mathematics are also theorems of intuitionistic mathematics. It is named after its discoverer, Harvey Friedman.

## 27.1   Definition

Let $A$ and $B$ be intuitionistic formulas, where no free variable of $B$ is quantified in $A$. The translation $A^B$ is defined by replacing each atomic subformula $C$ of $A$ by $C \vee B$. For purposes of the translation, $\bot$ is considered to be an atomic formula as well, hence it is replaced with $\bot \vee B$ (which is equivalent to $B$). Note that $\neg A$ is defined as an abbreviation for $A \rightarrow \bot$, hence $(\neg A)^B = A^B \rightarrow B$.

## 27.2   Application

The Friedman translation can be used to show the closure of many intuitionistic theories under the Markov rule, and to obtain partial conservativity results. The key condition is that the $\Delta^0_0$ sentences of the logic be decidable, allowing the unquantified theorems of the intuitionistic and classical theories to coincide.

For example, if $A$ is provable in Heyting arithmetic (HA), then $A^B$ is also provable in HA.[1] Moreover, if $A$ is a $\Sigma^0_1$-formula, then $A^B$ is in HA equivalent to $A \vee B$. This implies that:

- Heyting arithmetic is closed under the primitive recursive Markov rule (MPPR): if the formula $\neg\neg A$ is provable in HA, where $A$ is a $\Sigma^0_1$-formula, then $A$ is also provable in HA.

- Peano arithmetic is $\Pi^0_2$-conservative over Heyting arithmetic: if Peano arithmetic proves a $\Pi^0_2$-formula $A$, then $A$ is already provable in HA.

## 27.3   See also

- Gödel–Gentzen negative translation

## 27.4   Notes

[1] Harvey Friedman. Classically and Intuitionistically Provably Recursive Functions. In Scott, D. S. and Muller, G. H. Editors, Higher Set Theory, Volume 699 of Lecture Notes in Mathematics, Springer Verlag (1978), pp. 21–28. doi:10.1007/BFb0103100

# Chapter 28

# Gentzen's consistency proof

**Gentzen's consistency proof** is a result of proof theory in mathematical logic. It "reduces" the consistency of a simplified part of mathematics, not to something that could be proved in that same simplified part of mathematics (which would contradict the basic results of Kurt Gödel), but rather to a simpler logical principle.

## 28.1   Gentzen's theorem

In 1936 Gerhard Gentzen proved the consistency of first-order arithmetic using combinatorial methods. Gentzen's proof shows much more than merely that first-order arithmetic is consistent. Gentzen showed that the consistency of first-order arithmetic is provable, over the base theory of primitive recursive arithmetic with the additional principle of quantifier-free transfinite induction up to the ordinal $\varepsilon_0$. Informally, this additional principle means that there is a well-ordering on the set of finite rooted trees.

The principle of quantifier-free transfinite induction up to $\varepsilon_0$ says that for any formula $A(x)$ with no bound variables transfinite induction up to $\varepsilon_0$ holds. $\varepsilon_0$ is the first ordinal $\alpha$, such that $\omega^\alpha = \alpha$, i.e. the limit of the sequence:

$$\omega, \ \omega^\omega, \ \omega^{\omega^\omega}, \ \dots$$

To express ordinals in the language of arithmetic an ordinal notation is needed, i.e. a way to assign natural numbers to ordinals less than $\varepsilon_0$. This can be done in various ways, one example provided by Cantor's normal form theorem. That transfinite induction holds for a formula $A(x)$ means that $A$ does not define an infinite descending sequence of ordinals smaller than $\varepsilon_0$ (in which case $\varepsilon_0$ would not be well-ordered). Gentzen assigned ordinals smaller than $\varepsilon_0$ to proofs in first-order arithmetic and showed that if there is a proof of contradiction, then there is an infinite descending sequence of ordinals $< \varepsilon_0$ produced by a primitive recursive operation on proofs corresponding to a quantifier-free formula.[1]

## 28.2   Relation to Gödel's theorem

Gentzen's proof also highlights one commonly missed aspect of Gödel's second incompleteness theorem. It is sometimes claimed that the consistency of a theory can only be proved in a stronger theory. The theory obtained by adding quantifier-free transfinite induction to primitive recursive arithmetic proves the consistency of first-order arithmetic but is not stronger than first-order arithmetic. For example, it does not prove ordinary mathematical induction for all formulae, while first-order arithmetic does (it has this as an axiom schema). The resulting theory is not weaker than first-order arithmetic either, since it can prove a number-theoretical fact - the consistency of first-order arithmetic - that first-order arithmetic cannot. The two theories are simply incomparable.

Gentzen's proof is the first example of what is called proof-theoretical ordinal analysis. In ordinal analysis one gauges the strength of theories by measuring how large the (constructive) ordinals are that can be proven to be well-ordered, or equivalently for how large a (constructive) ordinal can transfinite induction be proven. A constructive ordinal is the order type of a recursive well-ordering of natural numbers.

Laurence Kirby and Jeff Paris proved in 1982 that Goodstein's theorem cannot be proven in Peano arithmetic based on Gentzen's theorem.

Hermann Weyl made the following comment in 1946 regarding the significance of Gentzen's consistency result following the devastating impact of Gödel's 1931 incompleteness result on Hilbert's plan to prove the consistency of mathematics.[2]

> It is likely that all mathematicians ultimately would have accepted Hilbert's approach had he been able to carry it out successfully. The first steps were inspiring and promising. But then Gödel dealt it a terrific blow (1931), from which it has not yet recovered. Gödel enumerated the symbols, formulas, and sequences of formulas in Hilbert's formalism in a certain way, and thus transformed the assertion of consistency into an arithmetic proposition. He could show that this proposition can neither be proved nor disproved within the formalism. This can mean only two things: either the reasoning by which a proof of consistency is given must contain some argument that has no formal counterpart within the system, i.e., we have not succeeded in completely formalizing the procedure of mathematical induction; or hope for a strictly "finitistic" proof of consistency must be given up altogether. When G. Gentzen finally succeeded in proving the consistency of arithmetic he trespassed those limits indeed by claiming as evident a type of reasoning that penetrates into Cantor's "second class of ordinal numbers."

Kleene (2009, p. 479) made the following comment in 1952 on the significance of Gentzen's result, particularly in the context of the formalist program which was initiated by Hilbert.

> The original proposals of the formalists to make classical mathematics secure by a consistency proof did not contemplate that such a method as transfinite induction up to $\varepsilon_0$ would have to be used. To what extent the Gentzen proof can be accepted as securing classical number theory in the sense of that problem formulation is in the present state of affairs a matter for individual judgement, depending on how ready one is to accept induction up to $\varepsilon_0$ as a finitary method.

## 28.3   Notes

[1] See Kleene (2009, pp. 476–499) for a full presentation of Gentzen's proof and various comments on the historic and philosophical significance of the result.

[2] Weyl (2012, p. 144).

## 28.4   References

- Gentzen, Gerhard (1936), "Die Widerspruchsfreiheit der reinen Zahlentheorie", *Mathematische Annalen* **112**: 493–565, doi:10.1007/BF01565428 - Translated as 'The consistency of arithmetic', in (Gentzen & Szabo 1969).

- Gentzen, Gerhard (1938), "Neue Fassung des Widerspruchsfreiheitsbeweises für die reine Zahlentheorie", *Forschungen zur Logik und zur Grundlegung der exakten Wissenschaften* **4**: 19–44 - Translated as 'New version of the consistency proof for elementary number theory', in (Gentzen & Szabo 1969).

- Gentzen, Gerhard (1969), M. E., Szabo, ed., *Collected Papers of Gerhard Gentzen*, Studies in logic and the foundations of mathematics (Hardcover ed.), Amsterdam: North-Holland, ISBN 0-7204-2254-X - an English translation of papers.

- Gödel, K. (2001) [1938], "Lecture at Zilsel's", in Feferman, Solomon, *Kurt Gödel: Collected Works*, vol.III Unpublished Essays and Lectures (Paperback ed.), Oxford University Press Inc., pp. 87–113, ISBN 0-19-514722-7

- Jervell, Herman Ruge (1999), *A course in proof theory* (textbook draft ed.)

- Kirby, L.; Paris, J. (1982), "Accessible independence results for Peano arithmetic" (PDF), *Bull. London Math. Soc.* (LMS) **14**: 285–293, doi:10.1112/blms/14.4.285

- Kleene, Stephen Cole (2009) [1952]. *Introduction to metamathematics.* Ishi Press International. ISBN 978-0-923891-57-2.

- Tait, W. W. (2005), "Gödel's reformulation of Gentzen's first consistency proof for arithmetic: the no-counterexample interpretation" (PDF), *The Bulletin of Symbolic Logic* (ASL) **11** (2): 225–238, doi:10.2178/bsl/1120231632, ISSN 1079-8986

- Weyl, Hermann (2012). *Levels of infinity: Selected writings on mathematics and philosophy.* New York: Dover Publications. ISBN 978-0-486-48903-2.

# Chapter 29

# Geometry of interaction

The **Geometry of Interaction** (GoI) was introduced by Jean-Yves Girard shortly after his work on Linear logic. In linear logic, proofs can be seen as various kinds of networks as opposed to the flat tree structures of sequent calculus. To distinguish the real proof nets from all the possible networks, Girard devised a criterium involving trips in the network. Trips can in fact be seen as some kind of operator acting on the proof. Drawing from this observation, Girard described directly this operator from the proof and has given a formula, the so-called *execution formula*, encoding the process of cut elimination at the level of operators.

One of the first significant applications of GoI was a better analysis[1] of Lamping's algorithm[2] for optimal reduction for the lambda calculus. GoI had a strong influence on game semantics for linear logic and PCF.

GoI has been applied to deep compiler optimisation for lambda calculi.[3] A bounded version of GoI dubbed **the Geometry of Synthesis** has been used to compile higher-order programming languages directly into static circuits.[4]

## 29.1 References

[1] Gonthier, G.; Abadi, M. N.; Lévy, J. J. (1992). "The geometry of optimal lambda reduction". *Proceedings of the 19th ACM SIGPLAN-SIGACT symposium on Principles of programming languages - POPL '92*. p. 15. doi:10.1145/143165.143172. ISBN 0897914538.

[2] Lamping, J. (1990). "An algorithm for optimal lambda calculus reduction". *Proceedings of the 17th ACM SIGPLAN-SIGACT symposium on Principles of programming languages - POPL '90*. p. 16. doi:10.1145/96709.96711. ISBN 0897913434.

[3] Mackie, I. (1995). "The geometry of interaction machine". *Proceedings of the 22nd ACM SIGPLAN-SIGACT symposium on Principles of programming languages - POPL '95*. p. 198. doi:10.1145/199448.199483. ISBN 0897916921.

[4] Dan R. Ghica. Function Interface Models for Hardware Compilation. MEMOCODE 2011.

## 29.2 Further reading

- GoI tutorial given at Siena 07 by Laurent Regnier, in the Linear Logic workshop,

# Chapter 30

# Original proof of Gödel's completeness theorem

The proof of Gödel's completeness theorem given by Kurt Gödel in his doctoral dissertation of 1929 (and a rewritten version of the dissertation, published as an article in 1930) is not easy to read today; it uses concepts and formalism that are no longer used and terminology that is often obscure. The version given below attempts to represent all the steps in the proof and all the important ideas faithfully, while restating the proof in the modern language of mathematical logic. This outline should not be considered a rigorous proof of the theorem.

## 30.1 Definitions and assumptions

We work with first-order predicate calculus. Our languages allow constant, function and relation symbols. Structures consist of (non-empty) domains and interpretations of the relevant symbols as constant members, functions or relations over that domain.

We fix some axiomatization of the predicate calculus: logical axioms and rules of inference. Any of the several well-known axiomatisations will do; we assume without proof all the basic well-known results about our formalism (such as the normal form theorem or the soundness theorem) that we need.

We axiomatize predicate calculus *without equality*, i.e. there are no special axioms expressing the properties of equality as a special relation symbol. After the basic form of the theorem is proved, it will be easy to extend it to the case of predicate calculus *with equality*.

## 30.2 Statement of the theorem and its proof

In the following, we state two equivalent forms of the theorem, and show their equivalence.

Later, we prove the theorem. This is done in the following steps:

1. Reducing the theorem to sentences (formulas with no free variables) in prenex form, i.e. with all quantifiers ($\forall$ and $\exists$) at the beginning. Furthermore, we reduce it to formulas whose first quantifier is $\forall$. This is possible because for every sentence, there is an equivalent one in prenex form whose first quantifier is $\forall$.

2. Reducing the theorem to sentences of the form $\forall x_1 \forall x_2 ... \forall x_k \exists y_1 \exists y_2 ... \exists y_m \, \varphi(x_1...x_k, y_1...y_m)$. While we cannot do this by simply rearranging the quantifiers, we show that it is yet enough to prove the theorem for sentences of that form.

3. Finally we prove the theorem for sentences of that form.

- This is done by first noting that a sentence such as B = $\exists x_1 \exists x_2 ... \exists x_k \, \exists y_1 \exists y_2 ... \exists y_m \, \varphi(x_1...x_k, y_1...y_m)$ is either refutable or has some model in which it holds; this model is simply assigning truth values to the subpropositions from which B is built. The reason for that is the completeness of propositional logic, with the existential quantifiers playing no role.

- We extend this result to more and more complex and lengthy sentences, $D_n$ (n=1,2...), built out from B, so that either any of them is refutable and therefore so is $\varphi$, or all of them are not refutable and therefore each holds in some model.

- We finally use the models in which the $D_n$ hold (in case all are not refutable) in order to build a model in which $\varphi$ holds.

### 30.2.1   Theorem 1. Every valid formula (true in all structures) is provable.

This is the most basic form of the completeness theorem. We immediately restate it in a form more convenient for our purposes:

### 30.2.2   Theorem 2. Every formula $\varphi$ is either refutable or satisfiable in some structure.

"$\varphi$ is refutable" means *by definition* "$\neg\varphi$ is provable".

### 30.2.3   Equivalence of both theorems

To see the equivalence, note first that if **Theorem 1** holds, and $\varphi$ is not satisfiable in any structure, then $\neg\varphi$ is valid in all structures and therefore provable, thus $\varphi$ is refutable and **Theorem 2** holds. If on the other hand **Theorem 2** holds and $\varphi$ is valid in all structures, then $\neg\varphi$ is not satisfiable in any structure and therefore refutable; then $\neg\neg\varphi$ is provable and then so is $\varphi$, thus **Theorem 1** holds.

### 30.2.4   Proof of theorem 2: first step

We approach the proof of **Theorem 2** by successively restricting the class of all formulas $\varphi$ for which we need to prove "$\varphi$ is either refutable or satisfiable". At the beginning we need to prove this for all possible formulas $\varphi$ in our language. However, suppose that for every formula $\varphi$ there is some formula $\psi$ taken from a more restricted class of formulas **C**, such that "$\psi$ is either refutable or satisfiable" → "$\varphi$ is either refutable or satisfiable". Then, once this claim (expressed in the previous sentence) is proved, it will suffice to prove "$\varphi$ is either refutable or satisfiable" only for $\varphi$'s belonging to the class **C**. Note also that if $\varphi$ is provably equivalent to $\psi$ (*i.e.*, ($\varphi\equiv\psi$) is provable), then it is indeed the case that "$\psi$ is either refutable or satisfiable" → "$\varphi$ is either refutable or satisfiable" (the soundness theorem is needed to show this).

There are standard techniques for rewriting an arbitrary formula into one that does not use function or constant symbols, at the cost of introducing additional quantifiers; we will therefore assume that all formulas are free of such symbols. Gödel's paper uses a version of first-order predicate calculus that has no function or constant symbols to begin with.

Next we consider a generic formula $\varphi$ (which no longer uses function or constant symbols) and apply the prenex form theorem to find a formula $\psi$ in *normal form* such that $\varphi\equiv\psi$ ($\psi$ being in *normal form* means that all the quantifiers in $\psi$, if there are any, are found at the very beginning of $\psi$). It follows now that we need only prove **Theorem 2** for formulas $\varphi$ in normal form.

Next, we eliminate all free variables from $\varphi$ by quantifying them existentially: if, say, $\mathbf{x_1...x_n}$ are free in $\varphi$, we form $\psi = \exists x_1...\exists x_n \phi$. If $\psi$ is satisfiable in a structure M, then certainly so is $\varphi$ and if $\psi$ is refutable, then $\neg\psi = \forall x_1...\forall x_n \neg\phi$ is provable, and then so is $\neg\varphi$, thus $\varphi$ is refutable. We see that we can restrict $\varphi$ to be a *sentence*, that is, a formula with no free variables.

Finally, we would like, for reasons of technical convenience, that the *prefix* of $\varphi$ (that is, the string of quantifiers at the beginning of $\varphi$, which is in normal form) begin with a universal quantifier and end with an existential quantifier. To achieve this for a generic $\varphi$ (subject to restrictions we have already proved), we take some one-place relation symbol **F** unused in φ, and two new variables **y** and **z**.. If φ = **(P)Φ**, where (P) stands for the prefix of φ and Φ for the *matrix* (the remaining, quantifier-free part of φ) we form $\psi = \forall y(P)\exists z(\Phi \wedge [F(y) \vee \neg F(z)])$ . Since $\forall y \exists z(F(y) \vee \neg F(z))$ is clearly provable, it is easy to see that $\phi = \psi$ is provable.

## 30.2.5 Reducing the theorem to formulas of degree 1

Our generic formula φ now is a sentence, in normal form, and its prefix starts with a universal quantifier and ends with an existential quantifier. Let us call the class of all such formulas **R**. We are faced with proving that every formula in **R** is either refutable or satisfiable. Given our formula φ, we group strings of quantifiers of one kind together in blocks:

$$\phi = (\forall x_1 ... \forall x_{k_1})(\exists x_{k_1+1} ... \exists x_{k_2}).......(\forall x_{k_{n-2}+1} ... \forall x_{k_{n-1}})(\exists x_{k_{n-1}+1} ... \exists x_{k_n})(\Phi)$$

We define the **degree** of $\phi$ to be the number of universal quantifier blocks, separated by existential quantifier blocks as shown above, in the prefix of $\phi$ . The following lemma, which Gödel adapted from Skolem's proof of the Löwenheim-Skolem theorem, lets us sharply reduce the complexity of the generic formula $\phi$ we need to prove the theorem for:

**Lemma.** Let **k>=1**. If every formula in **R** of degree **k** is either refutable or satisfiable, then so is every formula in **R** of degree **k+1**.

> **Comment**: Take a formula φ of degree k+1 of the form $\phi = (\forall x)(\exists y)(\forall u)(\exists v)(P)\psi$ , where $(P)\psi$ is the remainder of $\phi$ (it is thus of degree **k-1**). φ states that for every x there is a y such that... (something). It would have been nice to have a predicate $Q'$ so that for every x, $Q'(x,y)$ would be true if and only if y is the required one to make (something) true. Then we could have written a formula of degree k, which is equivalent to φ, namely $(\forall x')(\forall x)(\forall y)(\forall u)(\exists v)(\exists y')(P)Q'(x',y') \wedge (Q'(x,y) \rightarrow \psi)$ . This formula is indeed equivalent to φ because it states that for every x, if there is a y thatsatisfies Q'(x,y), then (something) holds, and furthermore, we know that there is such a y, because for every x', there is a y' that satisfies Q'(x',y'). Therefore φ follows from this formula. It is also easy to show that if the formula is false, then so is φ. **Unfortunately**, in general there is no such predicate Q'. However, this idea can be understood as a basis for the following proof of the Lemma.

**Proof.** Let φ be a formula of degree **k+1**; then we can write it as

$$\phi = (\forall x)(\exists y)(\forall u)(\exists v)(P)\psi$$

where **(P)** is the remainder of the prefix of $\phi$ (it is thus of degree **k-1**) and $\psi$ is the quantifier-free matrix of $\phi$ . **x, y, u** and **v** denote here *tuples* of variables rather than single variables; *e.g.* $(\forall x)$ really stands for $\forall x_1 \forall x_2 ... \forall x_n$ where $x_1 ... x_n$ are some distinct variables.

Let now **x'** and **y'** be tuples of previously unused variables of the same length as **x** and **y** respectively, and let **Q** be a previously unused relation symbol that takes as many arguments as the sum of lengths of **x** and y; we consider the formula

$$\Phi = (\forall x')(\exists y')Q(x',y') \wedge (\forall x)(\forall y)(Q(x,y) \rightarrow (\forall u)(\exists v)(P)\psi)$$

Clearly, $\Phi \rightarrow \phi$ is provable.

Now since the string of quantifiers $(\forall u)(\exists v)(P)$ does not contain variables from **x** or **y**, the following equivalence is easily provable with the help of whatever formalism we're using:

$$(Q(x,y) \rightarrow (\forall u)(\exists v)(P)\psi) \equiv (\forall u)(\exists v)(P)(Q(x,y) \rightarrow \psi)$$

And since these two formulas are equivalent, if we replace the first with the second inside Φ, we obtain the formula Φ' such that Φ≡Φ':

$$\Phi' = (\forall x')(\exists y')Q(x',y') \wedge (\forall x)(\forall y)(\forall u)(\exists v)(P)(Q(x,y) \rightarrow \psi)$$

Now $\Phi'$ has the form $(S)\rho \wedge (S')\rho'$ , where **(S)** and **(S')** are some quantifier strings, $\rho$ and $\rho'$ are quantifier-free, and, **furthermore**, no variable of **(S)** occurs in $\rho'$ and no variable of **(S')** occurs in $\rho$. Under such conditions every formula of the form $(T)(\rho \wedge \rho')$ , where **(T)** is a string of quantifiers containing all quantifiers in (S) and (S') interleaved among themselves in any fashion, but maintaining the relative order inside (S) and (S'), will be equivalent to the original formula $\Phi'$(this is yet another basic result in first-order predicate calculus that we rely on). To wit, we form $\Psi$ as follows:

$$\Psi = (\forall x')(\forall x)(\forall y)(\forall u)(\exists y')(\exists v)(P)Q(x',y') \wedge (Q(x,y) \rightarrow \psi)$$

and we have $\Phi' \equiv \Psi$ .

Now $\Psi$ is a formula of degree **k** and therefore by assumption either refutable or satisfiable. If $\Psi$ is satisfiable in a structure **M**, then, considering $\Psi \equiv \Phi' \equiv \Phi \wedge \Phi \rightarrow \phi$ , we see that $\phi$ is satisfiable as well. If $\Psi$ is refutable, then so is $\Phi$ , which is equivalent to it; thus $\neg\Phi$ is provable. Now we can replace all occurrences of **Q** inside the provable formula $\neg\Phi$ by some other formula dependent on the same variables, and we will still get a provable formula. (*This is yet another basic result of first-order predicate calculus. Depending on the particular formalism adopted for the calculus, it may be seen as a simple application of a "functional substitution" rule of inference, as in Gödel's paper, or it may be proved by considering the formal proof of $\neg\Phi$ , replacing in it all occurrences of Q by some other formula with the same free variables, and noting that all logical axioms in the formal proof remain logical axioms after the substitution, and all rules of inference still apply in the same way.*)

In this particular case, we replace Q(x',y') in $\neg\Phi$ with the formula $(\forall u)(\exists v)(P)\psi(x,y|x',y')$ . Here (x,y|x',y') means that instead of $\psi$ we are writing a different formula, in which x and y are replaced with x' and y'. Note that Q(x,y) is simply replaced by $(\forall u)(\exists v)(P)\psi$ .

$\neg\Phi$ then becomes

$$\neg((\forall x')(\exists y')(\forall u)(\exists v)(P)\psi(x,y|x',y') \wedge (\forall x)(\forall y)((\forall u)(\exists v)(P)\psi \rightarrow (\forall u)(\exists v)(P)\psi))$$

and this formula is provable; since the part under negation and after the $\wedge$ sign is obviously provable, and the part under negation and before the $\wedge$ sign is obviously φ, just with **x** and **y** replaced by **x'** and **y'**, we see that $\neg\phi$ is provable, and φ is refutable. We have proved that φ is either satisfiable or refutable, and this concludes the proof of the **Lemma**.

Notice that we could not have used $(\forall u)(\exists v)(P)\psi(x,y|x',y')$ instead of Q(x',y') from the beginning, because $\Psi$ would not have been a well-formed formula in that case. This is why we cannot naively use the argument appearing at the comment that precedes the proof.

### 30.2.6  Proving the theorem for formulas of degree 1

As shown by the **Lemma** above, we only need to prove our theorem for formulas φ in **R** of degree 1. φ cannot be of degree 0, since formulas in R have no free variables and don't use constant symbols. So the formula φ has the general form:

$$(\forall x_1...x_k)(\exists y_1...y_m)\phi(x_1...x_k, y_1...y_m).$$

Now we define an ordering of the k-tuples of natural numbers as follows: $(x_1...x_k) < (y_1...y_k)$ should hold if either $\Sigma_k(x_1...x_k) < \Sigma_k(y_1...y_k)$ , or $\Sigma_k(x_1...x_k) = \Sigma_k(y_1...y_k)$ , and $(x_1...x_k)$ precedes $(y_1...y_k)$ in lexicographic order. [Here $\Sigma_k(x_1...x_k)$ denotes the sum of the terms of the tuple.] Denote the nth tuple in this order by $(a_1^n...a_k^n)$ .

Set the formula $B_n$ as $\phi(z_{a_1^n}...z_{a_k^n}, z_{(n-1)m+2}, z_{(n-1)m+3}...z_{nm+1})$ . Then put $D_n$ as

$(\exists z_1...z_{nm+1})(B_1 \wedge B_2... \wedge B_n)$.

**Lemma**: For every $n$, $\varphi \to D_n$ .

**Proof**: By induction on n; we have

$$D_k \Leftarrow D_{k-1} \wedge (\forall z_1...z_{(n-1)m+1})(\exists z_{(n-1)m+2}...z_{nm+1})B_n \Leftarrow D_{k-1} \wedge (\forall z_{a_1^n}...z_{a_k^n})(\exists y_1...y_m)\phi(z_{a_1^n}...z_{a_k^n}, y_1...y_m)$$

, where the latter implication holds by variable substitution, since the ordering of the tuples is such that $(\forall k)(a_1^n...a_k^n) < (n-1)m + 2$ . But the last formula is equivalent to $D_{k-1} \wedge \varphi$.

For the base case, $D_1 \equiv (\exists z_1...z_{m+1})\phi(z_{a_1^1}...z_{a_k^1}, z_2, z_3...z_{m+1}) \equiv (\exists z_1...z_{m+1})\phi(z_1...z_1, z_2, z_3...z_{m+1})$ is obviously a corollary of $\varphi$ as well. So the **Lemma** is proven.

Now if $D_n$ is refutable for some $n$, it follows that $\varphi$ is refutable. On the other hand, suppose that $D_n$ is not refutable for any $n$. Then for each $n$ there is some way of assigning truth values to the distinct subpropositions $E_h$ (ordered by their first appearance in $D_n$ ; "distinct" here means either distinct predicates, or distinct bound variables) in $B_k$ , such that $D_n$ will be true when each proposition is evaluated in this fashion. This follows from the completeness of the underlying propositional logic.

We will now show that there is such an assignment of truth values to $E_h$ , so that all $D_n$ will be true: The $E_h$ appear in the same order in every $D_n$ ; we will inductively define a general assignment to them by a sort of "majority vote": Since there are infinitely many assignments (one for each $D_n$ ) affecting $E_1$ , either infinitely many make $E_1$ true, or infinitely many make it false and only finitely many make it true. In the former case, we choose $E_1$ to be true in general; in the latter we take it to be false in general. Then from the infinitely many $n$ for which $E_1$ through $E_{h-1}$ are assigned the same truth value as in the general assignment, we pick a general assignment to $E_h$ in the same fashion.

This general assignment must lead to every one of the $B_k$ and $D_k$ being true, since if one of the $B_k$ were false under the general assignment, $D_n$ would also be false for every $n > k$. But this contradicts the fact that for the finite collection of general $E_h$ assignments appearing in $D_k$ , there are infinitely many $n$ where the assignment making $D_n$ true matches the general assignment.

From this general assignment, which makes all of the $D_k$ true, we construct an interpretation of the language's predicates that makes $\varphi$ true. The universe of the model will be the natural numbers. Each i-ary predicate $\Psi$ should be true of the naturals $(u_1...u_i)$ precisely when the proposition $\Psi(z_{u_1}...z_{u_i})$ is either true in the general assignment, or not assigned by it (because it never appears in any of the $D_k$ ).

In this model, each of the formulas $(\exists y_1...y_m)\phi(a_1^n...a_k^n, y_1...y_m)$ is true by construction. But this implies that $\varphi$ itself is true in the model, since the $a^n$ range over all possible k-tuples of natural numbers. So $\varphi$ is satisfiable, and we are done.

**Intuitive explanation**

We may write each $B_i$ as $\Phi(x_1...x_k, y_1...y_m)$ for some x-s, which we may call "first arguments" and y-s that we may call "last arguments".

Take $B_1$ for example. Its "last arguments" are $z_2, z_3...z_{m+1}$, and for every possible combination of k of these variables there is some j so that they appear as "first arguments" in $B_j$. Thus for large enough $n_1$, $D_{n1}$ has the property that the "last arguments" of $B_1$ appear, in every possible combinations of k of them, as "first arguments" in other $B_j$-s within $D_n$. For every $B_i$ there is a $D_{ni}$ with the corresponding property.

Therefore in a model that satisfies all the $D_n$-s, there are objects corresponding to $z_1, z_2...$ and each combination of k of these appear as "first arguments" in some $B_j$, meaning that for every k of these objects $z_{p1}...z_{pk}$ there are $z_{q1}...z_{qm}$, which makes $\Phi(z_{p1}...z_{pk}, z_{q1}...z_{qm})$ satisfied. By taking a submodel with only these $z_1, z_2...$ objects, we have a model satisfying $\varphi$.

# 30.3 Extensions

### 30.3.1 Extension to first-order predicate calculus with equality

Gödel reduced a formula containing instances of the equality predicate to ones without it in an extended language. His method involves replacing a formula $\varphi$ containing some instances of equality with the formula

$$(\forall x)Eq(x,x) \wedge (\forall x,y,z)[Eq(x,y) \rightarrow (Eq(x,z) \rightarrow Eq(y,z))] \wedge (\forall x,y,z)[Eq(x,y) \rightarrow (Eq(z,x) \rightarrow$$
$$Eq(z,y))] \wedge (\forall x_1...x_k, y_1...y_k)[(Eq(x_1,y_1) \wedge ... \wedge Eq(x_k,y_k)) \rightarrow (A(x_1...x_k) \equiv A(y_1...y_k))] \wedge ... \wedge$$
$$(\forall x_1...x_m, y_1...y_m)[(Eq(x_1,y_1) \wedge ... \wedge Eq(x_m,y_m)) \rightarrow (Z(x_1...x_m) \equiv Z(y_1...y_m))] \wedge \varphi'.$$

Here $A...Z$ denote the predicates appearing in $\varphi$ (with $k...m$ their respective arities), and $\varphi'$ is the formula $\varphi$ with all occurrences of equality replaced with the new predicate $Eq$. If this new formula is refutable, the original $\varphi$ was as well; the same is true of satisfiability, since we may take a quotient of satisfying model of the new formula by the equivalence relation representing $Eq$. This quotient is well-defined with respect to the other predicates, and therefore will satisfy the original formula $\varphi$.

### 30.3.2 Extension to countable sets of formulas

Gödel also considered the case where there are a countably infinite collection of formulas. Using the same reductions as above, he was able to consider only those cases where each formula is of degree 1 and contains no uses of equality. For a countable collection of formulas $\phi^i$ of degree 1, we may define $B_k^i$ as above; then define $D_k$ to be the closure of $B_1^1...B_k^1, ..., B_1^k...B_k^k$. The remainder of the proof then went through as before.

### 30.3.3 Extension to arbitrary sets of formulas

When there is an uncountably infinite collection of formulas, the Axiom of Choice (or at least some weak form of it) is needed. Using the full AC, one can well-order the formulas, and prove the uncountable case with the same argument as the countable one, except with transfinite induction. Other approaches can be used to prove that the completeness theorem in this case is equivalent to the Boolean prime ideal theorem, a weak form of AC.

## 30.4 References

- Gödel, K (1929). "Über die Vollständigkeit des Logikkalküls". Doctoral dissertation. University Of Vienna. The first proof of the completeness theorem.

- Gödel, K (1930). "Die Vollständigkeit der Axiome des logischen Funktionenkalküls". *Monatshefte für Mathematik* (in German) **37** (1): 349–360. doi:10.1007/BF01696781. JFM 56.0046.04. The same material as the dissertation, except with briefer proofs, more succinct explanations, and omitting the lengthy introduction.

## 30.5 External links

- Stanford Encyclopedia of Philosophy: "Kurt Gödel"—by Juliette Kennedy.

- MacTutor biography: Kurt Gödel.

# Chapter 31

# Gödel's speed-up theorem

In mathematics, **Gödel's speed-up theorem**, proved by Gödel (1936), shows that there are theorems whose proofs can be drastically shortened by working in more powerful axiomatic systems.

Kurt Gödel showed how to find explicit examples of statements in formal systems that are provable in that system but whose shortest proof is absurdly long. For example, the statement:

"This statement cannot be proved in Peano arithmetic in fewer than a googolplex symbols"

is provable in Peano arithmetic (PA) but the shortest proof has at least a googolplex symbols, by an argument similar to the proof of Gödel's first incompleteness theorem: PA (if consistent) cannot prove the statement in fewer than a googolplex symbols, because the existence of such a proof would itself be a theorem of PA, that would contradict the statement which PA supposedly proved. But simply enumerating all strings of length up to a googolplex and checking that each such string is not a proof (in PA) of the statement, yields a proof of the statement that is necessarily longer than a googolplex symbols.

The statement has a short proof in a more powerful system: in fact it is easily provable in Peano arithmetic together with the statement that Peano arithmetic is consistent (which, per the incompleteness theorem, cannot be proved in Peano arithmetic).

In this argument, Peano arithmetic can be replaced by any more powerful consistent system, and a googolplex can be replaced by any number that can be described concisely in the system.

Harvey Friedman found some explicit natural examples of this phenomenon, giving some explicit statements in Peano arithmetic and other formal systems whose shortest proofs are ridiculously long (Smoryński 1982).

## 31.1 See also

- Blum's speedup theorem

- Blum's size theorem ("On the Size of Machines", Blum 1967)

- List of long proofs

## 31.2 References

- Buss, Samuel R. (1994), "On Gödel's theorems on lengths of proofs. I. Number of lines and speedup for arithmetics", *The Journal of Symbolic Logic* **59** (3): 737–756, doi:10.2307/2275906, ISSN 0022-4812, MR 1295967

- Buss, Samuel R. (1995), "On Gödel's theorems on lengths of proofs. II. Lower bounds for recognizing k symbol provability", in Clote, Peter; Remmel, Jeffrey, *Feasible mathematics, II (Ithaca, NY, 1992)*, Progr. Comput. Sci. Appl. Logic **13**, Boston, MA: Birkhäuser Boston, pp. 57–90, ISBN 978-0-8176-3675-3, MR 1322274

- Gödel, Kurt (1936), "Über die Länge von Beweisen", *Ergebinisse eines mathematischen Kolloquiums* (in German) **7**: 23–24, Reprinted with English translation in volume 1 of his collected works.

- Smoryński, C. (1982), "The varieties of arboreal experience", *Math. Intelligencer* **4** (4): 182–189, doi:10.1007/bf03023553, MR 0685558

# Chapter 32

# Double-negation translation

In proof theory, a discipline within mathematical logic, **double-negation translation**, sometimes called **negative translation**, is a general approach for embedding classical logic into intuitionistic logic, typically by translating formulas to formulas which are classically equivalent but intuitionistically inequivalent. Particular instances of double-negation translation include **Glivenko's translation** for propositional logic, and the **Gödel–Gentzen translation** and **Kuroda's translation** for first-order logic.

## 32.1   Propositional logic

The easiest double-negation translation to describe comes from **Glivenko's theorem**, proved by Valery Glivenko in 1929. It maps each classical formula $\varphi$ to its double negation $\neg\neg\varphi$.

Glivenko's theorem states:

> If $\varphi$ is a propositional formula, then $\varphi$ is a classical tautology if and only if $\neg\neg\varphi$ is an intuitionistic tautology.

Glivenko's theorem implies the more general statement:

> If $T$ is a set of propositional formulas, $T^*$ a set consisting of the double negated formulas of $T$, and $\varphi$ a propositional formula, then $T \vdash \varphi$ in classical logic if and only if $T^* \vdash \neg\neg\varphi$ in intuitionistic logic.

In particular, a set of propositional formulas is intuitionistically consistent if and only if it is classically satisfiable.

## 32.2   First-order logic

The *Gödel–Gentzen translation* (named after Kurt Gödel and Gerhard Gentzen) associates with each formula $\varphi$ in a first-order language another formula $\varphi^N$, which is defined inductively:

- If $\phi$ is atomic, then $\phi^N$ is $\neg\neg\phi$

- $(\phi \wedge \theta)^N$ is $\phi^N \wedge \theta^N$

- $(\phi \vee \theta)^N$ is $\neg(\neg\phi^N \wedge \neg\theta^N)$

- $(\phi \to \theta)^N$ is $\phi^N \to \theta^N$

- $(\neg\phi)^N$ is $\neg\phi^N$

- $(\forall\, x\, \phi)^N$ is $\forall\, x\, \phi^N$

- $(\exists\, x\, \phi)^N$ is $\neg\forall\, x\, \neg\phi^N$

Notice that $\varphi^N$ is classically equivalent to $\varphi$.

The fundamental soundness theorem states:[1]

> If $T$ is a set of axioms and $\varphi$ a formula, then $T$ proves $\varphi$ using classical logic if and only if $T^N$ proves $\varphi^N$ using intuitionistic logic.

Here $T^N$ consists of the double-negation translations of the formulas in $T$.

Note that $\varphi$ need not imply its negative translation $\varphi^N$ in intuitionistic first-order logic. Troelsta and Van Dalen[2] give a description (due to Leivant) of formulas which do imply their Gödel–Gentzen translation.

## 32.3   Variants

There are several alternative definitions of the negative translation. They are all provably equivalent in intuitionistic logic, but may be easier to apply in particular contexts.

One possibility is to change the clauses for disjunction and existential quantifier to

- $(\varphi \vee \theta)^N$ is $\neg\neg(\varphi^N \vee \theta^N)$

- $(\exists x\, \varphi)^N$ is $\neg\neg\exists x\, \varphi^N$

Then the translation can be succinctly described as: prefix $\neg\neg$ to every atomic formula, disjunction, and existential quantifier.

Another possibility (known as Kuroda's translation) is to construct $\varphi^N$ from $\varphi$ by putting $\neg\neg$ before the whole formula and after every universal quantifier. Notice that this reduces to the simple $\neg\neg\varphi$ translation if $\varphi$ is propositional.

It is also possible to define $\varphi^N$ by prefixing $\neg\neg$ before every subformula of $\varphi$, as done by Kolmogorov. Such a translation is the logical counterpart to the call-by-name continuation-passing style translation of functional programming languages along the lines of the Curry–Howard correspondence between proofs and programs.

## 32.4   Results

The double-negation translation was used by Gödel (1933) to study the relationship between classical and intuitionistic theories of the natural numbers ("arithmetic"). He obtains the following result:

> If a formula $\varphi$ is provable from the axioms of Peano arithmetic then $\varphi^N$ is provable from the axioms of intuitionistic Heyting arithmetic.

This result shows that if Heyting arithmetic is consistent then so is Peano arithmetic. This is because a contradictory formula $\theta \wedge \neg\theta$ is interpreted as $\theta^N \wedge \neg\theta^N$, which is still contradictory. Moreover, the proof of the relationship is entirely constructive, giving a way to transform a proof of $\theta \wedge \neg\theta$ in Peano arithmetic into a proof of $\theta^N \wedge \neg\theta^N$ in Heyting arithmetic. (By combining the double-negation translation with the Friedman translation, it is in fact possible to prove that Peano arithmetic is $\Pi^0_2$-conservative over Heyting arithmetic.)

The propositional mapping of $\varphi$ to $\neg\neg\varphi$ does not extend to a sound translation of first-order logic, because $\forall x\, \neg\neg\varphi(x) \rightarrow \neg\neg\forall x\, \varphi(x)$ is not a theorem of intuitionistic predicate logic. This explains why $\varphi^N$ has to be defined in a more complicated way in the first-order case.

## 32.5   See also

- Dialectica interpretation

## 32.6   Notes

[1] Avigad and Feferman 1998, p. 342; Buss 1998 p. 66

[2] Troelsta, van Dalen 1988, Ch. 2, Sec. 3)

## 32.7   References

- J. Avigad and S. Feferman (1998), "Gödel's Functional ("Dialectica") Interpretation", *Handbook of Proof Theory"*, S. Buss, ed., Elsevier. ISBN 0-444-89840-9

- S. Buss (1998), "Introduction to Proof Theory", *Handbook of Proof Theory*, S. Buss, ed., Elsevier. ISBN 0-444-89840-9

- G. Gentzen (1936), "Die Widerspruchfreiheit der reinen Zahlentheorie", *Mathematische Annalen*, v. 112, pp. 493–565 (German). Reprinted in English translation as "The consistency of arithmetic" in *The collected papers of Gerhard Gentzen*, M. E. Szabo, ed.

- V. Glivenko (1929), *Sur quelques points de la logique de M. Brouwer*, Bull. Soc. Math. Belg. 15, 183-188

- K. Gödel (1933), "Zur intuitionistischen Arithmetik und Zahlentheorie", *Ergebnisse eines mathematischen Kolloquiums*, v. 4, pp. 34–38 (German). Reprinted in English translation as "On intuitionistic arithmetic and number theory" in *The Undecidable*, M. Davis, ed., pp. 75–81.

- A. N. Kolmogorov (1925), "O principe tertium non datur" (Russian). Reprinted in English translation as "On the principle of the excluded middle" in *From Frege to Gödel*, van Heijenoort, ed., pp. 414–447.

- A. S. Troelsta (1977), "Aspects of Constructive Mathematics", *Handbook of Mathematical Logic"*, J. Barwise, ed., North-Holland. *ISBN 0-7204-2285-X*

- A.S. Troelsta and D. van Dalen (1988), *Constructivism in Mathematics. An Introduction*, volumes 121, 123 of *Studies in Logic and the Foundations of Mathematics*, North–Holland.

## 32.8   External links

- "Intuitionistic logic", Stanford Encyclopedia of Philosophy.

# Chapter 33

# Gödel's completeness theorem

**Gödel's completeness theorem** is a fundamental theorem in mathematical logic that establishes a correspondence between semantic truth and syntactic provability in first-order logic. It makes a close link between model theory that deals with what is true in different models, and proof theory that studies what can be formally proven in particular formal systems.

It was first proved by Kurt Gödel in 1929. It was then simplified in 1947, when Leon Henkin observed in his Ph.D. thesis that the hard part of the proof can be presented as the Model Existence Theorem (published in 1949). Henkin's proof was simplified by Gisbert Hasenjaeger in 1953.

## 33.1   Statement of the theorem

### 33.1.1   Preliminaries

There are numerous deductive systems for first-order logic, including systems of natural deduction and Hilbert-style systems. Common to all deductive systems is the notion of a **formal deduction**. This is a sequence (or, in some cases, a finite tree) of formulas with a specially-designated **conclusion**. The definition of a deduction is such that it is finite and that it is possible to verify algorithmically (by a computer, for example, or by hand) that a given sequence (or tree) of formulas is indeed a deduction.

A first-order formula is called **logically valid** if it is true in every structure for the language of the formula (i.e. for any assignment of values to the variables of the formula). To formally state, and then prove, the completeness theorem, it is necessary to also define a deductive system. A deductive system is called **complete** if every logically valid formula is the conclusion of some formal deduction, and the completeness theorem for a particular deductive system is the theorem that it is complete in this sense. Thus, in a sense, there is a different completeness theorem for each deductive system. A converse to completeness is **soundness,** the fact that only logically valid formulas are provable in the deductive system.

If some specific deductive system of first-order logic is sound and complete, then it is "perfect" (a formula is provable if and only if it is a semantic consequence of the axioms), thus equivalent to any other deductive system with the same quality (any proof in one system can be converted into the other).

### 33.1.2   Gödel's original formulation

The completeness theorem says that if a formula is logically valid then there is a finite deduction (a formal proof) of the formula.

Gödel's completeness theorem says that a deductive system of first-order predicate calculus is "complete" in the sense that no additional inference rules are required to prove all the logically valid formulas. A converse to completeness is **soundness,** the fact that only logically valid formulas are provable in the deductive system. Together with soundness (whose verification is easy), this theorem implies that a formula is logically valid if and only if it is the conclusion of a formal deduction.

### 33.1.3 Model existence theorem

The simplest version of this theorem that suffices in practice for most needs, and has connections with the Löwenheim–Skolem theorem, says:

Every consistent, countable first-order theory has a finite or countable model

A more general version can be expressed as :

Every consistent first-order theory with a well-orderable language has a model.

Here, a consistent theory is defined as one in which, for no formula F, both F and ¬F can be proven. See Consistency, the syntactic definition; the semantic definition would be tautological in this context.

This theorem by Henkin is the most directly obtained version of the completeness theorem in its simplest proof.

Given Henkin's theorem, the proof of the completeness theorem is as follows: If $\models A$ is valid, then $\neg A$ does not have models. By the contrapositive of Henkin's, then $\neg A$ is an inconsistent formula. But, by the definition of consistency, if $\neg A$ is inconsistent then it's possible to build a proof of $\vdash A$ .

### 33.1.4 More general form

It says that for any first-order theory $T$ with a well-orderable language, and any sentence $s$ in the language of the theory, there is a formal proof of $s$ in $T$ if and only if $s$ is satisfied by every model of $T$ ($s$ is a semantic consequence of $T$).

This more general theorem is used implicitly, for example, when a sentence is shown to be provable from the axioms of group theory by considering an arbitrary group and showing that the sentence is satisfied by that group. It is deduced from the model existence theorem as follows: if there is no formal proof of a formula then adding its negation to the axioms gives a consistent theory, which has thus a model, so that the formula is not a semantic consequence of the initial theory.

Gödel's original formulation is deduced by taking the particular case of a theory without any axiom.

### 33.1.5 As a theorem of arithmetic

The Model Existence Theorem and its proof can be formalized in the framework of Peano arithmetic. Precisely, we can systematically define a model of any consistent effective first-order theory $T$ in Peano arithmetic by interpreting each symbol of $T$ by an arithmetical formula whose free variables are the arguments of the symbol. However, the definition expressed by this formula is not recursive.

## 33.2 Consequences

An important consequence of the completeness theorem is that it is possible to recursively enumerate the semantic consequences of any effective first-order theory, by enumerating all the possible formal deductions from the axioms of the theory, and use this to produce an enumeration of their conclusions.

This comes in contrast with the direct meaning of the notion of semantic consequence, that quantifies over all structures in a particular language, which is clearly not a recursive definition.

Also, it makes the concept of "provability," and thus of "theorem," a clear concept that only depends on the chosen system of axioms of the theory, and not on the choice of a proof system.

## 33.3   Relationship to the incompleteness theorem

Gödel's incompleteness theorem, another celebrated result, shows that there are inherent limitations in what can be achieved with formal proofs in mathematics. The name for the incompleteness theorem refers to another meaning of *complete* (see model theory – Using the compactness and completeness theorems).

It shows that in any consistent effective theory $T$ containing Peano arithmetic (PA), the formula $CT$ expressing the consistency of $T$ cannot be proven within $T$.

Applying the completeness theorem to this result, gives the existence of a model of $T$ where the formula $CT$ is false. Such a model (precisely, the set of "natural numbers" it contains) is necessarily non-standard, as it contains the code number of a proof of a contradiction of $T$. But $T$ is consistent when viewed from the outside. Thus this code number of a proof of contradiction of $T$ must be a non-standard number.

In fact, the model of *any* theory containing PA obtained by the systematic construction of the arithmetical model existence theorem, is *always* non-standard with a non-equivalent provability predicate and a non-equivalent way to interpret its own construction, so that this construction is non-recursive (as recursive definitions would be unambiguous).

Also, there is no recursive non-standard model of PA.

## 33.4   Relationship to the compactness theorem

The completeness theorem and the compactness theorem are two cornerstones of first-order logic. While neither of these theorems can be proven in a completely effective manner, each one can be effectively obtained from the other.

The compactness theorem says that if a formula $\varphi$ is a logical consequence of a (possibly infinite) set of formulas $\Gamma$ then it is a logical consequence of a finite subset of $\Gamma$. This is an immediate consequence of the completeness theorem, because only a finite number of axioms from $\Gamma$ can be mentioned in a formal deduction of $\varphi$, and the soundness of the deduction system then implies $\varphi$ is a logical consequence of this finite set. This proof of the compactness theorem is originally due to Gödel.

Conversely, for many deductive systems, it is possible to prove the completeness theorem as an effective consequence of the compactness theorem.

The ineffectiveness of the completeness theorem can be measured along the lines of reverse mathematics. When considered over a countable language, the completeness and compactness theorems are equivalent to each other and equivalent to a weak form of choice known as weak König's lemma, with the equivalence provable in RCA$_0$ (a second-order variant of Peano arithmetic restricted to induction over $\Sigma^0{}_1$ formulas). Weak König's lemma is provable in ZF, the system of Zermelo–Fraenkel set theory without axiom of choice, and thus the completeness and compactness theorems for countable languages are provable in ZF. However the situation is different when the language is of arbitrary large cardinality since then, though the completeness and compactness theorems remain provably equivalent to each other in ZF, they are also provably equivalent to a weak form of the axiom of choice known as the ultrafilter lemma. In particular, no theory extending ZF can prove either the completeness or compactness theorems over arbitrary (possibly uncountable) languages without also proving the ultrafilter lemma on a set of same cardinality, knowing that on countable sets, the ultrafilter lemma becomes equivalent to weak König's lemma.

## 33.5   Completeness in other logics

The completeness theorem is a central property of first-order logic that does not hold for all logics. Second-order logic, for example, does not have a completeness theorem for its standard semantics (but does have the completeness property for Henkin semantics), and the same is true of all higher-order logics. It is possible to produce sound deductive systems for higher-order logics, but no such system can be complete. The set of logically-valid formulas in second-order logic is not enumerable.

Lindström's theorem states that first-order logic is the strongest (subject to certain constraints) logic satisfying both compactness and completeness.

A completeness theorem can be proved for modal logic or intuitionistic logic with respect to Kripke semantics.

## 33.6  Proofs

Gödel's original proof of the theorem proceeded by reducing the problem to a special case for formulas in a certain syntactic form, and then handling this form with an *ad hoc* argument.

In modern logic texts, Gödel's completeness theorem is usually proved with Henkin's proof, rather than with Gödel's original proof. Henkin's proof directly constructs a term model for any consistent first-order theory. James Margetson (2004) developed a computerized formal proof using the Isabelle theorem prover. Other proofs are also known.

## 33.7  See also

- Gödel's incompleteness theorems

- Original proof of Gödel's completeness theorem

## 33.8  Further reading

- Gödel, K (1929). "Über die Vollständigkeit des Logikkalküls". Doctoral dissertation. University Of Vienna. The first proof of the completeness theorem.

- Gödel, K (1930). "Die Vollständigkeit der Axiome des logischen Funktionenkalküls". *Monatshefte für Mathematik* (in German) **37** (1): 349–360. doi:10.1007/BF01696781. JFM 56.0046.04. The same material as the dissertation, except with briefer proofs, more succinct explanations, and omitting the lengthy introduction.

## 33.9  External links

- Stanford Encyclopedia of Philosophy: "Kurt Gödel"—by Juliette Kennedy.

- MacTutor biography: Kurt Gödel.

- Detlovs, Vilnis, and Podnieks, Karlis, "Introduction to mathematical logic."

# Chapter 34

# Gödel's incompleteness theorems

**Gödel's incompleteness theorems** are two theorems of mathematical logic that establish inherent limitations of all but the most trivial axiomatic systems capable of doing arithmetic. The theorems, proven by Kurt Gödel in 1931, are important both in mathematical logic and in the philosophy of mathematics. The two results are widely, but not universally, interpreted as showing that Hilbert's program to find a complete and consistent set of axioms for all mathematics is impossible, giving a negative answer to Hilbert's second problem.

The first incompleteness theorem states that no consistent system of axioms whose theorems can be listed by an "effective procedure" (i.e., any sort of algorithm) is capable of proving all truths about the relations of the natural numbers (arithmetic). For any such system, there will always be statements about the natural numbers that are true, but that are unprovable within the system. The second incompleteness theorem, an extension of the first, shows that such a system cannot demonstrate its own consistency.

## 34.1 Background

Because statements of a formal theory are written in symbolic form, it is possible to verify mechanically that a formal proof from a finite set of axioms is valid. This task, known as automatic proof verification, is closely related to automated theorem proving. The difference is that instead of constructing a new proof, the proof verifier simply checks that a provided formal proof (or, in instructions that can be followed to create a formal proof) is correct. This process is not merely hypothetical; systems such as Isabelle and Coq are used today to formalize proofs and then check their validity.

Many theories of interest include an infinite set of axioms, however. To verify a formal proof when the set of axioms is infinite, it must be possible to determine whether a statement that is claimed to be an axiom is actually an axiom. This issue arises in first order theories of arithmetic, such as Peano arithmetic, because the principle of mathematical induction is expressed as an infinite set of axioms (an axiom schema).

A formal theory is said to be *effectively generated* if its set of axioms is a recursively enumerable set. This means that there is a computer program that, in principle, could enumerate all the axioms of the theory without listing any statements that are not axioms. This is equivalent to the existence of a program that enumerates all the theorems of the theory without enumerating any statements that are not theorems. Examples of effectively generated theories with infinite sets of axioms include Peano arithmetic and Zermelo–Fraenkel set theory.

In choosing a set of axioms, one goal is to be able to prove as many correct results as possible, without proving any incorrect results. A set of axioms is complete if, for any statement in the axioms' language, that statement or its negation is provable from the axioms. A set of axioms is (simply) consistent if there is no statement such that both the statement and its negation are provable from the axioms. In the standard system of first-order logic, an inconsistent set of axioms will prove every statement in its language (this is sometimes called the principle of explosion), and is thus automatically complete. A set of axioms that is both complete and consistent, however, proves a maximal set of non-contradictory theorems. Gödel's incompleteness theorems show that in certain cases, it is not possible to obtain a theory that is effectively generated and complete and consistent.

# 34.2 First incompleteness theorem

**Gödel's first incompleteness theorem** first appeared as "Theorem VI" in Gödel's 1931 paper *On Formally Undecidable Propositions of Principia Mathematica and Related Systems I.*[1]

The formal theorem is written in highly technical language. It may be paraphrased in English as:

> Any effectively generated theory capable of expressing elementary arithmetic cannot be both consistent and complete. In particular, for any consistent, effectively generated formal theory that proves certain basic arithmetic truths, there is an arithmetical statement that is true,[2] but not provable in the theory.

The true but unprovable statement referred to by the theorem is often referred to as "the Gödel sentence" for the theory. The proof constructs a specific Gödel sentence for each consistent effectively generated theory, but there are infinitely many statements in the language of the theory that share the property of being true but unprovable. For example, the conjunction of the Gödel sentence and any logically valid sentence will have this property.

For each consistent formal theory $T$ having the required small amount of number theory, the corresponding Gödel sentence $G$ asserts: "$G$ cannot be proved within the theory $T$". This interpretation of $G$ leads to the following informal analysis. If $G$ were provable under the axioms and rules of inference of $T$, then $T$ would have a theorem, $G$, which effectively contradicts itself, and thus the theory $T$ would be inconsistent. This means that if the theory $T$ is consistent then $G$ cannot be proved within it, and so the theory $T$ is incomplete. Moreover, the claim $G$ makes about its own unprovability is correct. In this sense $G$ is not only unprovable but true, and provability-within-the-theory-$T$ is not the same as truth. This informal analysis can be formalized to make a rigorous proof of the incompleteness theorem, as described in the section "Proof sketch for the first theorem" below. The formal proof reveals exactly the hypotheses required for the theory $T$ in order for the self-contradictory nature of $G$ to lead to a genuine contradiction.

Each effectively generated theory has its own Gödel statement. It is possible to define a larger theory $T'$ that contains the whole of $T$, plus $G$ as an additional axiom. This will not result in a complete theory, because Gödel's theorem will also apply to $T'$, and thus $T'$ cannot be complete. In this case, $G$ is indeed a theorem in $T'$, because it is an axiom. Since $G$ states only that it is not provable in $T$, no contradiction is presented by its provability in $T'$. However, because the incompleteness theorem applies to $T'$, there will be a new Gödel statement $G'$ for $T'$, showing that $T'$ is also incomplete. $G'$ will differ from $G$ in that $G'$ will refer to $T'$, rather than $T$.

To prove the first incompleteness theorem, Gödel represented statements by numbers. Then the theory at hand, which is assumed to prove certain facts about numbers, also proves facts about its own statements, provided that it is effectively generated. Questions about the provability of statements are represented as questions about the properties of numbers, which would be decidable by the theory if it were complete. In these terms, the Gödel sentence states that no natural number exists with a certain, strange property. A number with this property would encode a proof of the inconsistency of the theory. If there were such a number then the theory would be inconsistent, contrary to the consistency hypothesis. So, under the assumption that the theory is consistent, there is no such number.

## 34.2.1 Meaning of the first incompleteness theorem

Gödel's first incompleteness theorem shows that any consistent effectively generated formal system that includes enough of the theory of the natural numbers is incomplete: there are true statements expressible in its language that are unprovable within the system. Thus no formal system (satisfying the hypotheses of the theorem) that aims to characterize the natural numbers can actually do so, as there will be true number-theoretical statements that that system cannot prove. This fact is sometimes thought to have severe consequences for the program of logicism proposed by Gottlob Frege and Bertrand Russell, which aimed to define the natural numbers in terms of logic (Hellman 1981, p. 451–468). Bob Hale and Crispin Wright argue that it is not a problem for logicism because the incompleteness theorems apply equally to first order logic as they do to arithmetic. They argue that only those who believe that the natural numbers are to be defined in terms of first order logic have this problem.

The existence of an incomplete formal system is, in itself, not particularly surprising. A system may be incomplete simply because not all the necessary axioms have been discovered. For example, Euclidean geometry without the parallel postulate is incomplete; it is not possible to prove or disprove the parallel postulate from the remaining axioms.

Gödel's theorem shows that, in theories that include a small portion of number theory, a complete and consistent finite list of axioms can *never* be created: each time a new statement is added as an axiom, there are other true statements that still cannot be proved, even with the new axiom. If an axiom is ever added that makes the system complete, it does so at the cost of making the system inconsistent. It is not even possible for there to be an infinite list of axioms that is complete, consistent, and can be enumerated by a computer program.

There *are* complete and consistent lists of axioms for arithmetic that *cannot* be enumerated by a computer program. For example, one might take all true statements about the natural numbers to be axioms (and no false statements), which gives the theory known as "true arithmetic". The difficulty is that there is no mechanical way to decide, given a statement about the natural numbers, whether it is an axiom of this theory, and thus there is no effective way to verify a formal proof in this theory.

Many logicians believe that Gödel's incompleteness theorems struck a fatal blow to David Hilbert's second problem, which asked for a finitary consistency proof for mathematics. The second incompleteness theorem, in particular, is often viewed as making the problem impossible. Not all mathematicians agree with this analysis, however, and the status of Hilbert's second problem is not yet decided (see "Modern viewpoints on the status of the problem").

## 34.2.2   Relation to the liar paradox

The liar paradox is the sentence "This sentence is false." An analysis of the liar sentence shows that it cannot be true (for then, as it asserts, it is false), nor can it be false (for then, it is true). A Gödel sentence $G$ for a theory $T$ makes a similar assertion to the liar sentence, but with truth replaced by provability: $G$ says "$G$ is not provable in the theory $T$." The analysis of the truth and provability of $G$ is a formalized version of the analysis of the truth of the liar sentence.

It is not possible to replace "not provable" with "false" in a Gödel sentence because the predicate "$Q$ is the Gödel number of a false formula" cannot be represented as a formula of arithmetic. This result, known as Tarski's undefinability theorem, was discovered independently by both Gödel, when he was working on the proof of the incompleteness theorem, and by the theorem's namesake, Alfred Tarski.

## 34.2.3   Extensions of Gödel's original result

Gödel demonstrated the incompleteness of the theory of *Principia Mathematica*, a particular theory of arithmetic, but a parallel demonstration could be given for any effective theory of a certain expressiveness. Gödel commented on this fact in the introduction to his paper, but restricted the proof to one system for concreteness. In modern statements of the theorem, it is common to state the effectiveness and expressiveness conditions as hypotheses for the incompleteness theorem, so that it is not limited to any particular formal theory. The terminology used to state these conditions was not yet developed in 1931 when Gödel published his results.

Gödel's original statement and proof of the incompleteness theorem requires the assumption that the theory is not just consistent but *ω-consistent*. A theory is **ω-consistent** if it is not ω-inconsistent, and is ω-inconsistent if there is a predicate $P$ such that for every specific natural number $m$ the theory proves $\sim P(m)$, and yet the theory also proves that there exists a natural number $n$ such that $P(n)$. That is, the theory says that a number with property $P$ exists while denying that it has any specific value. The ω-consistency of a theory implies its consistency, but consistency does not imply ω-consistency. J. Barkley Rosser (1936) strengthened the incompleteness theorem by finding a variation of the proof (Rosser's trick) that only requires the theory to be consistent, rather than ω-consistent. This is mostly of technical interest, since all true formal theories of arithmetic (theories whose axioms are all true statements about natural numbers) are ω-consistent, and thus Gödel's theorem as originally stated applies to them. The stronger version of the incompleteness theorem that only assumes consistency, rather than ω-consistency, is now commonly known as Gödel's incompleteness theorem and as the Gödel–Rosser theorem.

# 34.3 Second incompleteness theorem

**Gödel's second incompleteness theorem** first appeared as "Theorem XI" in Gödel's 1931 paper *On Formally Undecidable Propositions in Principia Mathematica and Related Systems I*.

Like with the first incompleteness theorem, Gödel wrote this theorem in highly technical formal mathematics. It may be paraphrased in English as:

> For any formal effectively generated theory $T$ including basic arithmetical truths and also certain truths about formal provability, if $T$ includes a statement of its own consistency then $T$ is inconsistent.

This strengthens the first incompleteness theorem, because the statement constructed in the first incompleteness theorem does not directly express the consistency of the theory. The proof of the second incompleteness theorem is obtained by formalizing the proof of the first incompleteness theorem within the theory itself.

A technical subtlety in the second incompleteness theorem is how to express the consistency of $T$ as a formula in the language of $T$. There are many ways to do this, and not all of them lead to the same result. In particular, different formalizations of the claim that $T$ is consistent may be inequivalent in $T$, and some may even be provable. For example, first-order Peano arithmetic (PA) can prove that the largest consistent subset of PA is consistent. But since PA is consistent, the largest consistent subset of PA is just PA, so in this sense PA "proves that it is consistent". What PA does not prove is that the largest consistent subset of PA is, in fact, the whole of PA. (The term "largest consistent subset of PA" is technically ambiguous, but what is meant here is the largest consistent initial segment of the axioms of PA ordered according to specific criteria; i.e., by "Gödel numbers", the numbers encoding the axioms as per the scheme used by Gödel mentioned above).

For Peano arithmetic, or any familiar explicitly axiomatized theory $T$, it is possible to canonically define a formula $\text{Con}(T)$ expressing the consistency of $T$; this formula expresses the property that "there does not exist a natural number coding a sequence of formulas, such that each formula is either one of the axioms of $T$, a logical axiom, or an immediate consequence of preceding formulas according to the rules of inference of first-order logic, and such that the last formula is a contradiction".

The formalization of $\text{Con}(T)$ depends on two factors: formalizing the notion of a sentence being derivable from a set of sentences and formalizing the notion of being an axiom of $T$. Formalizing derivability can be done in canonical fashion: given an arithmetical formula $A(x)$ defining a set of axioms, one can canonically form a predicate $\text{ProvA}(P)$, which expresses that a sentence $P$ is provable from the set of axioms defined by $A(x)$.

In addition, the standard proof of the second incompleteness theorem assumes that $\text{ProvA}(P)$ satisfies the Hilbert–Bernays provability conditions. Letting $\#(P)$ represent the Gödel number of a formula $P$, the derivability conditions say:

1. If $T$ proves $P$, then T proves $\text{ProvA}(\#(P))$.

2. $T$ proves 1.; that is, $T$ proves that if $T$ proves $P$, then $T$ proves $\text{ProvA}(\#(P))$. In other words, $T$ proves that $\text{ProvA}(\#(P))$ implies $\text{ProvA}(\#(\text{ProvA}(\#(P))))$.

3. $T$ proves that if $T$ proves that $(P \to Q)$ and $T$ proves $P$ then $T$ proves $Q$. In other words, $T$ proves that $\text{ProvA}(\#(P \to Q))$ and $\text{ProvA}(\#(P))$ imply $\text{ProvA}(\#(Q))$.

## 34.3.1 Implications for consistency proofs

Gödel's second incompleteness theorem also implies that a theory $T_1$ satisfying the technical conditions outlined above cannot prove the consistency of any theory $T_2$ that proves the consistency of $T_1$. This is because such a theory $T_1$ can prove that if $T_2$ proves the consistency of $T_1$, then $T_1$ is in fact consistent. For the claim that $T_1$ is consistent has form "for all numbers $n$, $n$ has the decidable property of not being a code for a proof of contradiction in $T_1$". If $T_1$ were in fact inconsistent, then $T_2$ would prove for some $n$ that $n$ is the code of a contradiction in $T_1$. But if $T_2$ also proved that $T_1$ is consistent (that is, that there is no such n), then it would itself be inconsistent. This reasoning can be formalized in $T_1$ to show that if $T_2$ is consistent, then $T_1$ is consistent. Since, by second incompleteness theorem, $T_1$ does not prove its consistency, it cannot prove the consistency of $T_2$ either.

This corollary of the second incompleteness theorem shows that there is no hope of proving, for example, the consistency of Peano arithmetic using any finitistic means that can be formalized in a theory the consistency of which is provable in Peano arithmetic. For example, the theory of primitive recursive arithmetic (PRA), which is widely accepted as an accurate formalization of finitistic mathematics, is provably consistent in PA. Thus PRA cannot prove the consistency of PA. This fact is generally seen to imply that Hilbert's program, which aimed to justify the use of "ideal" (infinitistic) mathematical principles in the proofs of "real" (finitistic) mathematical statements by giving a finitistic proof that the ideal principles are consistent, cannot be carried out.

The corollary also indicates the epistemological relevance of the second incompleteness theorem. It would actually provide no interesting information if a theory $T$ proved its consistency. This is because inconsistent theories prove everything, including their consistency. Thus a consistency proof of $T$ in $T$ would give us no clue as to whether $T$ really is consistent; no doubts about the consistency of $T$ would be resolved by such a consistency proof. The interest in consistency proofs lies in the possibility of proving the consistency of a theory $T$ in some theory $T'$ that is in some sense less doubtful than $T$ itself, for example weaker than $T$. For many naturally occurring theories $T$ and $T'$, such as $T$ = Zermelo–Fraenkel set theory and $T'$ = primitive recursive arithmetic, the consistency of $T'$ is provable in $T$, and thus $T'$ can't prove the consistency of $T$ by the above corollary of the second incompleteness theorem.

The second incompleteness theorem does not rule out consistency proofs altogether, only consistency proofs that could be formalized in the theory that is proved consistent. For example, Gerhard Gentzen proved the consistency of Peano arithmetic (PA) in a different theory that includes an axiom asserting that the ordinal called $\varepsilon_0$ is wellfounded; see Gentzen's consistency proof. Gentzen's theorem spurred the development of ordinal analysis in proof theory.

## 34.4   Examples of undecidable statements

See also: List of statements undecidable in ZFC

There are two distinct senses of the word "undecidable" in mathematics and computer science. The first of these is the proof-theoretic sense used in relation to Gödel's theorems, that of a statement being neither provable nor refutable in a specified deductive system. The second sense, which will not be discussed here, is used in relation to computability theory and applies not to statements but to decision problems, which are countably infinite sets of questions each requiring a yes or no answer. Such a problem is said to be undecidable if there is no computable function that correctly answers every question in the problem set (see undecidable problem).

Because of the two meanings of the word undecidable, the term independent is sometimes used instead of undecidable for the "neither provable nor refutable" sense. The usage of "independent" is also ambiguous, however. Some use it to mean just "not provable", leaving open whether an independent statement might be refuted.

Undecidability of a statement in a particular deductive system does not, in and of itself, address the question of whether the truth value of the statement is well-defined, or whether it can be determined by other means. Undecidability only implies that the particular deductive system being considered does not prove the truth or falsity of the statement. Whether there exist so-called "absolutely undecidable" statements, whose truth value can never be known or is ill-specified, is a controversial point in the philosophy of mathematics.

The combined work of Gödel and Paul Cohen has given two concrete examples of undecidable statements (in the first sense of the term): The continuum hypothesis can neither be proved nor refuted in ZFC (the standard axiomatization of set theory), and the axiom of choice can neither be proved nor refuted in ZF (which is all the ZFC axioms *except* the axiom of choice). These results do not require the incompleteness theorem. Gödel proved in 1940 that neither of these statements could be disproved in ZF or ZFC set theory. In the 1960s, Cohen proved that neither is provable from ZF, and the continuum hypothesis cannot be proven from ZFC.

In 1973, the Whitehead problem in group theory was shown to be undecidable, in the first sense of the term, in standard set theory.

Gregory Chaitin produced undecidable statements in algorithmic information theory and proved another incompleteness theorem in that setting. Chaitin's incompleteness theorem states that for any theory that can represent enough arithmetic, there is an upper bound $c$ such that no specific number can be proven in that theory to have Kolmogorov complexity greater than $c$. While Gödel's theorem is related to the liar paradox, Chaitin's result is related to Berry's paradox.

### 34.4.1 Undecidable statements provable in larger systems

These are natural mathematical equivalents of the Gödel "true but undecidable" sentence. They can be proved in a larger system which is generally accepted as a valid form of reasoning, but are undecidable in a more limited system such as Peano Arithmetic.

In 1977, Paris and Harrington proved that the Paris-Harrington principle, a version of the Ramsey theorem, is undecidable in the first-order axiomatization of arithmetic called Peano arithmetic, but can be proven in the larger system of second-order arithmetic. Kirby and Paris later showed Goodstein's theorem, a statement about sequences of natural numbers somewhat simpler than the Paris-Harrington principle, to be undecidable in Peano arithmetic.

Kruskal's tree theorem, which has applications in computer science, is also undecidable from Peano arithmetic but provable in set theory. In fact Kruskal's tree theorem (or its finite form) is undecidable in a much stronger system codifying the principles acceptable based on a philosophy of mathematics called predicativism. The related but more general graph minor theorem (2003) has consequences for computational complexity theory.

## 34.5 Limitations of Gödel's theorems

The conclusions of Gödel's theorems are only proven for the formal theories that satisfy the necessary hypotheses. Not all axiom systems satisfy these hypotheses, even when these systems have models that include the natural numbers as a subset. For example, there are first-order axiomatizations of Euclidean geometry, of real closed fields, and of arithmetic in which multiplication is not *provably* total; none of these meet the hypotheses of Gödel's theorems. The key fact is that these axiomatizations are not expressive enough to define the set of natural numbers or develop basic properties of the natural numbers. Regarding the third example, Dan Willard (2001) has studied many weak systems of arithmetic which do not satisfy the hypotheses of the second incompleteness theorem, and which are consistent and capable of proving their own consistency (see self-verifying theories).

Gödel's theorems only apply to effectively generated (that is, recursively enumerable) theories. If all true statements about natural numbers are taken as axioms for a theory, then this theory is a consistent, complete extension of Peano arithmetic (called true arithmetic) for which none of Gödel's theorems apply in a meaningful way, because this theory is not recursively enumerable.

The second incompleteness theorem only shows that the consistency of certain theories cannot be proved from the axioms of those theories themselves. It does not show that the consistency cannot be proved from other (consistent) axioms. For example, the consistency of the Peano arithmetic can be proved in Zermelo–Fraenkel set theory (ZFC), or in theories of arithmetic augmented with transfinite induction, as in Gentzen's consistency proof.

## 34.6 Relationship with computability

The incompleteness theorem is closely related to several results about undecidable sets in recursion theory.

Stephen Cole Kleene (1943) presented a proof of Gödel's incompleteness theorem using basic results of computability theory. One such result shows that the halting problem is undecidable: there is no computer program that can correctly determine, given any program $P$ as input, whether $P$ eventually halts when run with a particular given input. Kleene showed that the existence of a complete effective theory of arithmetic with certain consistency properties would force the halting problem to be decidable, a contradiction. This method of proof has also been presented by Shoenfield (1967, p. 132); Charlesworth (1980); and Hopcroft and Ullman (1979).

Franzén (2005, p. 73) explains how Matiyasevich's solution to Hilbert's 10th problem can be used to obtain a proof to Gödel's first incompleteness theorem. Matiyasevich proved that there is no algorithm that, given a multivariate polynomial $p(x_1, x_2,...,x_k)$ with integer coefficients, determines whether there is an integer solution to the equation $p = 0$. Because polynomials with integer coefficients, and integers themselves, are directly expressible in the language of arithmetic, if a multivariate integer polynomial equation $p = 0$ does have a solution in the integers then any sufficiently strong theory of arithmetic $T$ will prove this. Moreover, if the theory $T$ is $\omega$-consistent, then it will never prove that a particular polynomial equation has a solution when in fact there is no solution in the integers. Thus, if $T$ were complete and $\omega$-consistent, it would be possible to determine algorithmically whether a polynomial equation has a solution by merely enumerating proofs of $T$ until either "$p$ has a solution" or "$p$ has no solution" is found, in contradiction to Matiyasevich's theorem. Moreover, for each consistent effectively generated theory $T$, it is possible to effectively generate a multivariate polynomial $p$ over the integers such that the equation $p = 0$ has no solutions over the integers, but the lack of solutions cannot be proved in $T$ (Davis 2006:416, Jones 1980).

Smorynski (1977, p. 842) shows how the existence of recursively inseparable sets can be used to prove the first incompleteness theorem. This proof is often extended to show that systems such as Peano arithmetic are essentially undecidable (see Kleene 1967, p. 274).

Chaitin's incompleteness theorem gives a different method of producing independent sentences, based on Kolmogorov complexity. Like the proof presented by Kleene that was mentioned above, Chaitin's theorem only applies to theories with the additional property that all their axioms are true in the standard model of the natural numbers. Gödel's incompleteness theorem is distinguished by its applicability to consistent theories that nonetheless include statements that are false in the standard model; these theories are known as $\omega$-inconsistent.

## 34.7    Proof sketch for the first theorem

Main article: Proof sketch for Gödel's first incompleteness theorem

The proof by contradiction has three essential parts. To begin, choose a formal system that meets the proposed criteria:

1. Statements in the system can be represented by natural numbers (known as Gödel numbers). The significance of this is that properties of statements—such as their truth and falsehood—will be equivalent to determining whether their Gödel numbers have certain properties, and that properties of the statements can therefore be demonstrated by examining their Gödel numbers. This part culminates in the construction of a formula expressing the idea that *"statement S is provable in the system"* (which can be applied to any statement "S" in the system).

2. In the formal system it is possible to construct a number whose matching statement, when interpreted, is self-referential and essentially says that it (i.e. the statement itself) is unprovable. This is done using a technique called "diagonalization" (so-called because of its origins as Cantor's diagonal argument).

3. Within the formal system this statement permits a demonstration that it is neither provable nor disprovable in the system, and therefore the system cannot in fact be $\omega$-consistent. Hence the original assumption that the proposed system met the criteria is false.

### 34.7.1    Arithmetization of syntax

The main problem in fleshing out the proof described above is that it seems at first that to construct a statement $p$ that is equivalent to "$p$ cannot be proved", $p$ would somehow have to contain a reference to $p$, which could easily give rise to an infinite regress. Gödel's ingenious technique is to show that statements can be matched with numbers (often called the arithmetization of syntax) in such a way that *"proving a statement"* can be replaced with *"testing whether a number has a given property"*. This allows a self-referential formula to be constructed in a way that avoids any infinite regress of definitions. The same technique was later used by Alan Turing in his work on the Entscheidungsproblem.

In simple terms, a method can be devised so that every formula or statement that can be formulated in the system gets a unique number, called its **Gödel number**, in such a way that it is possible to mechanically convert back and forth between formulas and Gödel numbers. The numbers involved might be very long indeed (in terms of number of digits), but this is not a barrier; all that matters is that such numbers can be constructed. A simple example is the way in which English is stored as a sequence of numbers in computers using ASCII or Unicode:

- The word **HELLO** is represented by 72-69-76-76-79 using decimal ASCII, ie the number 7269767679.

- The logical statement **x=y => y=x** is represented by 120-061-121-032-061-062-032-121-061-120 using octal ASCII, ie the number 120061121032061062032121061120.

In principle, proving a statement true or false can be shown to be equivalent to proving that the number matching the statement does or doesn't have a given property. Because the formal system is strong enough to support reasoning about *numbers in general*, it can support reasoning about *numbers that represent formulae and statements* as well. Crucially, because the system can support reasoning about *properties of numbers*, the results are equivalent to reasoning about *provability of their equivalent statements*.

### 34.7.2 Construction of a statement about "provability"

Having shown that in principle the system can indirectly make statements about provability, by analyzing properties of those numbers representing statements it is now possible to show how to create a statement that actually does this.

A formula $F(x)$ that contains exactly one free variable $x$ is called a *statement form* or *class-sign*. As soon as $x$ is replaced by a specific number, the statement form turns into a *bona fide* statement, and it is then either provable in the system, or not. For certain formulas one can show that for every natural number n, F(n) is true if and only if it can be proven (the precise requirement in the original proof is weaker, but for the proof sketch this will suffice). In particular, this is true for every specific arithmetic operation between a finite number of natural numbers, such as "2×3=6".

Statement forms themselves are not statements and therefore cannot be proved or disproved. But every statement form $F(x)$ can be assigned a Gödel number denoted by $G(F)$. The choice of the free variable used in the form $F(x)$ is not relevant to the assignment of the Gödel number $G(F)$.

The notion of provability itself can also be encoded by Gödel numbers, in the following way: since a proof is a list of statements which obey certain rules, the Gödel number of a proof can be defined. Now, for every statement $p$, one may ask whether a number $x$ is the Gödel number of its proof. The relation between the Gödel number of $p$ and $x$, the potential Gödel number of its proof, is an arithmetical relation between two numbers. Therefore there is a statement form Bew($y$) that uses this arithmetical relation to state that a Gödel number of a proof of $y$ exists:

> Bew($y$) = $\exists x$ ( $y$ is the Gödel number of a formula and $x$ is the Gödel number of a proof of the formula encoded by $y$).

The name **Bew** is short for *beweisbar*, the German word for "provable"; this name was originally used by Gödel to denote the provability formula just described. Note that "Bew($y$)" is merely an abbreviation that represents a particular, very long, formula in the original language of $T$; the string "Bew" itself is not claimed to be part of this language.

An important feature of the formula Bew($y$) is that if a statement $p$ is provable in the system then Bew($G(p)$) is also provable. This is because any proof of $p$ would have a corresponding Gödel number, the existence of which causes Bew($G(p)$) to be satisfied.

### 34.7.3 Diagonalization

The next step in the proof is to obtain a statement that says it is unprovable. Although Gödel constructed this statement directly, the existence of at least one such statement follows from the diagonal lemma, which says that for any sufficiently strong formal system and any statement form $F$ there is a statement $p$ such that the system proves

> $p \leftrightarrow F(G(p))$.

By letting $F$ be the negation of Bew($x$), we obtain the theorem

> $p \leftrightarrow \sim Bew(G(p))$

and the $p$ defined by this roughly states that its own Gödel number is the Gödel number of an unprovable formula.

The statement $p$ is not literally equal to ~Bew($\mathbf{G}(p)$); rather, $p$ states that if a certain calculation is performed, the resulting Gödel number will be that of an unprovable statement. But when this calculation is performed, the resulting Gödel number turns out to be the Gödel number of $p$ itself. This is similar to the following sentence in English:

", when preceded by itself in quotes, is unprovable.", when preceded by itself in quotes, is unprovable.

This sentence does not directly refer to itself, but when the stated transformation is made the original sentence is obtained as a result, and thus this sentence asserts its own unprovability. The proof of the diagonal lemma employs a similar method.

Now, assume that the axiomatic system is $\omega$-consistent, and let $p$ be the statement obtained in the previous section.

If $p$ were provable, then Bew($\mathbf{G}(p)$) would be provable, as argued above. But $p$ asserts the negation of Bew($\mathbf{G}(p)$). Thus the system would be inconsistent, proving both a statement and its negation. This contradiction shows that $p$ cannot be provable.

If the negation of $p$ were provable, then Bew($\mathbf{G}(p)$) would be provable (because $p$ was constructed to be equivalent to the negation of Bew($\mathbf{G}(p)$)). However, for each specific number $x$, $x$ cannot be the Gödel number of the proof of $p$, because $p$ is not provable (from the previous paragraph). Thus on one hand the system proves there is a number with a certain property (that it is the Gödel number of the proof of $p$), but on the other hand, for every specific number $x$, we can prove that it does not have this property. This is impossible in an $\omega$-consistent system. Thus the negation of $p$ is not provable.

Thus the statement $p$ is undecidable in our axiomatic system: it can neither be proved nor disproved within the system.

In fact, to show that $p$ is not provable only requires the assumption that the system is consistent. The stronger assumption of $\omega$-consistency is required to show that the negation of $p$ is not provable. Thus, if $p$ is constructed for a particular system:

- If the system is $\omega$-consistent, it can prove neither $p$ nor its negation, and so $p$ is undecidable.

- If the system is consistent, it may have the same situation, or it may prove the negation of $p$. In the later case, we have a statement ("not $p$") which is false but provable, and the system is not $\omega$-consistent.

If one tries to "add the missing axioms" to avoid the incompleteness of the system, then one has to add either $p$ or "not $p$" as axioms. But then the definition of "being a Gödel number of a proof" of a statement changes. which means that the formula Bew($x$) is now different. Thus when we apply the diagonal lemma to this new Bew, we obtain a new statement $p$, different from the previous one, which will be undecidable in the new system if it is $\omega$-consistent.

## 34.7.4   Proof via Berry's paradox

George Boolos (1989) sketches an alternative proof of the first incompleteness theorem that uses Berry's paradox rather than the liar paradox to construct a true but unprovable formula. A similar proof method was independently discovered by Saul Kripke (Boolos 1998, p. 383). Boolos's proof proceeds by constructing, for any computably enumerable set $S$ of true sentences of arithmetic, another sentence which is true but not contained in $S$. This gives the first incompleteness theorem as a corollary. According to Boolos, this proof is interesting because it provides a "different sort of reason" for the incompleteness of effective, consistent theories of arithmetic (Boolos 1998, p. 388).

## 34.7.5   Formalized proofs

Formalized proofs of versions of the incompleteness theorem have been developed by Natarajan Shankar in 1986 using Nqthm (Shankar 1994) and by Russell O'Connor in 2003 using Coq (O'Connor 2005).

## 34.8 Proof sketch for the second theorem

The main difficulty in proving the second incompleteness theorem is to show that various facts about provability used in the proof of the first incompleteness theorem can be formalized within the system using a formal predicate for provability. Once this is done, the second incompleteness theorem follows by formalizing the entire proof of the first incompleteness theorem within the system itself.

Let $p$ stand for the undecidable sentence constructed above, and assume that the consistency of the system can be proven from within the system itself. The demonstration above shows that if the system is consistent, then $p$ is not provable. The proof of this implication can be formalized within the system, and therefore the statement "$p$ is not provable", or "not $P(p)$" can be proven in the system.

But this last statement is equivalent to $p$ itself (and this equivalence can be proven in the system), so $p$ can be proven in the system. This contradiction shows that the system must be inconsistent.

## 34.9 Discussion and implications

The incompleteness results affect the philosophy of mathematics, particularly versions of formalism, which use a single system of formal logic to define their principles. One can paraphrase the first theorem as saying the following:

> An all-encompassing axiomatic system can never be found that is able to prove *all* mathematical truths, but no falsehoods.

On the other hand, from a strict formalist perspective this paraphrase would be considered meaningless because it presupposes that mathematical "truth" and "falsehood" are well-defined in an absolute sense, rather than relative to each formal system.

The following rephrasing of the second theorem is even more unsettling to the foundations of mathematics:

> If an axiomatic system can be proven to be consistent from within itself, then it is inconsistent.

Therefore, to establish the consistency of a system S, one needs to use some other system T, but a proof in T is not completely convincing unless T's consistency has already been established without using S.

Theories such as Peano arithmetic, for which any computably enumerable consistent extension is incomplete, are called essentially undecidable or **essentially incomplete**.

### 34.9.1 Minds and machines

Main article: Mechanism (philosophy) § Gödelian arguments

Authors including the philosopher J. R. Lucas and physicist Roger Penrose have debated what, if anything, Gödel's incompleteness theorems imply about human intelligence. Much of the debate centers on whether the human mind is equivalent to a Turing machine, or by the Church–Turing thesis, any finite machine at all. If it is, and if the machine is consistent, then Gödel's incompleteness theorems would apply to it.

Hilary Putnam (1960) suggested that while Gödel's theorems cannot be applied to humans, since they make mistakes and are therefore inconsistent, it may be applied to the human faculty of science or mathematics in general. Assuming that it is consistent, either its consistency cannot be proved or it cannot be represented by a Turing machine.

Avi Wigderson (2010) has proposed that the concept of mathematical "knowability" should be based on computational complexity rather than logical decidability. He writes that "when *knowability* is interpreted by modern standards, namely via computational complexity, the Gödel phenomena are very much with us."

### 34.9.2  Paraconsistent logic

Although Gödel's theorems are usually studied in the context of classical logic, they also have a role in the study of paraconsistent logic and of inherently contradictory statements (*dialetheia*). Graham Priest (1984, 2006) argues that replacing the notion of formal proof in Gödel's theorem with the usual notion of informal proof can be used to show that naive mathematics is inconsistent, and uses this as evidence for dialetheism. The cause of this inconsistency is the inclusion of a truth predicate for a theory within the language of the theory (Priest 2006:47). Stewart Shapiro (2002) gives a more mixed appraisal of the applications of Gödel's theorems to dialetheism.

### 34.9.3  Appeals to the incompleteness theorems in other fields

Appeals and analogies are sometimes made to the incompleteness theorems in support of arguments that go beyond mathematics and logic. Several authors have commented negatively on such extensions and interpretations, including Torkel Franzén (2005); Alan Sokal and Jean Bricmont (1999); and Ophelia Benson and Jeremy Stangroom (2006). Bricmont and Stangroom (2006, p. 10), for example, quote from Rebecca Goldstein's comments on the disparity between Gödel's avowed Platonism and the anti-realist uses to which his ideas are sometimes put. Sokal and Bricmont (1999, p. 187) criticize Régis Debray's invocation of the theorem in the context of sociology; Debray has defended this use as metaphorical (ibid.).

### 34.9.4  Role of self-reference

Torkel Franzén (2005, p. 46) observes:

> Gödel's proof of the first incompleteness theorem and Rosser's strengthened version have given many the impression that the theorem can only be proved by constructing self-referential statements [...] or even that only strange self-referential statements are known to be undecidable in elementary arithmetic. To counteract such impressions, we need only introduce a different kind of proof of the first incompleteness theorem.

He then proposes the proofs based on computability, or on information theory, as described earlier in this article, as examples of proofs that should "counteract such impressions".

## 34.10  History

After Gödel published his proof of the completeness theorem as his doctoral thesis in 1929, he turned to a second problem for his habilitation. His original goal was to obtain a positive solution to Hilbert's second problem (Dawson 1997, p. 63). At the time, theories of the natural numbers and real numbers similar to second-order arithmetic were known as "analysis", while theories of the natural numbers alone were known as "arithmetic".

Gödel was not the only person working on the consistency problem. Ackermann had published a flawed consistency proof for analysis in 1925, in which he attempted to use the method of ε-substitution originally developed by Hilbert. Later that year, von Neumann was able to correct the proof for a theory of arithmetic without any axioms of induction. By 1928, Ackermann had communicated a modified proof to Bernays; this modified proof led Hilbert to announce his belief in 1929 that the consistency of arithmetic had been demonstrated and that a consistency proof of analysis would likely soon follow. After the publication of the incompleteness theorems showed that Ackermann's modified proof must be erroneous, von Neumann produced a concrete example showing that its main technique was unsound (Zach 2006, p. 418, Zach 2003, p. 33).

In the course of his research, Gödel discovered that although a sentence which asserts its own falsehood leads to paradox, a sentence that asserts its own non-provability does not. In particular, Gödel was aware of the result now called Tarski's indefinability theorem, although he never published it. Gödel announced his first incompleteness theorem to Carnap, Feigel and Waismann on August 26, 1930; all four would attend a key conference in Königsberg the following week.

## 34.10.1 Announcement

The 1930 Königsberg conference was a joint meeting of three academic societies, with many of the key logicians of the time in attendance. Carnap, Heyting, and von Neumann delivered one-hour addresses on the mathematical philosophies of logicism, intuitionism, and formalism, respectively (Dawson 1996, p. 69). The conference also included Hilbert's retirement address, as he was leaving his position at the University of Göttingen. Hilbert used the speech to argue his belief that all mathematical problems can be solved. He ended his address by saying,

> For the mathematician there is no *Ignorabimus*, and, in my opinion, not at all for natural science either. ... The true reason why [no one] has succeeded in finding an unsolvable problem is, in my opinion, that there is no unsolvable problem. In contrast to the foolish *Ignoramibus*, our credo avers: We must know. We shall know!

This speech quickly became known as a summary of Hilbert's beliefs on mathematics (its final six words, "*Wir müssen wissen. Wir werden wissen!*", were used as Hilbert's epitaph in 1943). Although Gödel was likely in attendance for Hilbert's address, the two never met face to face (Dawson 1996, p. 72).

Gödel announced his first incompleteness theorem at a roundtable discussion session on the third day of the conference. The announcement drew little attention apart from that of von Neumann, who pulled Gödel aside for conversation. Later that year, working independently with knowledge of the first incompleteness theorem, von Neumann obtained a proof of the second incompleteness theorem, which he announced to Gödel in a letter dated November 20, 1930 (Dawson 1996, p. 70). Gödel had independently obtained the second incompleteness theorem and included it in his submitted manuscript, which was received by *Monatshefte für Mathematik* on November 17, 1930.

Gödel's paper was published in the *Monatshefte* in 1931 under the title *Über formal unentscheidbare Sätze der Principia Mathematica und verwandter Systeme I* (On Formally Undecidable Propositions in Principia Mathematica and Related Systems I). As the title implies, Gödel originally planned to publish a second part of the paper; it was never written.

## 34.10.2 Generalization and acceptance

Gödel gave a series of lectures on his theorems at Princeton in 1933–1934 to an audience that included Church, Kleene, and Rosser. By this time, Gödel had grasped that the key property his theorems required is that the theory must be effective (at the time, the term "general recursive" was used). Rosser proved in 1936 that the hypothesis of ω-consistency, which was an integral part of Gödel's original proof, could be replaced by simple consistency, if the Gödel sentence was changed in an appropriate way. These developments left the incompleteness theorems in essentially their modern form.

Gentzen published his consistency proof for first-order arithmetic in 1936. Hilbert accepted this proof as "finitary" although (as Gödel's theorem had already shown) it cannot be formalized within the system of arithmetic that is being proved consistent.

The impact of the incompleteness theorems on Hilbert's program was quickly realized. Bernays included a full proof of the incompleteness theorems in the second volume of *Grundlagen der Mathematik* (1939), along with additional results of Ackermann on the ε-substitution method and Gentzen's consistency proof of arithmetic. This was the first full published proof of the second incompleteness theorem.

## 34.10.3 Criticisms

### Finsler

Paul Finsler (1926) used a version of Richard's paradox to construct an expression that was false but unprovable in a particular, informal framework he had developed. Gödel was unaware of this paper when he proved the incompleteness theorems (Collected Works Vol. IV., p. 9). Finsler wrote to Gödel in 1931 to inform him about this paper, which Finsler felt had priority for an incompleteness theorem. Finsler's methods did not rely on formalized provability, and had only a superficial resemblance to Gödel's work (van Heijenoort 1967:328). Gödel read the paper but found it deeply flawed, and his response to Finsler laid out concerns about the lack of formalization (Dawson:89). Finsler continued to argue for his philosophy of mathematics, which eschewed formalization, for the remainder of his career.

### Zermelo

In September 1931, Ernst Zermelo wrote Gödel to announce what he described as an "essential gap" in Gödel's argument (Dawson:76). In October, Gödel replied with a 10-page letter (Dawson:76, Grattan-Guinness:512-513). But Zermelo did not relent and published his criticisms in print with "a rather scathing paragraph on his young competitor" (Grattan-Guinness:513). Gödel decided that to pursue the matter further was pointless, and Carnap agreed (Dawson:77). Much of Zermelo's subsequent work was related to logics stronger than first-order logic, with which he hoped to show both the consistency and categoricity of mathematical theories.

### Wittgenstein

Ludwig Wittgenstein wrote several passages about the incompleteness theorems that were published posthumously in his 1953 *Remarks on the Foundations of Mathematics*. Gödel was a member of the Vienna Circle during the period in which Wittgenstein's early ideal language philosophy and Tractatus Logico-Philosophicus dominated the circle's thinking. Writings in Gödel's Nachlass express the belief that Wittgenstein deliberately misread his ideas.

Multiple commentators have read Wittgenstein as misunderstanding Gödel (Rodych 2003), although Juliet Floyd and Hilary Putnam (2000), as well as Graham Priest (2004) have provided textual readings arguing that most commentary misunderstands Wittgenstein. On their release, Bernays, Dummett, and Kreisel wrote separate reviews on Wittgenstein's remarks, all of which were extremely negative (Berto 2009:208). The unanimity of this criticism caused Wittgenstein's remarks on the incompleteness theorems to have little impact on the logic community. In 1972, Gödel, stated: "Has Wittgenstein lost his mind? Does he mean it seriously?" (Wang 1996:197), and wrote to Karl Menger that Wittgenstein's comments demonstrate a willful misunderstanding of the incompleteness theorems writing:

> "It is clear from the passages you cite that Wittgenstein did "not" understand [the first incompleteness theorem] (or pretended not to understand it). He interpreted it as a kind of logical paradox, while in fact is just the opposite, namely a mathematical theorem within an absolutely uncontroversial part of mathematics (finitary number theory or combinatorics)." (Wang 1996:197)

Since the publication of Wittgenstein's *Nachlass* in 2000, a series of papers in philosophy have sought to evaluate whether the original criticism of Wittgenstein's remarks was justified. Floyd and Putnam (2000) argue that Wittgenstein had a more complete understanding of the incompleteness theorem than was previously assumed. They are particularly concerned with the interpretation of a Gödel sentence for an $\omega$-inconsistent theory as actually saying "I am not provable", since the theory has no models in which the provability predicate corresponds to actual provability. Rodych (2003) argues that their interpretation of Wittgenstein is not historically justified, while Bays (2004) argues against Floyd and Putnam's philosophical analysis of the provability predicate. Berto (2009) explores the relationship between Wittgenstein's writing and theories of paraconsistent logic.

## 34.11   See also

- Gödel's completeness theorem

- Gödel's speed-up theorem

- Löb's Theorem

- *Minds, Machines and Gödel*

- Münchhausen trilemma

- Non-standard model of arithmetic

- Provability logic

- Tarski's undefinability theorem

- Third Man Argument

## 34.12 Notes

[1] The Roman numeral "I" indicates that Gödel intended to publish a sequel but "The prompt acceptance of his results was one of the reasons that made him change his plan", cf the text and its footnote 68a in van Heijenoort 1967:616

[2] The word "true" is used disquotationally here: the Gödel sentence is true in this sense because it "asserts its own unprovability and it is indeed unprovable" (Smoryński 1977 p. 825; also see Franzén 2005 pp. 28–33). It is also possible to read "$GT$ is true" in the formal sense that primitive recursive arithmetic proves the implication $\mathrm{Con}(T) \rightarrow GT$, where $\mathrm{Con}(T)$ is a canonical sentence asserting the consistency of $T$ (Smoryński 1977 p. 840, Kikuchi and Tanaka 1994 p. 403). However, the arithmetic statement in question is *false* in some nonstandard models of arithmetic.

## 34.13 References

### 34.13.1 Articles by Gödel

- "Über formal unentscheidbare Sätze der Principia Mathematica und verwandter Systeme, I.". *Monatshefte für Mathematik und Physik* **38**: 173–98.

- 1931, *Über formal unentscheidbare Sätze der Principia Mathematica und verwandter Systeme, I.* and *On formally undecidable propositions of Principia Mathematica and related systems I* in Solomon Feferman, ed., 1986. *Kurt Gödel Collected works, Vol. I*. Oxford University Press: 144-195. The original German with a facing English translation, preceded by a very illuminating introductory note by Kleene.

    - Hirzel, Martin, 2000, *On formally undecidable propositions of Principia Mathematica and related systems I.*. A modern translation by Hirzel.

- 1951, *Some basic theorems on the foundations of mathematics and their implications* in Solomon Feferman, ed., 1995. *Kurt Gödel Collected works, Vol. III*. Oxford University Press: 304-23.

### 34.13.2 Translations, during his lifetime, of Gödel's paper into English

None of the following agree in all translated words and in typography. The typography is a serious matter, because Gödel expressly wished to emphasize "those metamathematical notions that had been defined in their usual sense before . . ." (van Heijenoort 1967:595). Three translations exist. Of the first John Dawson states that: "The Meltzer translation was seriously deficient and received a devastating review in the *Journal of Symbolic Logic*; "Gödel also complained about Braithwaite's commentary (Dawson 1997:216). "Fortunately, the Meltzer translation was soon supplanted by a better one prepared by Elliott Mendelson for Martin Davis's anthology *The Undecidable* . . . he found the translation "not quite so good" as he had expected . . . [but because of time constraints he] agreed to its publication" (ibid). (In a footnote Dawson states that "he would regret his compliance, for the published volume was marred throughout by sloppy typography and numerous misprints" (ibid)). Dawson states that "The translation that Gödel favored was that by Jean van Heijenoort" (ibid). For the serious student another version exists as a set of lecture notes recorded by Stephen Kleene and J. B. Rosser "during lectures given by Gödel at to the Institute for Advanced Study during the spring of 1934" (cf commentary by Davis 1965:39 and beginning on p. 41); this version is titled "On Undecidable Propositions of Formal Mathematical Systems". In their order of publication:

- B. Meltzer (translation) and R. B. Braithwaite (Introduction), 1962. *On Formally Undecidable Propositions of Principia Mathematica and Related Systems*, Dover Publications, New York (Dover edition 1992), ISBN 0-486-66980-7 (pbk.) This contains a useful translation of Gödel's German abbreviations on pp. 33–34. As noted above, typography, translation and commentary is suspect. Unfortunately, this translation was reprinted with all its suspect content by

    - Stephen Hawking editor, 2005. *God Created the Integers: The Mathematical Breakthroughs That Changed History*, Running Press, Philadelphia, ISBN 0-7624-1922-9. Gödel's paper appears starting on p. 1097, with Hawking's commentary starting on p. 1089.

- Martin Davis editor, 1965. *The Undecidable: Basic Papers on Undecidable Propositions, Unsolvable problems and Computable Functions*, Raven Press, New York, no ISBN. Gödel's paper begins on page 5, preceded by one page of commentary.

- Jean van Heijenoort editor, 1967, 3rd edition 1967. *From Frege to Gödel: A Source Book in Mathematical Logic, 1879-1931*, Harvard University Press, Cambridge Mass., ISBN 0-674-32449-8 (pbk).[1] van Heijenoort did the translation. He states that "Professor Gödel approved the translation, which in many places was accommodated to his wishes." (p. 595). Gödel's paper begins on p. 595; van Heijenoort's commentary begins on p. 592.

- Martin Davis editor, 1965, ibid. "On Undecidable Propositions of Formal Mathematical Systems." A copy with Gödel's corrections of errata and Gödel's added notes begins on page 41, preceded by two pages of Davis's commentary. Until Davis included this in his volume this lecture existed only as mimeographed notes.

## Citation

[1]  van Heijenoort, Jean. "From Frege to Gödel: A Source Book in Mathematical Logic, 1879-1931". *This link goes to the Google Books page for the text.* The original print book was published by Harvard University Press in 1977 and is widely available from booksellers. Retrieved 9 April 2014.

## 34.13.3   Articles by others

- George Boolos, 1989, "A New Proof of the Gödel Incompleteness Theorem", *Notices of the American Mathematical Society* v. 36, pp. 388–390 and p. 676, reprinted in Boolos, 1998, *Logic, Logic, and Logic*, Harvard Univ. Press. ISBN 0-674-53766-1

- Bernd Buldt, "The Scope of Gödel's First Incompleteness Theorem", *Logica Universalis* 8,2014,499–552. doi:10-014-0107-3 "Free preprint"

- Arthur Charlesworth, 1980, "A Proof of Godel's Theorem in Terms of Computer Programs," *Mathematics Magazine*, v. 54 n. 3, pp. 109–121. JStor

- Martin Davis, "The Incompleteness Theorem", in Notices of the AMS vol. 53 no. 4 (April 2006), p. 414.

- Jean van Heijenoort, 1963. "Gödel's Theorem" in Edwards, Paul, ed., *Encyclopedia of Philosophy, Vol. 3*. Macmillan: 348-57.

- Geoffrey Hellman, *How to Gödel a Frege-Russell: Gödel's Incompleteness Theorems and Logicism*. Noûs, Vol. 15, No. 4, Special Issue on Philosophy of Mathematics. (Nov., 1981), pp. 451–468.

- David Hilbert, 1900, "Mathematical Problems." English translation of a lecture delivered before the International Congress of Mathematicians at Paris, containing Hilbert's statement of his Second Problem.

- Kikuchi, Makoto; Tanaka, Kazuyuki (1994), "On formalization of model-theoretic proofs of Gödel's theorems", *Notre Dame Journal of Formal Logic* **35** (3): 403–412, doi:10.1305/ndjfl/1040511346, ISSN 0029-4527, MR 1326122

- Stephen Cole Kleene, 1943, "Recursive predicates and quantifiers," reprinted from *Transactions of the American Mathematical Society*, v. 53 n. 1, pp. 41–73 in Martin Davis 1965, *The Undecidable* (loc. cit.) pp. 255–287.

- John Barkley Rosser, 1936, "Extensions of some theorems of Gödel and Church," reprinted from the *Journal of Symbolic Logic* vol. 1 (1936) pp. 87–91, in Martin Davis 1965, *The Undecidable* (loc. cit.) pp. 230–235.

- John Barkley Rosser, 1939, "An Informal Exposition of proofs of Gödel's Theorem and Church's Theorem", Reprinted from the *Journal of Symbolic Logic*, vol. 4 (1939) pp. 53–60, in Martin Davis 1965, *The Undecidable* (loc. cit.) pp. 223–230

- C. Smoryński, "The incompleteness theorems", in J. Barwise, ed., *Handbook of Mathematical Logic*, North-Holland 1982 ISBN 978-0-444-86388-1, pp. 821–866.

- Dan E. Willard (2001), "Self-Verifying Axiom Systems, the Incompleteness Theorem and Related Reflection Principles", *Journal of Symbolic Logic*, v. 66 n. 2, pp. 536–596. doi:10.2307/2695030

- Richard Zach (2003), "The Practice of Finitism: Epsilon Calculus and Consistency Proofs in Hilbert's Program" (PDF), *Synthese* (Berlin, New York: Springer-Verlag) **137** (1): 211–259, doi:10.1023/A:1026247421383, ISSN 0039-7857

- Richard Zach (2005), "Paper on the incompleteness theorems", in Ivor Grattan-Guinness, *Landmark Writings in Western Mathematics*, Elsevier, pp. 917–25, doi:10.1016/B978-044450871-3/50152-2

## 34.13.4  Books about the theorems

- Francesco Berto. *There's Something about Gödel: The Complete Guide to the Incompleteness Theorem* John Wiley and Sons. 2010.

- Domeisen, Norbert, 1990. *Logik der Antinomien*. Bern: Peter Lang. 142 S. 1990. ISBN 3-261-04214-1. Zentralblatt MATH

- Torkel Franzén, 2005. *Gödel's Theorem: An Incomplete Guide to its Use and Abuse*. A.K. Peters. ISBN 1-56881-238-8 MR 2007d:03001

- Douglas Hofstadter, 1979. *Gödel, Escher, Bach: An Eternal Golden Braid*. Vintage Books. ISBN 0-465-02685-0. 1999 reprint: ISBN 0-465-02656-7. MR 80j:03009

- Douglas Hofstadter, 2007. *I Am a Strange Loop*. Basic Books. ISBN 978-0-465-03078-1. ISBN 0-465-03078-5. MR 2008g:00004

- Stanley Jaki, OSB, 2005. *The drama of the quantities*. Real View Books.

- Per Lindström, 1997, *Aspects of Incompleteness*, Lecture Notes in Logic v. 10.

- J.R. Lucas, FBA, 1970. *The Freedom of the Will*. Clarendon Press, Oxford, 1970.

- Ernest Nagel, James Roy Newman, Douglas Hofstadter, 2002 (1958). *Gödel's Proof*, revised ed. ISBN 0-8147-5816-9. MR 2002i:03001

- Rudy Rucker, 1995 (1982). *Infinity and the Mind: The Science and Philosophy of the Infinite*. Princeton Univ. Press. MR 84d:03012

- Smith, Peter, 2007. *An Introduction to Gödel's Theorems*. Cambridge University Press. MathSciNet

- N. Shankar, 1994. *Metamathematics, Machines and Gödel's Proof*, Volume 38 of Cambridge tracts in theoretical computer science. ISBN 0-521-58533-3

- Raymond Smullyan, 1991. *Godel's Incompleteness Theorems*. Oxford Univ. Press.

- —, 1994. *Diagonalization and Self-Reference*. Oxford Univ. Press. MR 96c:03001

- Hao Wang, 1997. *A Logical Journey: From Gödel to Philosophy*. MIT Press. ISBN 0-262-23189-1 MR 97m:01090

### 34.13.5   Miscellaneous references

- Francesco Berto. "The Gödel Paradox and Wittgenstein's Reasons" *Philosophia Mathematica* (III) 17. 2009.

- John W. Dawson, Jr., 1997. *Logical Dilemmas: The Life and Work of Kurt Gödel*, A. K. Peters, Wellesley Mass, ISBN 1-56881-256-6.

- Goldstein, Rebecca, 2005, *Incompleteness: the Proof and Paradox of Kurt Gödel*, W. W. Norton & Company. ISBN 0-393-05169-2

- Juliet Floyd and Hilary Putnam, 2000, "A Note on Wittgenstein's 'Notorious Paragraph' About the Gödel Theorem", *Journal of Philosophy* v. 97 n. 11, pp. 624–632.

- David Hilbert and Paul Bernays, *Grundlagen der Mathematik*, Springer-Verlag.

- John Hopcroft and Jeffrey Ullman 1979, *Introduction to Automata Theory, Languages, and Computation*, Addison-Wesley, ISBN 0-201-02988-X

- James P. Jones, *Undecidable Diophantine Equations*, Bulletin of the American Mathematical Society v. 3 n. 2, 1980, pp. 859–862.

- Stephen Cole Kleene, 1967, *Mathematical Logic*. Reprinted by Dover, 2002. ISBN 0-486-42533-9

- Russell O'Connor, 2005, "Essential Incompleteness of Arithmetic Verified by Coq", Lecture Notes in Computer Science v. 3603, pp. 245–260.

- Graham Priest, 2006, *In Contradiction: A Study of the Transconsistent*, Oxford University Press, ISBN 0-19-926329-9

- Graham Priest, 2004, *Wittgenstein's Remarks on Gödel's Theorem* in Max Kölbel, ed., *Wittgenstein's lasting significance*, Psychology Press, pp. 207–227.

- Graham Priest, 1984, "Logic of Paradox Revisited", *Journal of Philosophical Logic*, v. 13,` n. 2, pp. 153–179

- Hilary Putnam, 1960, *Minds and Machines* in Sidney Hook, ed., *Dimensions of Mind: A Symposium*. New York University Press. Reprinted in Anderson, A. R., ed., 1964. *Minds and Machines*. Prentice-Hall: 77.

- Rautenberg, Wolfgang (2010), *A Concise Introduction to Mathematical Logic* (3rd ed.), New York: Springer Science+Business Media, doi:10.1007/978-1-4419-1221-3, ISBN 978-1-4419-1220-6.

- Victor Rodych, 2003, "Misunderstanding Gödel: New Arguments about Wittgenstein and New Remarks by Wittgenstein", *Dialectica* v. 57 n. 3, pp. 279–313. doi:10.1111/j.1746-8361.2003.tb00272.x

- Stewart Shapiro,2002, "Incompleteness and Inconsistency",*Mind*,v. 111,pp817–32. doi:10.1093/mind/111.4447

- Alan Sokal and Jean Bricmont, 1999, *Fashionable Nonsense: Postmodern Intellectuals' Abuse of Science*, Picador. ISBN 0-312-20407-8

- Joseph R. Shoenfield (1967), *Mathematical Logic*. Reprinted by A.K. Peters for the Association for Symbolic Logic, 2001. ISBN 978-1-56881-135-2

- Jeremy Stangroom and Ophelia Benson, *Why Truth Matters*, Continuum. ISBN 0-8264-9528-1

- George Tourlakis, *Lectures in Logic and Set Theory, Volume 1, Mathematical Logic*, Cambridge University Press, 2003. ISBN 978-0-521-75373-9

- Wigderson, Avi (2010), "The Gödel Phenomena in Mathematics: A Modern View" (PDF), *Kurt Gödel and the Foundations of Mathematics: Horizons of Truth*, Cambridge University Press

- Hao Wang, 1996, *A Logical Journey: From Gödel to Philosophy*, The MIT Press, Cambridge MA, ISBN 0-262-23189-1.

- Richard Zach, 2006, "Hilbert's program then and now", in *Philosophy of Logic*, Dale Jacquette (ed.), Handbook of the Philosophy of Science, v. 5., Elsevier, pp. 411–447.

## 34.14  External links

- 

- Godel's Incompleteness Theorems on *In Our Time* at the BBC. (listen now)

- Kurt Gödel entry by Juliette Kennedy in the *Stanford Encyclopedia of Philosophy*, July 5, 2011

- Gödel's Incompleteness Theorems entry by Panu Raatikainen in the *Stanford Encyclopedia of Philosophy*, November 11, 2013

- MacTutor biographies:

    - Kurt Gödel.

    - Gerhard Gentzen.

    - What is Mathematics:Gödel'{}s Theorem and Around by *Karlis Podnieks*. An online free book.

- World's shortest explanation of Gödel's theorem using a printing machine as an example.

- October 2011 RadioLab episode about/including Gödel's Incompleteness theorem

- Hazewinkel, Michiel, ed. (2001), "Gödel incompleteness theorem", *Encyclopedia of Mathematics*, Springer, ISBN 978-1-55608-010-4

# Chapter 35

# Hardy hierarchy

In computability theory, computational complexity theory and proof theory, the **Hardy hierarchy**, named after G. H. Hardy, is an ordinal-indexed family of functions $h\alpha$: $\mathbf{N} \to \mathbf{N}$ (where $\mathbf{N}$ is the set of natural numbers, $\{0, 1, ...\}$). It is related to the fast-growing hierarchy and slow-growing hierarchy. The hierarchy was first described in Hardy's 1904 paper, "A theorem concerning the infinite cardinal numbers".

## 35.1  Definition

Let $\mu$ be a large countable ordinal such that a fundamental sequence is assigned to every limit ordinal less than $\mu$. The **Hardy hierarchy** of functions $h\alpha$: $\mathbf{N} \to \mathbf{N}$, for $\alpha < \mu$, is then defined as follows:

- $h_0(n) = n$,

- $h_{\alpha+1}(n) = h_\alpha(n+1)$,

- $h_\alpha(n) = h_{\alpha[n]}(n)$ if $\alpha$ is a limit ordinal.

Here $\alpha[n]$ denotes the $n^{\text{th}}$ element of the fundamental sequence assigned to the limit ordinal $\alpha$. A standardized choice of fundamental sequence for all $\alpha \leq \varepsilon_0$ is described in the article on the fast-growing hierarchy.

Caicedo (2007) defines a modified Hardy hierarchy of functions $H_\alpha$ by using the standard fundamental sequences, but with $\alpha[n+1]$ (instead of $\alpha[n]$) in the third line of the above definition.

## 35.2  Relation to fast-growing hierarchy

The Wainer hierarchy of functions $f\alpha$ and the Hardy hierarchy of functions $h\alpha$ are related by $f\alpha = h\omega^\alpha$ for all $\alpha < \varepsilon_0$. Thus, for any $\alpha < \varepsilon_0$, $h\alpha$ grows much more slowly than does $f\alpha$. However, the Hardy hierarchy "catches up" to the Wainer hierarchy at $\alpha = \varepsilon_0$, such that $f\varepsilon_0$ and $h\varepsilon_0$ have the same growth rate, in the sense that $f\varepsilon_0(n-1) \leq h\varepsilon_0(n) \leq f\varepsilon_0(n+1)$ for all $n \geq 1$. (Gallier 1991)

## 35.3  References

- Hardy, G.H. (1904), "A theorem concerning the infinite cardinal numbers", *Quarterly Journal of Mathematics* **35**: 87–94

- Gallier, Jean H. (1991), "What's so special about Kruskal's theorem and the ordinal $\Gamma_0$? A survey of some results in proof theory", *Ann. Pure Appl. Logic* **53** (3): 199–260, doi:10.1016/0168-0072(91)90022-E, MR 1129778 PDF's: part 1 2 3. (In particular part 3, Section 12, pp. 59–64, "A Glimpse at Hierarchies of Fast and Slow Growing Functions".)

- Caicedo, A. (2007), "Goodstein's function" (PDF), *Revista Colombiana de Matemáticas* **41** (2): 381–391.

# Chapter 36

# Herbrand's theorem

Not to be confused with Herbrand–Ribet theorem or Herbrand's theorem on ramification groups.

**Herbrand's theorem** is a fundamental result of mathematical logic obtained by Jacques Herbrand (1930).[1] It essentially allows a certain kind of reduction of first-order logic to propositional logic. Although Herbrand originally proved his theorem for arbitrary formulas of first-order logic,[2] the simpler version shown here, restricted to formulas in prenex form containing only existential quantifiers became more popular.

Let

$$(\exists y_1, \ldots, y_n) F(y_1, \ldots, y_n)$$

be a formula of first-order logic with

$$F(y_1, \ldots, y_n)$$

Then

$$(\exists y_1, \ldots, y_n) F(y_1, \ldots, y_n)$$

is valid if and only if there exists a finite sequence of terms $t_{ij}$, possibly in an expansion of the language, with

$$1 \leq i \leq k \text{ and } 1 \leq j \leq n,$$

such that

$$F(t_{11}, \ldots, t_{1n}) \vee \ldots \vee F(t_{k1}, \ldots, t_{kn})$$

is valid. If it is valid,

$$F(t_{11}, \ldots, t_{1n}) \vee \ldots \vee F(t_{k1}, \ldots, t_{kn})$$

is called a *Herbrand disjunction* for

$(\exists y_1, \ldots, y_n) F(y_1, \ldots, y_n).$

Informally: a formula $A$ in prenex form containing existential quantifiers only is provable (valid) in first-order logic if and only if a disjunction composed of substitution instances of the quantifier-free subformula of $A$ is a tautology (propositionally derivable).

The restriction to formulas in prenex form containing only existential quantifiers does not limit the generality of the theorem, because formulas can be converted to prenex form and their universal quantifiers can be removed by Herbrandization. Conversion to prenex form can be avoided, if *structural* Herbrandization is performed. Herbrandization can be avoided by imposing additional restrictions on the variable dependencies allowed in the Herbrand disjunction.

## 36.1 Proof Sketch

A proof of the non-trivial direction of the theorem can be constructed according to the following steps:

1. If the formula $(\exists y_1, \ldots, y_n) F(y_1, \ldots, y_n)$ is valid, then by completeness of cut-free sequent calculus, which follows from Gentzen's cut-elimination theorem, there is a cut-free proof of $\vdash (\exists y_1, \ldots, y_n) F(y_1, \ldots, y_n)$.

2. Starting from above downwards, remove the inferences that introduce existential quantifiers.

3. Remove contraction-inferences on previously existentially quantified formulas, since the formulas (now with terms substituted for previously quantified variables) might not be identical anymore after the removal of the quantifier inferences.

4. The removal of contractions accumulates all the relevant substitution instances of $F(y_1, \ldots, y_n)$ in the right side of the sequent, thus resulting in a proof of $\vdash F(t_{11}, \ldots, t_{1n}), \ldots, F(t_{k1}, \ldots, t_{kn})$, from which the Herbrand disjunction can be obtained.

However, sequent calculus and cut-elimination were not known at the time of Herbrand's theorem, and Herbrand had to prove his theorem in a more complicated way.

## 36.2 Generalizations of Herbrand's Theorem

- Herbrand's theorem has been extended to arbitrary higher-order logics by using expansion-tree proofs.[3] The deep representation of expansion-tree proofs correspond to Herbrand disjunctions, when restricted to first-order logic.

- Herbrand disjunctions and expansion-tree proofs have been extended with a notion of cut. Due to the complexity of cut-elimination, herbrand disjunctions with cuts can be non-elementarily smaller than a standard herbrand disjunction.

- Herbrand disjunctions have been generalized to Herbrand sequents, allowing Herbrand's theorem to be stated for sequents: "a skolemized sequent is derivable iff it has a Herbrand sequent".

## 36.3 See also

- Herbrand structure
- Herbrand interpretation
- Herbrand universe
- Compactness theorem

## 36.4   Notes

[1]  J. Herbrand: Recherches sur la theorie de la demonstration. Travaux de la Societe des Sciences et des Lettres de Varsovie, Class III, Sciences Mathematiques et Physiques, 33, 1930.

[2]  Samuel R. Buss: "Handbook of Proof Theory". Chapter 1, "An Introduction to Proof Theory". Elsevier, 1998.

[3]  Dale Miller: A Compact Representation of Proofs. Studia Logica, 46(4), pp. 347-–370, 1987.

## 36.5   References

• Buss, Samuel R. (1995), "On Herbrand's Theorem", in Maurice, Daniel; Leivant, Raphaël, *Logic and Computational Complexity*, Lecture Notes in Computer Science, Berlin, New York: Springer-Verlag, pp. 195–209, ISBN 978-3-540-60178-4.

# Chapter 37

# Hilbert system

*In mathematical physics,* Hilbert system *is an infrequently used term for a physical system described by a C\*-algebra.*

In logic, especially mathematical logic, a **Hilbert system**, sometimes called **Hilbert calculus** or **Hilbert–Ackermann system**, is a type of system of formal deduction attributed to Gottlob Frege[1] and David Hilbert. These deductive systems are most often studied for first-order logic, but are of interest for other logics as well.

Most variants of Hilbert systems take a characteristic tack in the way they balance a trade-off between logical axioms and rules of inference.[1] Hilbert systems can be characterised by the choice of a large number of schemes of logical axioms and a small set of rules of inference. Systems of natural deduction take the opposite tack, including many deduction rules but very few or no axiom schemes. The most commonly studied Hilbert systems have either just one rule of inference — modus ponens, for propositional logics — or two — with generalisation, to handle predicate logics, as well — and several infinite axiom schemes. Hilbert systems for propositional modal logics, sometimes called Hilbert-Lewis systems, are generally axiomatised with two additional rules, the necessitation rule and the uniform substitution rule.

A characteristic feature of the many variants of Hilbert systems is that the *context* is not changed in any of their rules of inference, while both natural deduction and sequent calculus contain some context-changing rules. Thus, if we are interested only in the derivability of tautologies, no hypothetical judgments, then we can formalize the Hilbert system in such a way that its rules of inference contain only judgments of a rather simple form. The same cannot be done with the other two deductions systems : as context is changed in some of their rules of inferences, they cannot be formalized so that hypothetical judgments could be avoided — not even if we want to use them just for proving derivability of tautologies.

## 37.1 Formal deductions

In a Hilbert-style deduction system, a **formal deduction** is a finite sequence of formulas in which each formula is either an axiom or is obtained from previous formulas by a rule of inference. These formal deductions are meant to mirror natural-language proofs, although they are far more detailed.

Suppose $\Gamma$ is a set of formulas, considered as **hypotheses**. For example $\Gamma$ could be a set of axioms for group theory or set theory. The notation $\Gamma \vdash \phi$ means that there is a deduction that ends with $\phi$ using as axioms only **logical axioms** and elements of $\Gamma$. Thus, informally, $\Gamma \vdash \phi$ means that $\phi$ is provable assuming all the formulas in $\Gamma$.

Hilbert-style deduction systems are characterized by the use of numerous schemes of **logical axioms**. An axiom scheme is an infinite set of axioms obtained by substituting all formulas of some form into a specific pattern. The set of logical axioms includes not only those axioms generated from this pattern, but also any generalization of one of those axioms. A generalization of a formula is obtained by prefixing zero or more universal quantifiers on the formula; thus

$$\forall y (\forall x Pxy \rightarrow Pty)$$

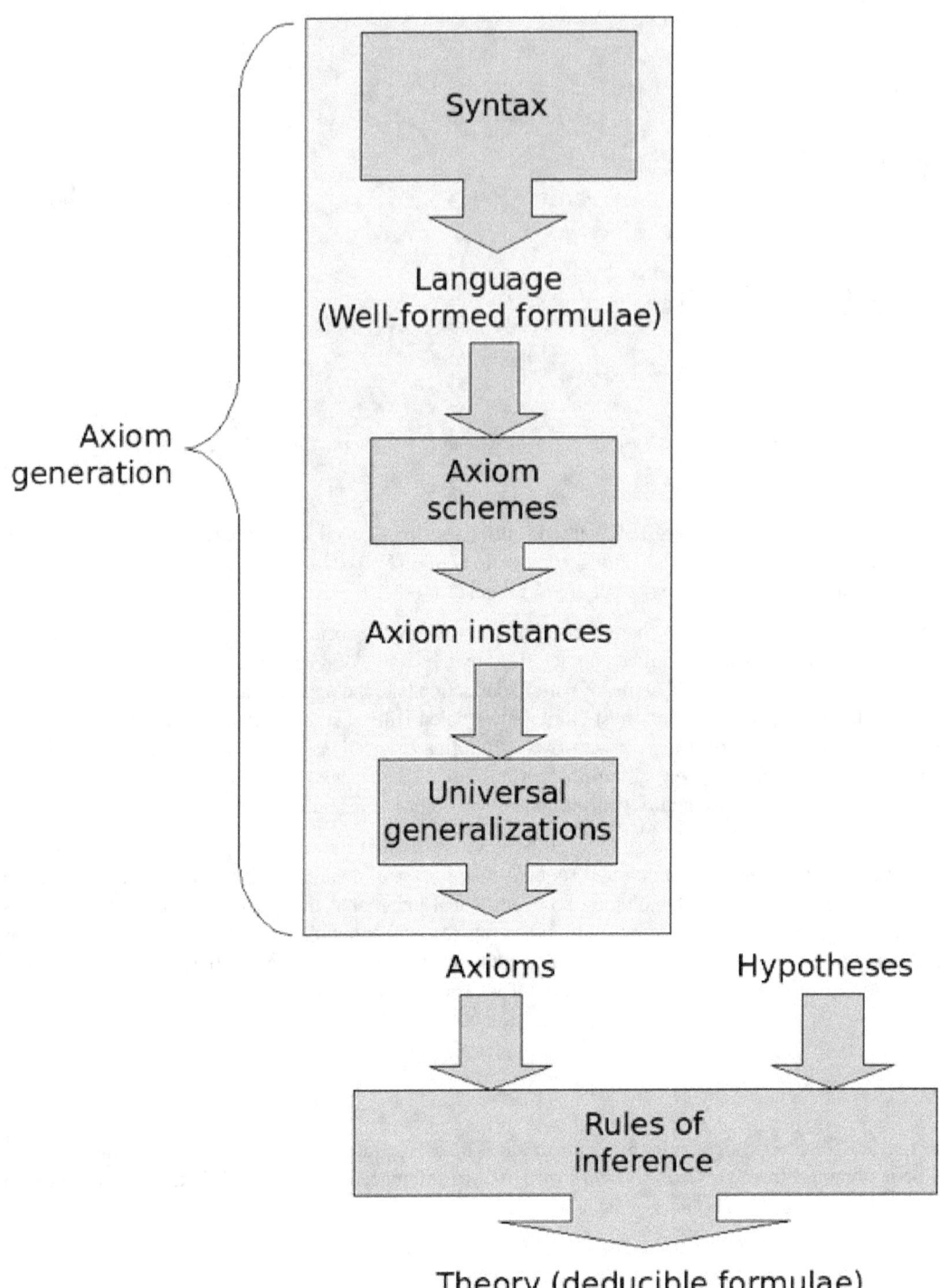

*A graphic representation of the deduction system*

is a generalization of $\forall x P x y \to P t y$ .

## 37.1.1 Logical axioms

There are several variant axiomatisations of predicate logic, since for any logic there is freedom in choosing axioms and rules that characterise that logic. We describe here a Hilbert system with nine axioms and just the rule modus ponens, which we call the one-rule axiomatisation and which describes classical equational logic. We deal with a minimal language for this logic, where formulas use only the connectives $\neg$ and $\rightarrow$ and only the quantifier $\forall$. Later we show how the system can be extended to include additional logical connectives, such as $\wedge$ and $\vee$, without enlarging the class of deducible formulas.

The first four logical axiom schemes allow (together with modus ponens) for the manipulation of logical connectives.

$$\phi \rightarrow \phi$$

$$\phi \rightarrow (\psi \rightarrow \phi)$$

$$(\phi \rightarrow (\psi \rightarrow \xi)) \rightarrow ((\phi \rightarrow \psi) \rightarrow (\phi \rightarrow \xi))$$

$$(\neg\phi \rightarrow \neg\psi) \rightarrow (\psi \rightarrow \phi)$$

The axiom P1 is redundant, as it follows from P3, P2 and modus ponens. These axioms describe classical propositional logic; without axiom P4 we get positive implicational logic. Minimal logic is achieved either by adding instead the axiom P4m, or by defining $\neg\phi$ as $\phi \rightarrow \bot$.

$$(\phi \rightarrow \psi) \rightarrow ((\phi \rightarrow \neg\psi) \rightarrow \neg\phi)$$

Intuitionistic logic is achieved by adding axioms P4i and P5i to positive implicational logic, or by adding axiom P5i to minimal logic. Both P4i and P5i are theorems of classical propositional logic.

$$(\phi \rightarrow \neg\phi) \rightarrow \neg\phi$$

$$\neg\phi \rightarrow (\phi \rightarrow \psi)$$

Note that these are axiom schemes, which represent infinitely many specific instances of axioms. For example, P1 might represent the particular axiom instance $p \rightarrow p$, or it might represent $(p \rightarrow q) \rightarrow (p \rightarrow q)$ : the $\phi$ is a place where any formula can be placed. A variable such as this that ranges over formulae is called a 'schematic variable'.

With a second rule of uniform substitution (US), we can change each of these axiom schemes into a single axiom, replacing each schematic variable by some propositional variable that isn't mentioned in any axiom to get what we call the substitutional axiomatisation. Both formalisations have variables, but where the one-rule axiomatisation has schematic variables that are outside the logic's language, the substitutional axiomatisation uses propositional variables that do the same work by expressing the idea of a variable ranging over formulae with a rule that uses substitution.

> US. Let $\phi(p)$ be a formula with one or more instances of the propositional variable $p$, and let $\psi$ be another formula. Then from $\phi(p)$, infer $\phi(\psi)$.

The next three logical axiom schemes provide ways to add, manipulate, and remove universal quantifiers.

> Q5. $\forall x\,(\phi) \rightarrow \phi[x := t]$ where $t$ may be substituted for $x$ in $\phi$
>
> Q6. $\forall x\,(\phi \rightarrow \psi) \rightarrow (\forall x\,(\phi) \rightarrow \forall x\,(\psi))$
>
> Q7. $\phi \rightarrow \forall x\,(\phi)$ where $x$ is not a free variable of $\phi$.

These three additional rules extend the propositional system to axiomatise classical predicate logic. Likewise, these three rules extend system for intuitionstic propositional logic (with P1-3 and P4i and P5i) to intuitionistic predicate logic.

Universal quantification is often given an alternative axiomatisation using an extra rule of generalisation (see the section on Metatheorems), in which case the rules Q5 and Q6 are redundant.

The final axiom schemes are required to work with formulas involving the equality symbol.

I8. $x = x$ for every variable $x$.

I9. $(x = y) \rightarrow (\phi[z := x] \rightarrow \phi[z := y])$

## 37.2  Conservative extensions

It is common to include in a Hilbert-style deduction system only axioms for implication and negation. Given these axioms, it is possible to form conservative extensions of the deduction theorem that permit the use of additional connectives. These extensions are called conservative because if a formula $\varphi$ involving new connectives is rewritten as a logically equivalent formula $\theta$ involving only negation, implication, and universal quantification, then $\varphi$ is derivable in the extended system if and only if $\theta$ is derivable in the original system. When fully extended, a Hilbert-style system will resemble more closely a system of natural deduction.

### 37.2.1  Existential quantification

- Introduction

$$\forall x(\phi \rightarrow \exists y(\phi[x := y]))$$

- Elimination

$$\forall x(\phi \rightarrow \psi) \rightarrow \exists x(\phi) \rightarrow \psi \text{ where } x \text{ is not a free variable of } \psi.$$

### 37.2.2  Conjunction and Disjunction

- Conjunction introduction and elimination

$$\alpha \rightarrow \beta \rightarrow \alpha \wedge \beta$$

$$\alpha \wedge \beta \rightarrow \alpha$$

$$\alpha \wedge \beta \rightarrow \beta$$

- Disjunction introduction and elimination

$$\alpha \rightarrow \alpha \vee \beta$$

$$\beta \rightarrow \alpha \vee \beta$$

$$(\alpha \rightarrow \gamma) \rightarrow (\beta \rightarrow \gamma) \rightarrow \alpha \vee \beta \rightarrow \gamma$$

## 37.3 Metatheorems

Because Hilbert-style systems have very few deduction rules, it is common to prove **metatheorems** that show that additional deduction rules add no deductive power, in the sense that a deduction using the new deduction rules can be converted into a deduction using only the original deduction rules.

Some common metatheorems of this form are:

- The **deduction theorem**: $\Gamma; \phi \vdash \psi$ if and only if $\Gamma \vdash \phi \to \psi$.

- $\Gamma \vdash \phi \leftrightarrow \psi$ if and only if $\Gamma \vdash \phi \to \psi$ and $\Gamma \vdash \psi \to \phi$.

- Contraposition: If $\Gamma; \phi \vdash \psi$ then $\Gamma; \neg\psi \vdash \neg\phi$.

- Generalization: If $\Gamma \vdash \phi$ and $x$ does not occur free in any formula of $\Gamma$ then $\Gamma \vdash \forall x \phi$.

## 37.4 Alternative axiomatizations

Further information: List of logic systems

The axiom 3 above is credited to Łukasiewicz.[2] The original system by Frege had axioms P2 and P3 but four other axioms instead of axiom P4 (see Frege's propositional calculus). Russell and Whitehead also suggested a system with five propositional axioms.

## 37.5 Further connections

Axioms P1, P2 and P3, with the deduction rule modus ponens (formalising intuitionistic propositional logic), correspond to combinatory logic base combinators **I**, **K** and **S** with the application operator. Proofs in the Hilbert system then correspond to combinator terms in combinatory logic. See also Curry-Howard correspondence.

## 37.6 Notes

[1] Máté & Ruzsa 1997:129

[2] A. Tarski, Logic, semantics, metamathematics, Oxford, 1956

## 37.7 References

- Curry, Haskell B.; Robert Feys (1958). *Combinatory Logic Vol. I* **1**. Amsterdam: North Holland.

- Monk, J. Donald (1976). *Mathematical Logic*. Graduate Texts in Mathematics. Berlin, New York: Springer-Verlag. ISBN 978-0-387-90170-1.

- Ruzsa, Imre; Máté, András (1997). *Bevezetés a modern logikába* (in Hungarian). Budapest: Osiris Kiadó.

- Tarski, Alfred (1990). *Bizonyítás és igazság* (in Hungarian). Budapest: Gondolat. It is a Hungarian translation of Alfred Tarski's selected papers on semantic theory of truth.

- David Hilbert (1927) "The foundations of mathematics", translated by Stephan Bauer-Menglerberg and Dagfinn Føllesdal (pp. 464–479). in:

- van Heijenoort, Jean (1967). *From Frege to Gödel: A Source Book in Mathematical Logic, 1879–1931* (3rd printing 1976 ed.). Cambridge MA: Harvard University Press. ISBN 0-674-32449-8.

   Hilbert's 1927, Based on an earlier 1925 "foundations" lecture (pp. 367–392), presents his 17 axioms -- axioms of implication #1-4, axioms about & and V #5-10, axioms of negation #11-12, his logical ε-axiom #13, axioms of equality #14-15, and axioms of number #16-17 -- along with the other necessary elements of his Formalist "proof theory" -- e.g. induction axioms, recursion axioms, etc; he also offers up a spirited defense against L.E.J. Brouwer's Intuitionism. Also see Hermann Weyl's (1927) comments and rebuttal (pp. 480–484), Paul Bernay's (1927) appendix to Hilbert's lecture (pp. 485–489) and Luitzen Egbertus Jan Brouwer's (1927) response (pp. 490–495)

- Kleene, Stephen Cole (1952). *Introduction to Metamathematics* (10th impression with 1971 corrections ed.). Amsterdam NY: North Holland Publishing Company. ISBN 0-7204-2103-9.

   See in particular Chapter IV Formal System (pp. 69–85) wherein Kleene presents subchapters §16 Formal symbols, §17 Formation rules, §18 Free and bound variables (including substitution), §19 Transformation rules (e.g. modus ponens) -- and from these he presents 21 "postulates" -- 18 axioms and 3 "immediate-consequence" relations divided as follows: Postulates for the propostional calculus #1-8, Additional postulates for the predicate calculus #9-12, and Additional postulates for number theory #13-21.

## 37.8   External links

- Gaifman, Haim. "A Hilbert Type Deductive System for Sentential Logic, Completeness and Compactness." (pdf).

- Farmer, W. M. "Propositional logic" (pdf). It describes (among others) a part of the Hilbert-style deduction system (restricted to propositional calculus).

# Chapter 38

# Hilbert's program

In mathematics, **Hilbert's program**, formulated by German mathematician David Hilbert, was a proposed solution to the foundational crisis of mathematics, when early attempts to clarify the foundations of mathematics were found to suffer from paradoxes and inconsistencies. As a solution, Hilbert proposed to ground all existing theories to a finite, complete set of axioms, and provide a proof that these axioms were consistent. Hilbert proposed that the consistency of more complicated systems, such as real analysis, could be proven in terms of simpler systems. Ultimately, the consistency of all of mathematics could be reduced to basic arithmetic.

Gödel's incompleteness theorems, published in 1931, showed that Hilbert's program was unattainable for key areas of mathematics. In his first theorem, Gödel showed that any consistent system with a computable set of axioms which is capable of expressing arithmetic can never be complete: it is possible to construct a statement that can be shown to be true, but that cannot be derived from the formal rules of the system. In his second theorem, he showed that such a system could not prove its own consistency, so it certainly cannot be used to prove the consistency of anything stronger with certainty. This refuted Hilbert's assumption that a finitistic system could be used to prove the consistency of itself, and therefore anything else.

## 38.1 Statement of Hilbert's Program

The main goal of Hilbert's program was to provide secure foundations for all mathematics. In particular this should include:

- A formalization of all mathematics; in other words all mathematical statements should be written in a precise formal language, and manipulated according to well defined rules.

- Completeness: a proof that all true mathematical statements can be proved in the formalism.

- Consistency: a proof that no contradiction can be obtained in the formalism of mathematics. This consistency proof should preferably use only "finitistic" reasoning about finite mathematical objects.

- Conservation: a proof that any result about "real objects" obtained using reasoning about "ideal objects" (such as uncountable sets) can be proved without using ideal objects.

- Decidability: there should be an algorithm for deciding the truth or falsity of any mathematical statement.

## 38.2 Gödel's Incompleteness Theorems

Main article: Gödel's incompleteness theorems

Kurt Gödel showed that most of the goals of Hilbert's program were impossible to achieve, at least if interpreted in the most obvious way. His second incompleteness theorem stated that any consistent theory powerful enough to encode addition and multiplication of integers cannot prove its own consistency. This wipes out most of Hilbert's program as follows:

- It is not possible to formalize **all** of mathematics, as any attempt at such a formalism will omit some true mathematical statements.

- An easy consequence of Gödel's incompleteness theorem is that there is no complete consistent extension of even Peano arithmetic with a recursively enumerable set of axioms, so in particular most interesting mathematical theories are not complete.

- A theory such as Peano arithmetic cannot even prove its own consistency, so a restricted "finitistic" subset of it certainly cannot prove the consistency of more powerful theories such as set theory.

- There is no algorithm to decide the truth (or provability) of statements in any consistent extension of Peano arithmetic. (Strictly speaking this result only appeared a few years after Gödel's theorem, because at the time the notion of an algorithm had not been precisely defined.)

## 38.3   Hilbert's program after Gödel

Many current lines of research in mathematical logic, proof theory and reverse mathematics can be viewed as natural continuations of Hilbert's original program. Much of it can be salvaged by changing its goals slightly (Zach 2005), and with the following modifications some of it was successfully completed:

- Although it is not possible to formalize **all** mathematics, it is possible to formalize essentially all the mathematics that anyone uses. In particular Zermelo–Fraenkel set theory, combined with first-order logic, gives a satisfactory and generally accepted formalism for essentially all current mathematics.

- Although it is not possible to prove completeness for systems at least as powerful as Peano arithmetic (at least if they have a computable set of axioms), it is possible to prove forms of completeness for many interesting systems. The first big success was by Gödel himself (before he proved the incompleteness theorems) who proved the completeness theorem for first-order logic, showing that any logical consequence of a series of axioms is provable. An example of a non-trivial theory for which completeness has been proved is the theory of algebraically closed fields of given characteristic.

- The question of whether there are finitary consistency proofs of strong theories is difficult to answer, mainly because there is no generally accepted definition of a "finitary proof". Most mathematicians in proof theory seem to regard finitary mathematics as being contained in Peano arithmetic, and in this case it is not possible to give finitary proofs of reasonably strong theories. On the other hand Gödel himself suggested the possibility of giving finitary consistency proofs using finitary methods that cannot be formalized in Peano arithmetic, so he seems to have had a more liberal view of what finitary methods might be allowed. A few years later, Gentzen gave a consistency proof for Peano arithmetic. The only part of this proof that was not clearly finitary was a certain transfinite induction up to the ordinal $\varepsilon_0$. If this transfinite induction is accepted as a finitary method, then one can assert that there is a finitary proof of the consistency of Peano arithmetic. More powerful subsets of second order arithmetic have been given consistency proofs by Gaisi Takeuti and others, and one can again debate about exactly how finitary or constructive these proofs are. (The theories that have been proved consistent by these methods are quite strong, and include most "ordinary" mathematics.)

- Although there is no algorithm for deciding the truth of statements in Peano arithmetic, there are many interesting and non-trivial theories for which such algorithms have been found. For example, Tarski found an algorithm that can decide the truth of any statement in analytic geometry (more precisely, he proved that the theory of real closed fields is decidable). Given the Cantor–Dedekind axiom, this algorithm can be regarded as an algorithm to decide the truth of any statement in Euclidean geometry. This is substantial as few people would consider Euclidean geometry a trivial theory.

## 38.4 See also

- Grundlagen der Mathematik

- Foundational crisis of mathematics

- Atomism

## 38.5 References

- G. Gentzen, 1936/1969. Die Widerspruchfreiheit der reinen Zahlentheorie. *Mathematische Annalen* 112:493–565. Translated as 'The consistency of arithmetic', in *The collected papers of Gerhard Gentzen*, M. E. Szabo (ed.), 1969.

- D. Hilbert. 'Die Grundlagen Der Elementaren Zahlentheorie'. *Mathematische Annalen* 104:485–94. Translated by W. Ewald as 'The Grounding of Elementary Number Theory', pp. 266–273 in Mancosu (ed., 1998) *From Brouwer to Hilbert: The debate on the foundations of mathematics in the 1920s*, Oxford University Press. New York.

- S.G. Simpson, 1988. Partial realizations of Hilbert's program. *Journal of Symbolic Logic* 53:349–363.

- R. Zach, 2005. Hilbert's Program Then and Now. Manuscript, arXiv:math/0508572v1.

## 38.6 External links

- Entry on Hilbert's program by Richard Zach at the Stanford Encyclopedia of Philosophy.

# Chapter 39

# Independence (mathematical logic)

In mathematical logic, **independence** refers to the unprovability of a sentence from other sentences.

A sentence σ is **independent** of a given first-order theory $T$ if $T$ neither proves nor refutes σ; that is, it is impossible to prove σ from $T$, and it is also impossible to prove from $T$ that σ is false. Sometimes, σ is said (synonymously) to be *undecidable* from $T$; this is not the same meaning of "decidability" as in a decision problem.

A theory $T$ is **independent** if each axiom in $T$ is not provable from the remaining axioms in $T$. A theory for which there is an independent set of axioms is **independently axiomatizable**.

## 39.1   Usage note

Some authors say that σ is independent of $T$ if $T$ simply cannot prove σ, and do not necessarily assert by this that $T$ cannot refute σ. These authors will sometimes say "σ is independent of and consistent with $T$" to indicate that $T$ can neither prove nor refute σ.

## 39.2   Independence results in set theory

Many interesting statements in set theory are independent of Zermelo-Fraenkel set theory (ZF). The following statements in set theory are known to be independent of ZF, granting that ZF is consistent:

- The axiom of choice

- The continuum hypothesis and the generalised continuum hypothesis

- The Suslin conjecture

The following statements (none of which have been proved false) cannot be proved in ZFC to be independent of ZFC, even if the added hypothesis is granted that ZFC is consistent. However, they cannot be proved in ZFC (granting that ZFC is consistent), and few working set theorists expect to find a refutation of them in ZFC.

- The existence of strongly inaccessible cardinals

- The existence of large cardinals

- The non-existence of Kurepa trees

The following statements are inconsistent with the axiom of choice, and therefore with ZFC. However they are probably independent of ZF, in a corresponding sense to the above: They cannot be proved in ZF, and few working set theorists expect to find a refutation in ZF. However ZF cannot prove that they are independent of ZF, even with the added hypothesis that ZF is consistent.

- The Axiom of determinacy
- The axiom of real determinacy
- AD+

## 39.3   See also

- List of statements undecidable in ZFC
- Parallel postulate for an example in geometry
- Truth

## 39.4   References

- Mendelson, Elliott (1997), *An Introduction to Mathematical Logic* (4th ed.), London: Chapman & Hall, ISBN 978-0-412-80830-2

- Monk, J. Donald (1976), *Mathematical Logic*, Graduate Texts in Mathematics, Berlin, New York: Springer-Verlag, ISBN 978-0-387-90170-1

# Chapter 40

# Interpretability

In mathematical logic, **interpretability** is a relation between formal theories that expresses the possibility of interpreting or translating one into the other.

## 40.1 Informal definition

Assume T and S are formal theories. Slightly simplified, T is said to be *interpretable* in S if and only if the language of T can be translated into the language of S in such a way that S proves the translation of every theorem of T. Of course, there are some natural conditions on admissible translations here, such as the necessity for a translation to preserve the logical structure of formulas.

This concept, together with weak interpretability, was introduced by Alfred Tarski in 1953. Three other related concepts are cointerpretability, logical tolerance, and cotolerance, introduced by Giorgi Japaridze in 1992-1993.

## 40.2 References

- Japaridze, G., and De Jongh, D. (1998) "The logic of provability" in Buss, S., ed., *Handbook of Proof Theory*. North-Holland: 476-546.

- Alfred Tarski, Andrzej Mostowski, and Raphael Robinson (1953) *Undecidable Theories*. North-Holland.

## 40.3 See also

# Chapter 41

# Japaridze's polymodal logic

**Japaridze's polymodal logic (GLP)**, is a system of provability logic with infinitely many modal (provability) operators. This system has played an important role in some applications of provability algebras in proof theory, and has been extensively studied since the late 1980s. It is named after Giorgi Japaridze.

## 41.1 Language and axiomatization

The language of GLP extends that of the language of classical propositional logic by including the infinite series [0],[1],[2],... of "necessity" operators. Their dual "possibility" operators <0>,<1>,<2>,... are defined by $<n>p = \neg[n]\neg p$.

The axioms of GLP are all classical tautologies and all formulas of one of the following forms:

- $[n](p \rightarrow q) \rightarrow ([n]p \rightarrow [n]q)$

- $[n]([n]p \rightarrow p) \rightarrow [n]p$

- $[n]p \rightarrow [n+1]p$

- $<n>p \rightarrow [n+1]<n>p$

And the rules of inference are:

- From $p$ and $p \rightarrow q$ conclude $q$

- From $p$ conclude $[0]p$

## 41.2 Provability semantics

Consider a "sufficiently strong" first-order theory $T$ such as Peano Arithmetic **PA**. Define the series $T_0, T_1, T_2,...$ of theories as follows:

- $T_0$ is $T$

- $T_{n+1}$ is the extension of $Tn$ through the additional axioms $\forall x F(x)$ for each formula $F(x)$ such that $Tn$ proves all of the formulas $F(0), F(1), F(2),...$

For each $n$, let $\text{Pr}n(x)$ be a natural arithmetization of the predicate "$x$ is the Gödel number of a sentence provable in $Tn$.

A *realization* is a function $^*$ which sends each nonlogical atom $a$ of the language of GLP to a sentence a$^*$ of the language of $T$. It extends to all formulas of the language of GLP by stipulating that $^*$ commutes with the Boolean connectives, and that $([n]F)^*$ is $\text{Pr}\_n(`F^{*'})$, where '$F^{*'}$ stands for the (numeral for) the Gödel number of $F^*$.

An **arithmetical completeness theorem**[1] for GLP states that a formula $F$ is provable in GLP if and only if, for every interpretation $^*$, the sentence $F^*$ is provable in $T$.

The above understanding of the series $T_0, T_1, T_2,...$ of theories is not the only natural understanding yielding the soundness and completeness of GLP. For instance, each theory $Tn$ can be understood as $T$ augmented with all true $\prod_1$ sentences as additional axioms. George Boolos showed[2] that GLP remains sound and complete with analysis (second-order arithmetic) in the role of the base theory $T$.

## 41.3   Other semantics

GLP has been shown[3] to be incomplete with respect to any class of Kripke frames.

A natural topological semantics of GLP interprets modalities as derivative operators of a polytopological space. Such spaces are called GLP-spaces whenever they satisfy all the axioms of GLP. GLP is complete w.r.t. the class of all GLP-spaces.[4]

## 41.4   Computational complexity

The problem of being a theorem of GLP is PSPACE-complete. So is the same problem restricted to only variable-free formulas of GLP.[5]

## 41.5   History

GLP, under the name GP, was introduced by Giorgi Japaridze in his PhD thesis "Modal Logical Means of Investigating Provability" (Moscow State University, 1986) and first published in.[6] The completeness theorem for GLP with respect to its provability interpretation was also first proven in.[7] Later, in,[8] Beklemishev came up with a simpler proof of the same theorem. The non-existence of Kripke frames for GLP was shown in.[9] A more extensive study of Kripke models for GLP was conducted by Beklemishev in.[10] Topological models for GLP were studied by Beklemishev, Bezhanishvili, Icar and Gabelaia,.[11][12] The decidability of GLP in polynomial space was proven by I. Shapirovsky,[13] and the PSPACE-hardness of its variable-free fragment was proven by F.Pakhomov.[14] Among the most notable applications of GLP has been its use in proof-theoretically analyzing Peano arithmetic, elaborating a canonical way for recovering ordinal notation system up to $\varepsilon_0$ from the corresponding algebra, and constructing simple combinatorial independent statements (Beklemishev [15]).

An extensive survey of GLP in the context of provability logics in general was given by George Boolos in his book "The Logic of Provability".[16]

## 41.6   Literature

- L. Beklemishev, "Provability algebras and proof-theoretic ordinals, I". Annals of Pure and Applied Logic 128 (2004), pp. 103–123.

- L. Beklemishev, J. Joosten and M. Vervoort, "A finitary treatment of the closed fragment of Japaridze's provability logic". Journal of Logic and Computation 15 (2005), No 4, pp. 447–463.

- L. Beklemishev, "Kripke semantics for provability logic GLP". Annals of Pure and Applied Logic 161, 756–774 (2010).

- L. Beklemishev, G. Bezhanishvili and T. Icar, "On topological models of GLP". Ways of proof theory, Ontos Mathematical Logic, 2, eds. R. Schindler, Ontos Verlag, Frankfurt, 2010, pp. 133–153.

- L. Beklemishev, "On the Craig interpolation and the fixed point properties of GLP". Proofs, Categories and Computations. S. Feferman et al., eds., College Publications 2010. pp. 49–60.

- L. Beklemishev, "Ordinal completeness of bimodal provability logic GLB". Lecture Notes in Computer Science 6618 (2011), pp. 1–15.

- L. Beklemishev, "A simplified proof of arithmetical completeness theorem for provability logic GLP". Proceedings of the Steklov Institute of Mathematics 274 (2011), pp. 25–33.

- L. Beklemishev and D. Gabelaia, "Topological completeness of provability logic GLP". Annals of Pure and Applied Logic 164 (2013), pp. 1201–1223.

- G. Boolos, "The analytical completeness of Japaridze's polymodal logics". Annals of Pure and Applied Logic 61 (1993), pp. 95–111.

- G. Boolos, "The Logic of Provability". Cambridge University Press, 1993.

- E.V. Dashkov, "On the positive fragment of the polymodal provability logic GLP". Mathematical Notes 2012; 91:318–333.

- D. Fernandez-Duque and J.Joosten, "Well-orders in the transfinite Japaridze algebra". Logic Journal of the IGPL 22 (2014), pp. 933–963.

- G. Japaridze, "The polymodal logic of provability". Intensional Logics and Logical Structure of Theories. Metsniereba, Tbilisi, 1988, pp. 16–48 (Russian).

- F. Pakhomov, "On the complexity of the closed fragment of Japaridze's provability logic". Archive for Mathematical Logic 53 (2014), pp. 949–967.

- D.S. Shamkanov, "Interpolation properties for provability logics GL and GLP". Proceedings of the Steklov Institute of Mathematics 274 (2011), pp. 303–316.

- I. Shapirovsky, "PSPACE-decidability of Japaridze's polymodal logic". Advances in Modal Logic 7 (2008), pp. 289–304.

## 41.7 References

[1] G. Japaridze, "The polymodal logic of provability". Intensional Logics and Logical Structure of Theories. Metsniereba, Tbilisi, 1988, pp. 16–48 (Russian)

[2] G. Boolos, "The analytical completeness of Japaridze's polymodal logics". Annals of Pure and Applied Logic 61 (1993), pp. 95–111.

[3] G. Japaridze, "The polymodal logic of provability". Intensional Logics and Logical Structure of Theories. Metsniereba, Tbilisi, 1988, pp. 16–48 (Russian).

[4] L. Beklemishev and D. Gabelaia, "Topological completeness of provability logic GLP". Annals of Pure and Applied Logic 164 (2013), pp. 1201–1223.

[5] F. Pakhomov, "On the complexity of the closed fragment of Japaridze's provability logic". Archive for Mathematical Logic 53 (2014), pp. 949–967.

[6] G. Japaridze, "The polymodal logic of provability". Intensional Logics and Logical Structure of Theories. Metsniereba, Tbilisi, 1988, pp. 16–48 (Russian).

[7] G. Japaridze, "The polymodal logic of provability". Intensional Logics and Logical Structure of Theories. Metsniereba, Tbilisi, 1988, pp. 16–48 (Russian).

[8] L. Beklemishev, "A simplified proof of arithmetical completeness theorem for provability logic GLP". Proceedings of the Steklov Institute of Mathematics 274 (2011), pp. 25–33.

[9] G. Japaridze, "The polymodal logic of provability". Intensional Logics and Logical Structure of Theories. Metsniereba, Tbilisi, 1988, pp. 16–48 (Russian).

[10] L. Beklemishev, "Kripke semantics for provability logic GLP". Annals of Pure and Applied Logic 161, 756–774 (2010).

[11] L. Beklemishev, G. Bezhanishvili and T. Icar, "On topological models of GLP". Ways of proof theory, Ontos Mathematical Logic, 2, eds. R. Schindler, Ontos Verlag, Frankfurt, 2010, pp. 133–153.

[12] L. Beklemishev and D. Gabelaia, "Topological completeness of provability logic GLP". Annals of Pure and Applied Logic 164 (2013), pp. 1201–1223.

[13] I. Shapirovsky, "PSPACE-decidability of Japaridze's polymodal logic". Advances in Modal Logic 7 (2008), pp. 289-304.

[14] F. Pakhomov, "On the complexity of the closed fragment of Japaridze's provability logic". Archive for Mathematical Logic 53 (2014), pp. 949–967.

[15] L. Beklemishev, "Provability algebras and proof-theoretic ordinals, I". Annals of Pure and Applied Logic 128 (2004), pp. 103–123.

[16] G. Boolos, "The Logic of Provability". Cambridge University Press, 1993.

# Chapter 42

# Judgment (mathematical logic)

For other uses, see Judgment (disambiguation).

In mathematical logic, a **judgment** can be an assertion about occurrence of a free variable in an expression of the object language, or about provability of a proposition (either as a tautology or from a given context), but judgments can be also other inductively definable assertions in the metatheory. Judgments are used for example in formalizing deduction systems: a logical axiom expresses a judgment, premises of a rule of inference are formed as a sequence of judgments, and their conclusion is a judgment as well. Also the result of a proof expresses a judgment, and the used hypotheses are formed as a sequence of judgments.

A characteristic feature of the variants of Hilbert-style deduction systems is that the *context* is not changed in any of their rules of inference, while both natural deduction and sequent calculus contain some context-changing rules. Thus, if we are interested only in the derivability of tautologies, not hypothetical judgments, then we can formalize the Hilbert-style deduction system in such a way that its rules of inference contain only judgments of a rather simple form. The same cannot be done with the other two deductions systems: as context is changed in some of their rules of inferences, they cannot be formalized so that hypothetical judgments could be avoided—not even if we want to use them just for proving derivability of tautologies.

This basic diversity among the various calculi allows such difference, that the same basic thought (e.g. deduction theorem) must be proven as a metatheorem in Hilbert-style deduction system, while it can be declared explicitly as a rule of inference in natural deduction.

In type theory, some analogous notions are used as in mathematical logic (giving rise to connections between the two fields, e.g. Curry-Howard correspondence). The abstraction in the notion of *judgment* in mathematical logic can be exploited also in foundation of type theory as well.

## 42.1   See also

- Simply typed lambda calculus

- Mathematical logic

## 42.2   External links

- "Judgments in formal systems". *Everything*$_2$.

- Pfenning, Frank (Spring 2004). "Natural Deduction" (PDF). *15-815 Automated Theorem Proving*.

• Martin-Löf, Per (1983). "On the meaning of the logical constants and the justifications of the logical laws". *Siena Lectures*.

# Chapter 43

# Lambda-mu calculus

In mathematical logic and computer science, the **lambda-mu calculus** is an extension of the lambda calculus, and was introduced by M. Parigot.[1] It introduces two new operators: the mu operator (which is completely different both from the mu operator found in computability theory and from the μ operator of modal μ-calculus) and the bracket operator. Proof-theoretically, it provides a well-behaved formulation of classical natural deduction.

One of the main goals of this extended calculus is to be able to describe expressions corresponding to theorems in classical logic. According to the Curry–Howard isomorphism, lambda calculus on its own can express theorems in intuitionistic logic only, and several classical logical theorems can't be written at all. However with these new operators one is able to write terms that have the type of, for example, Peirce's law.

Semantically these operators correspond to continuations, found in some functional programming languages.

## 43.1 Formal definition

We can augment the definition of a lambda expression to gain one in the context of lambda-mu calculus. The three main expressions found in lambda calculus are as follows:

1. V, a *variable*, where V is any identifier.

2. λV.E, an *abstraction*, where $V$ is any identifier and $E$ is any lambda expression.

3. (E E′), an *application*, where E and E′ are any lambda expressions.

For details, see the corresponding article.

In addition to the traditional λ-variables, the lambda-mu calculus includes a distinct set of μ-variables. These μ-variables can be used to *name* or *freeze* arbitrary subterms, allowing us to later abstract on those names. The set of terms contains *unnamed* (all traditional lambda expressions are of this kind) and *named* terms. The terms that are added by the lambda-mu calculus are of the form:

1. [α]t is a named term, where α is a μ-variable and t is an unnamed term.

2. (μ α. E) is an unnamed term, where α is a μ-variable and E is a named term.

## 43.2 Reduction

The basic reduction rules used in the lambda-mu calculus are the following:

These rules cause the calculus to be confluent. Further reduction rules could be added to provide us with a stronger notion of normal form, though this would be at the expense of confluence.

## 43.3   See also

- Lambda Calculus

- Classical pure type systems for typed generalizations of lambda calculi with control

## 43.4   References

[1] Michel Parigot. *λμ-Calculus: An algorithmic interpretation of classical natural deduction. Lecture Notes in Computer Science*, Volume **624**, pages 190-201, 1992.

## 43.5   External links

- Lambda-mu relevant discussion on Lambda the Ultimate.

# Chapter 44

# Large countable ordinal

Main article: Ordinal number

In the mathematical discipline of set theory, there are many ways of describing specific countable ordinals. The smallest ones can be usefully and non-circularly expressed in terms of their Cantor normal forms. Beyond that, many ordinals of relevance to proof theory still have computable ordinal notations. However, it is not possible to decide effectively whether a given putative ordinal notation is a notation or not (for reasons somewhat analogous to the unsolvability of the halting problem); various more-concrete ways of defining ordinals that definitely have notations are available.

Since there are only countably many notations, all ordinals with notations are exhausted well below the first uncountable ordinal $\omega_1$; their supremum is called ***Church–Kleene $\omega_1$*** or $\omega_1^{CK}$ (not to be confused with the first uncountable ordinal, $\omega_1$), described below. Ordinal numbers below $\omega_1^{CK}$ are the **recursive** ordinals (see below). Countable ordinals larger than this may still be defined, but do not have notations.

Due to the focus on countable ordinals, ordinal arithmetic is used throughout, except where otherwise noted. The ordinals described here are not as large as the ones described in large cardinals, but they are large among those that have constructive notations (descriptions). Larger and larger ordinals can be defined, but they become more and more difficult to describe.

## 44.1 Generalities on recursive ordinals

Main article: Recursive ordinal

### 44.1.1 Ordinal notations

Main article: Ordinal notation

Recursive ordinals (or computable ordinals) are certain countable ordinals: loosely speaking those represented by a computable function. There are several equivalent definitions of this: the simplest is to say that a computable ordinal is the order-type of some recursive (i.e., computable) well-ordering of the natural numbers; so, essentially, an ordinal is recursive when we can present the set of smaller ordinals in such a way that a computer (Turing machine, say) can manipulate them (and, essentially, compare them).

A different definition uses Kleene's system of ordinal notations. Briefly, an ordinal notation is either the name zero (describing the ordinal 0), or the successor of an ordinal notation (describing the successor of the ordinal described by that notation), or a Turing machine (computable function) that produces an increasing sequence of ordinal notations (that describe the ordinal that is the limit of the sequence), and ordinal notations are (partially) ordered so as to make the successor of $o$ greater than $o$ and to make the limit greater than any term of the sequence (this order is computable;

however, the set **O** of ordinal notations itself is highly non-recursive, owing to the impossibility of deciding whether a given Turing machine does indeed produce a sequence of notations); a recursive ordinal is then an ordinal described by some ordinal notation.

Any ordinal smaller than a recursive ordinal is itself recursive, so the set of all recursive ordinals forms a certain (countable) ordinal, the Church-Kleene ordinal (see below).

It is tempting to forget about ordinal notations, and only speak of the recursive ordinals themselves: and some statements are made about recursive ordinals which, in fact, concern the notations for these ordinals. This leads to difficulties, however, as even the smallest infinite ordinal, $\omega$, has many notations, some of which cannot be proven to be equivalent to the obvious notation (the limit of the simplest program that enumerates all natural numbers).

### 44.1.2   Relationship to systems of arithmetic

There is a relation between computable ordinals and certain formal systems (containing arithmetic, that is, at least a reasonable fragment of Peano arithmetic).

Certain computable ordinals are so large that while they can be given by a certain ordinal notation $o$, a given formal system might not be sufficiently powerful to show that $o$ is, indeed, an ordinal notation: the system does not show transfinite induction for such large ordinals.

For example, the usual first-order Peano axioms do not prove transfinite induction for (or beyond) $\varepsilon_0$: while the ordinal $\varepsilon_0$ can easily be arithmetically described (it is countable), the Peano axioms are not strong enough to show that it is indeed an ordinal; in fact, transfinite induction on $\varepsilon_0$ proves the consistency of Peano's axioms (a theorem by Gentzen), so by Gödel's second incompleteness theorem, Peano's axioms cannot formalize that reasoning. (This is at the basis of the Kirby–Paris theorem on Goodstein sequences.) We say that $\varepsilon_0$ measures the proof-theoretic strength of Peano's axioms.

But we can do this for systems far beyond Peano's axioms. For example, the proof-theoretic strength of Kripke–Platek set theory is the Bachmann-Howard ordinal (see below), and, in fact, merely adding to Peano's axioms the axioms that state the well-ordering of all ordinals below the Bachmann–Howard ordinal is sufficient to obtain all arithmetical consequences of Kripke–Platek set theory.

## 44.2   Specific recursive ordinals

### 44.2.1   Predicative definitions and the Veblen hierarchy

Main article: Veblen function

We have already mentioned (see Cantor normal form) the ordinal $\varepsilon_0$, which is the smallest satisfying the equation $\omega^\alpha = \alpha$, so it is the limit of the sequence $0, 1, \omega, \omega^\omega, \omega^{\omega^\omega}$, etc. The next ordinal satisfying this equation is called $\varepsilon_1$: it is the limit of the sequence

$$\varepsilon_0 + 1, \qquad \omega^{\varepsilon_0 + 1} = \varepsilon_0 \cdot \omega, \qquad \omega^{\omega^{\varepsilon_0 + 1}} = (\varepsilon_0)^\omega, \qquad \text{etc.}$$

More generally, the $\iota$-th ordinal such that $\omega^\alpha = \alpha$ is called $\varepsilon_\iota$. We could define $\zeta_0$ as the smallest ordinal such that $\varepsilon_\alpha = \alpha$, but since the Greek alphabet does not have transfinitely many letters it is better to use a more robust notation: define ordinals $\varphi_\gamma(\beta)$ by transfinite induction as follows: let $\varphi_0(\beta) = \omega^\beta$ and let $\varphi_{\gamma+1}(\beta)$ be the $\beta$-th fixed point of $\varphi_\gamma$ (i.e., the $\beta$-th ordinal such that $\varphi_\gamma(\alpha) = \alpha$; so for example, $\varphi_1(\beta) = \varepsilon_\beta$), and when $\delta$ is a limit ordinal, define $\varphi_\delta(\alpha)$ as the $\alpha$-th common fixed point of the $\varphi_\gamma$ for all $\gamma < \delta$. This family of functions is known as the **Veblen hierarchy**. (There are inessential variations in the definition, such as letting, for $\delta$ a limit ordinal, $\varphi_\delta(\alpha)$ be the limit of the $\varphi_\gamma(\alpha)$ for $\gamma < \delta$: this essentially just shifts the indices by 1, which is harmless.) $\varphi_\gamma$ is called the $\gamma^{th}$ **Veblen function** (to the base $\omega$).

Ordering: $\varphi_\alpha(\beta) < \varphi_\gamma(\delta)$ if and only if either ( $\alpha = \gamma$ and $\beta < \delta$ ) or ( $\alpha < \gamma$ and $\beta < \varphi_\gamma(\delta)$ ) or ( $\alpha > \gamma$ and $\varphi_\alpha(\beta) < \delta$ ).

### 44.2.2 The Feferman–Schütte ordinal and beyond

The smallest ordinal such that $\varphi_\alpha(0) = \alpha$ is known as the Feferman–Schütte ordinal and generally written $\Gamma_0$ . It can be described as the set of all ordinals that can be written as finite expressions, starting from zero, using only the Veblen hierarchy and addition. The Feferman-Schütte ordinal is important because, in a sense that is complicated to make precise, it is the smallest (infinite) ordinal that cannot be ("predicatively") described using smaller ordinals. It measures the strength of such systems as "arithmetical transfinite recursion".

More generally, $\Gamma\alpha$ enumerates the ordinals that cannot be obtained from smaller ordinals using addition and the Veblen functions.

It is, of course, possible to describe ordinals beyond the Feferman-Schütte ordinal. One could continue to seek fixed points in more and more complicated manner: enumerate the fixed points of $\alpha \mapsto \Gamma_\alpha$ , then enumerate the fixed points of *that*, and so on, and then look for the first ordinal $\alpha$ such that $\alpha$ is obtained in $\alpha$ steps of this process, and continue diagonalizing in this *ad hoc* manner. This leads to the definition of the "small" and "large" Veblen ordinals.

### 44.2.3 Impredicative ordinals

Main article: Ordinal collapsing function

To go far beyond the Feferman-Schütte ordinal, one needs to introduce new methods. Unfortunately there is not yet any standard way to do this: every author in the subject seems to have invented their own system of notation, and it is quite hard to translate between the different systems. The first such system was introduced by Bachmann in 1950 (in an *ad hoc* manner), and different extensions and variations of it were described by Buchholz, Takeuti (ordinal diagrams), Feferman (θ systems), Aczel, Bridge, Schütte, and Pohlers. However most systems use the same basic idea, of constructing new countable ordinals by using the existence of certain uncountable ordinals. Here is an example of such a definition, described in much greater detail in the article on ordinal collapsing function:

- $\psi(\alpha)$ is defined to be the smallest ordinal that cannot be constructed by starting with 0, 1, ω and Ω, and repeatedly applying addition, multiplication and exponentiation, and $\psi$ to previously constructed ordinals (except that $\psi$ can only be applied to arguments less than $\alpha$, to ensure that it is well defined).

Here $\Omega = \omega_1$ is the first uncountable ordinal. It is put in because otherwise the function $\psi$ gets "stuck" at the smallest ordinal σ such that εσ=σ: in particular $\psi(\alpha)$=σ for any ordinal $\alpha$ satisfying σ≤α≤Ω. However the fact that we included Ω allows us to get past this point: $\psi(\Omega+1)$ is greater than σ. The key property of Ω that we used is that it is greater than any ordinal produced by $\psi$.

To construct still larger ordinals, we can extend the definition of $\psi$ by throwing in more ways of constructing uncountable ordinals. There are several ways to do this, described to some extent in the article on ordinal collapsing function.

The **Bachmann-Howard ordinal** (sometimes just called the **Howard ordinal**, $\psi(\varepsilon\Omega_{+1})$ with the notation above) is an important one, because it describes the proof-theoretic strength of Kripke-Platek set theory. Indeed, the main importance of these large ordinals, and the reason to describe them, is their relation to certain formal systems as explained above. However, such powerful formal systems as full second-order arithmetic, let alone Zermelo-Fraenkel set theory, seem beyond reach for the moment.

### 44.2.4 "Unrecursable" recursive ordinals

By dropping the requirement of having a useful description, even larger recursive countable ordinals can be obtained as the ordinals measuring the strengths of various strong theories; roughly speaking, these ordinals are the smallest ordinals that the theories cannot prove are well ordered. By taking stronger and stronger theories such as second-order arithmetic, Zermelo set theory, Zermelo-Fraenkel set theory, or Zermelo-Fraenkel set theory with various large cardinal axioms, one gets some extremely large recursive ordinals. (Strictly speaking it is not known that all of these really are ordinals: by construction, the ordinal strength of a theory can only be proven to be an ordinal from an even stronger theory. So for the large cardinal axioms this becomes quite unclear.)

## 44.3 Beyond recursive ordinals

### 44.3.1 The Church–Kleene ordinal

The set of recursive ordinals is an ordinal that is the smallest ordinal that *cannot* be described in a recursive way. (It is not the order type of any recursive well-ordering of the integers.) That ordinal is a countable ordinal called the Church–Kleene ordinal, $\omega_1^{CK}$. Thus, $\omega_1^{CK}$ is the smallest non-recursive ordinal, and there is no hope of precisely "describing" any ordinals from this point on — we can only *define* them. But it is still far less than the first uncountable ordinal, $\omega_1$. However, as its symbol suggests, it behaves in many ways rather like $\omega_1$.

### 44.3.2 Admissible ordinals

Main article: Admissible ordinal

The Church-Kleene ordinal is again related to Kripke-Platek set theory, but now in a different way: whereas the Bachmann-Howard ordinal (described above) was the smallest ordinal for which KP does not prove transfinite induction, the Church-Kleene ordinal is the smallest $\alpha$ such that the construction of the Gödel universe, $L$, up to stage $\alpha$, yields a model $L_\alpha$ of KP. Such ordinals are called **admissible**, thus $\omega_1^{CK}$ is the smallest admissible ordinal (beyond $\omega$ in case the axiom of infinity is not included in KP).

By a theorem of Sacks, the countable admissible ordinals are exactly those constructed in a manner similar to the Church-Kleene ordinal but for Turing machines with oracles. One sometimes writes $\omega_\alpha^{CK}$ for the $\alpha$-th ordinal that is either admissible or a limit of admissible.

### 44.3.3 Beyond admissible ordinals

An ordinal that is both admissible and a limit of admissibles, or equivalently such that $\alpha$ is the $\alpha$-th admissible ordinal, is called *recursively inaccessible*. There exists a theory of large ordinals in this manner that is highly parallel to that of (small) large cardinals. For example, we can define recursively *Mahlo ordinals*: these are the $\alpha$ such that every $\alpha$-recursive closed unbounded subset of $\alpha$ contains an admissible ordinal (a recursive analog of the definition of a Mahlo cardinal). But note that we are still talking about possibly countable ordinals here. (While the existence of inaccessible or Mahlo cardinals cannot be proved in Zermelo-Fraenkel set theory, that of recursively inaccessible or recursively Mahlo ordinals is a theorem of ZFC: in fact, any regular cardinal is recursively Mahlo and more, but even if we limit ourselves to countable ordinals, ZFC proves the existence of recursively Mahlo ordinals. They are, however, beyond the reach of Kripke-Platek set theory.)

An admissible ordinal $\alpha$ is called *nonprojectible* if there is no total $\alpha$-recursive injective function mapping $\alpha$ into a smaller ordinal. (This is trivially true for regular cardinals; however, we are mainly interested in countable ordinals.) Being nonprojectible is a much stronger condition than being admissible, recursively inaccessible, or even recursively Mahlo. It is equivalent to the statement that the Gödel universe, $L$, up to stage $\alpha$, yields a model $L_\alpha$ of KP + $\Sigma_1$-separation.

### 44.3.4 "Unprovable" ordinals

We can imagine even larger ordinals that are still countable. For example, if ZFC has a transitive model (a hypothesis stronger than the mere hypothesis of consistency, and implied by the existence of an inaccessible cardinal), then there exists a countable $\alpha$ such that $L_\alpha$ is a model of ZFC. Such ordinals are beyond the strength of ZFC in the sense that it cannot (by construction) prove their existence.

Even larger countable ordinals, called the *stable ordinals*, can be defined by indescribability conditions or as those $\alpha$ such that $L_\alpha$ is a 1-elementary submodel of $L$; the existence of these ordinals can be proven in ZFC,[1] and they are closely related to the nonprojectible ordinals.

# 44.4 A pseudo-well-ordering

Within the scheme of notations of Kleene some represent ordinals and some do not. One can define a recursive total ordering that is a subset of the Kleene notations and has an initial segment which is well-ordered with order-type $\omega_1^{CK}$. Every recursively enumerable (or even hyperarithmetic) nonempty subset of this total ordering has a least element. So it resembles a well-ordering in some respects. For example, one can define the arithmetic operations on it. Yet it is not possible to effectively determine exactly where the initial well-ordered part ends and the part lacking a least element begins.

# 44.5 References

Most books describing large countable ordinals are on proof theory, and unfortunately tend to be out of print.

## 44.5.1 On recursive ordinals

- Wolfram Pohlers, *Proof theory*, Springer 1989 ISBN 0-387-51842-8 (for Veblen hierarchy and some impredicative ordinals). This is probably the most readable book on large countable ordinals (which is not saying much).

- Gaisi Takeuti, *Proof theory*, 2nd edition 1987 ISBN 0-444-10492-5 (for ordinal diagrams)

- Kurt Schütte, *Proof theory*, Springer 1977 ISBN 0-387-07911-4 (for Veblen hierarchy and some impredicative ordinals)

- Craig Smorynski, *The varieties of arboreal experience* Math. Intelligencer 4 (1982), no. 4, 182–189; contains an informal description of the Veblen hierarchy.

- Hartley Rogers, Jr., *Theory of Recursive Functions and Effective Computability* McGraw-Hill (1967) ISBN 0-262-68052-1 (describes recursive ordinals and the Church–Kleene ordinal)

- Larry W. Miller, *Normal Functions and Constructive Ordinal Notations*, *The Journal of Symbolic Logic*, volume 41, number 2, June 1976, pages 439 to 459, JSTOR 2272243,

- Hilbert Levitz, *Transfinite Ordinals and Their Notations: For The Uninitiated*, expository article (8 pages, in PostScript)

- Herman Ruge Jervell, Truth and provability, manuscript in progress.

## 44.5.2 Beyond recursive ordinals

- Barwise, Jon (1976). *Admissible Sets and Structures: an Approach to Definability Theory*. Perspectives in Mathematical Logic. Springer-Verlag. ISBN 3-540-07451-1.

- Hinman, Peter G. (1978). *Recursion-theoretic hierarchies*. Perspectives in Mathematical Logic. Springer-Verlag.

### 44.5.3   Both recursive and nonrecursive ordinals

- Michael Rathjen, "The realm of ordinal analysis." in S. Cooper and J. Truss (eds.): *Sets and Proofs*. (Cambridge University Press, 1999) 219–279. At Postscript file.

### 44.5.4   Inline references

[1]  Barwise (1976), theorem 7.2.

# Chapter 45

# LowerUnits

In proof compression LowerUnits (**LU**) is an algorithm used to compress propositional logic resolution proofs. The main idea of LowerUnits is to exploit the following fact:[1]

**Theorem:** Let $\varphi$ be a potentially redundant proof, and $\eta$ be the redundant proof | redundant node. If $\eta$'s clause is a unit clause, then $\varphi$ is redundant.

The algorithm targets exactly the class of global redundancy stemming from multiple resolutions with unit clauses. The algorithm takes its name from the fact that, when this rewriting is done and the resulting proof is displayed as a DAG (directed acyclic graph), the unit node $\eta$ appears lower (i.e., closer to the root) than it used to appear in the original proof.

A naive implementation exploiting theorem would require the proof to be traversed and fixed after each unit node is lowered. It is possible, however, to do better by first collecting and removing all the unit nodes in a single traversal, and afterwards fixing the whole proof in a single second traversal. Finally, the collected and fixed unit nodes have to be reinserted at the bottom of the proof.

Care must be taken with cases when a unit node $\eta'$ occurs above in the subproof that derives another unit node $\eta$. In such cases, $\eta$ depends on $\eta'$. Let $\ell$ be the single literal of the unit clause of $\eta'$. Then any occurrence of $\overline{\ell}$ in the subproof above $\eta$ will not be cancelled by resolution inferences with $\eta'$ anymore. Consequently, $\overline{\ell}$ will be propagated downwards when the proof is fixed and will appear in the clause of $\eta$. Difficulties with such dependencies can be easily avoided if we reinsert the upper unit node $\eta'$ after reinserting the unit node $\eta$ (i.e. after reinsertion, $\eta'$ must appear below $\eta$, to cancel the extra literal $\overline{\ell}$ from $\eta$'s clause). This can be ensured by collecting the unit nodes in a queue during a bottom-up traversal of the proof and reinserting them in the order they were queued.

The algorithm for fixing a proof containing many roots performs a top-down traversal of the proof, recomputing the resolvents and replacing broken nodes (e.g. nodes having deletedNodeMarker as one of their parents) by their surviving parents (e.g. the other parent, in case one parent was deletedNodeMarker).

When unit nodes are collected and removed from a proof of a clause $\kappa$ and the proof is fixed, the clause $\kappa'$ in the root node of the new proof is not equal to $\kappa$ anymore, but contains (some of) the duals of the literals of the unit clauses that have been removed from the proof. The reinsertion of unit nodes at the bottom of the proof resolves $\kappa'$ with the clauses of (some of) the collected unit nodes, in order to obtain a proof of $\kappa$ again.

## 45.1   Algorithm

General structure of the algorithm

**Algorithm** LowerUnits Input: A proof $\psi$ Output: A proof $\psi'$ with no global redundancy with unit redundant node (unitsQueue, $\psi_b$) $\leftarrow$ collectUnits( $\psi$ ); $\psi_f \leftarrow$ fix( $\psi_b$ ); fixedUnitsQueue $\leftarrow$ fix(unitsQueue); $\psi' \leftarrow$ reinsertUnits( $\psi_f$, fixedUnitsQueue); **return** $\psi'$ ;

- "$\leftarrow$" is a shorthand for "changes to". For instance, "*largest* $\leftarrow$ *item*" means that the value of *largest* changes to the value of *item*.

- "**return**" terminates the algorithm and outputs the value that follows.

We collect the unit clauses as follow

**Algorithm** CollectUnits Input: A proof $\psi$ Output: A pair containing a queue of all unit nodes (unitsQueue) that are used more than once in $\psi$ and a broken proof $\psi_b$ $\psi_b \leftarrow \psi$ ; traverse $\psi_b$ bottom-up and **foreach** node $\eta$ in $\psi_b$ **do if** $\eta$ is unit and $\eta$ has more than one child **then** add $\eta$ to unitsQueue; remove $\eta$ from $\psi_b$ ; **end end return** (unitsQueue, $\psi_b$ );

- "$\leftarrow$" is a shorthand for "changes to". For instance, "*largest $\leftarrow$ item*" means that the value of *largest* changes to the value of *item*.

- "**return**" terminates the algorithm and outputs the value that follows.

Then we reinsert the units

**Algorithm** ReinsertUnits Input: A proof $\psi_f$ (with a single root) and a queue $q$ of root nodes Output: A proof $\psi'$ $\psi' \leftarrow \psi_f$ ; **while** $q \neq \emptyset$ **do** $\eta \leftarrow$ first element of $q$ ; $q \leftarrow$ tail of $q$ ; **if** $\eta$ is resolvable with root of $\psi'$ **then** $\psi' \leftarrow$ resolvent of $\eta$ with the root of $\psi'$ ; **end end return** $\psi'$ ;

- "$\leftarrow$" is a shorthand for "changes to". For instance, "*largest $\leftarrow$ item*" means that the value of *largest* changes to the value of *item*.

- "**return**" terminates the algorithm and outputs the value that follows.

## 45.2  Notes

[1] Fontaine, Pascal; Merz, Stephan; Woltzenlogel Paleo, Bruno. *Compression of Propositional Resolution Proofs via Partial Regularization.* 23rd International Conference on Automated Deduction, 2011.

# Chapter 46

# Mathematical fallacy

In mathematics, certain kinds of mistaken proof are often exhibited, and sometimes collected, as illustrations of a concept of **mathematical fallacy**. There is a distinction between a simple *mistake* and a *mathematical fallacy* in a proof: a mistake in a proof leads to an **invalid proof** just in the same way, but in the best-known examples of mathematical fallacies, there is some concealment in the presentation of the proof. For example, the reason validity fails may be a division by zero that is hidden by algebraic notation. There is a striking quality of the mathematical fallacy: as typically presented, it leads not only to an absurd result, but does so in a crafty or clever way.[1] Therefore, these fallacies, for pedagogic reasons, usually take the form of spurious proofs of obvious contradictions. Although the proofs are flawed, the errors, usually by design, are comparatively subtle, or designed to show that certain steps are conditional, and should not be applied in the cases that are the exceptions to the rules.

The traditional way of presenting a mathematical fallacy is to give an invalid step of deduction mixed in with valid steps, so that the meaning of fallacy is here slightly different from the logical fallacy. The latter applies normally to a form of argument that is not a genuine rule of logic, where the problematic mathematical step is typically a correct rule applied with a tacit wrong assumption. Beyond pedagogy, the resolution of a fallacy can lead to deeper insights into a subject (such as the introduction of Pasch's axiom of Euclidean geometry[2] and the five color theorem of graph theory). *Pseudaria*, an ancient lost book of false proofs, is attributed to Euclid.[3]

Mathematical fallacies exist in many branches of mathematics. In elementary algebra, typical examples may involve a step where division by zero is performed, where a root is incorrectly extracted or, more generally, where different values of a multiple valued function are equated. Well-known fallacies also exist in elementary Euclidean geometry and calculus.

## 46.1  Howlers

Examples exist of *mathematically **correct** results derived by **incorrect** lines of reasoning*. Such an argument, however true the conclusion, is mathematically invalid and is commonly known as a **howler**. Consider for instance the calculation (anomalous cancellation):

$$\frac{16}{64} = \frac{1\cancel{6}}{\cancel{6}4} = \frac{1}{4}.$$

Although the conclusion $\frac{16}{64} = \frac{1}{4}$ is correct, there is a fallacious, invalid cancellation in the middle step. Bogus proofs, calculations, or derivations constructed to produce a correct result in spite of incorrect logic or operations were termed *howlers* by Maxwell.[4] Outside the field of mathematics the term "*howler*" has various meanings, generally less specific.

## 46.2   Division by zero

The division-by-zero fallacy has many variants. The following example uses division by zero to "prove" that $2 = 1$, but can be modified to prove that any number equals any other number.

1. Let $a$ and $b$ be equal non-zero quantities

   $$a = b$$

2. Multiply by $a$

   $$a^2 = ab$$

3. Subtract $b^2$

   $$a^2 - b^2 = ab - b^2$$

4. Factor both sides; the left factors as a difference of squares, the right is factored through its greatest common divisor)

   $$(a - b)(a + b) = b(a - b)$$

5. Divide out $(a - b)$

   $$a + b = b$$

6. Observing that $a = b$

   $$b + b = b$$

7. Combine like terms on the left

   $$2b = b$$

8. Divide by the non-zero $b$

   $$2 = 1$$

$Q.E.D.$[5]

The fallacy is in line 5: the progression from line 4 to line 5 involves division by $a - b$, which is zero since $a$ equals $b$. Since division by zero is undefined, the argument is invalid.

## 46.3   Multivalued functions

Many functions do not have a unique inverse. For instance squaring a number gives a unique value, but there are two possible square roots of a positive number. The square root is multivalued. One value can be chosen by convention as the principal value, in the case of the square root the non-negative value is the principal value, but there is no guarantee that the square root function given by this principal value of the square of a number will be equal to the original number, e.g. the square root of the square of −2 is 2.

## 46.4  Calculus

Calculus as the mathematical study of infinitesimal change and limits can lead to mathematical fallacies if the properties of integrals and differentials are ignored. For instance, a naive use of integration by parts can be used to give a false proof that 0 = 1.[6] Letting $u = \frac{1}{\log x}$ and $dv = \frac{dx}{x}$, we may write:

$$\int \frac{1}{x \log x} dx = 1 + \int \frac{1}{x \log x} dx$$

after which the antiderivatives may be cancelled yielding 0 = 1. The problem is that antiderivatives are only defined up to a constant and shifting them by 1 or indeed any number is allowed. The error really comes to light when we introduce arbitrary integration limits $a$ and $b$.

$$\int_a^b \frac{1}{x \log x} dx = 1\big|_a^b + \int_a^b \frac{1}{x \log x} dx = 0 + \int_a^b \frac{1}{x \log x} dx = \int_a^b \frac{1}{x \log x} dx$$

Since the difference between two values of a constant function vanishes, the same definite integral appears on both sides of the equation.

## 46.5  Power and root

Fallacies involving disregarding the rules of elementary arithmetic through an incorrect manipulation of the radical. For complex numbers the failure of power and logarithm identities has led to many fallacies.

### 46.5.1  Positive and negative roots

Proof of

$5 = 4$

1. Start from

   $-20 = -20$

2. Write this as

   $25 - 45 = 16 - 36$

3. Rewrite as

   $5^2 - 5 * 9 = 4^2 - 4 * 9$

4. Add $81/4$ on both sides:

   $5^2 - 5 * 9 + 81/4 = 4^2 - 4 * 9 + 81/4$

5. These are perfect squares:

   $(5 - 9/2)^2 = (4 - 9/2)^2$

6. Take the square root of both sides:

$$5 - 9/2 = 4 - 9/2$$

7. Add $9/2$ on both sides:

$$5 = 4$$

*Q.E.D.*[7]

The fallacy is in line 6: $a^2=b^2$ only implies $a=b$ if $a$ and $b$ have the same sign, which not the case here. In this case it implies $a=-b$ and should read

$$5 - 9/2 = -(4 - 9/2).$$

### 46.5.2   Square roots of negative numbers

Invalid proofs utilizing powers and roots are often of the following kind:[8]

$$1 = \sqrt{1} = \sqrt{(-1)(-1)} = \sqrt{-1}\sqrt{-1} = i \cdot i = -1.$$

The fallacy is that the rule $\sqrt{xy} = \sqrt{x}\sqrt{y}$ is generally valid only if both $x$ and $y$ are positive (when dealing with real numbers), which is not the case here.

Although the fallacy is easily detected here, sometimes it is concealed more effectively in notation. For instance,[9] consider the equation

$$\cos^2 x = 1 - \sin^2 x$$

which holds as a consequence of the Pythagorean theorem. Then, by taking a square root,

$$\cos x = (1 - \sin^2 x)^{\frac{1}{2}}$$

so that

$$1 + \cos x = 1 + (1 - \sin^2 x)^{\frac{1}{2}}.$$

But evaluating this when $x = \pi$ implies

$$1 - 1 = 1 + (1 - 0)^{\frac{1}{2}}$$

or

$$0 = 2$$

which is incorrect.

The error in each of these examples fundamentally lies in the fact that any equation of the form

$$x^2 = a^2$$

has two solutions, provided $a \neq 0$,

$$x = \pm a$$

and it is essential to check which of these solutions is relevant to the problem at hand.[10] In the above fallacy, the square root that allowed the second equation to be deduced from the first is valid only when $\cos x$ is positive. In particular, when $x$ is set to $\pi$, the second equation is rendered invalid.

Another example of this kind of fallacy, where the error is immediately detectable, is the following invalid proof that $-2 = 2$. Letting $x = -2$, and then squaring gives

$$x^2 = 4$$

whereupon taking a square root implies

$$x = \sqrt{4} = 2,$$

so that $x = -2 = 2$, which is absurd. Clearly when the square root was extracted, it was the *negative* root $-2$, rather than the *positive* root, that was relevant for the particular solution in the problem.

Alternatively, imaginary roots are obfuscated in the following:

$$\sqrt{-1} = (-1)^{\frac{2}{4}} = ((-1)^2)^{\frac{1}{4}} = 1^{\frac{1}{4}} = 1$$

The error here lies in the last equality, where we are ignoring the other fourth roots of $1$,[11] which are $-1$, $i$ and $-i$ (where $i$ is the imaginary unit). Seeing as we have squared our figure and then taken roots, we cannot always assume that all the roots will be correct. So the correct fourth roots are $i$ and $-i$, which are the imaginary numbers defined to square to $-1$.

### 46.5.3 Complex exponents

When a number is raised to a complex power, the result is not uniquely defined (see Failure of power and logarithm identities). If this property is not recognized, then errors such as the following can result:

$$e^{2\pi i} = 1$$
$$(e^{2\pi i})^i = 1^i$$
$$e^{-2\pi} = 1$$

The error here is that the rule of multiplying exponents as when going to the third line does not apply unmodified with complex exponents, even if when putting both sides to the power $i$ only the principal value is chosen. When treated as multivalued functions, both sides produce the same set of values, being $\{e^{2\pi n} \mid n \in \mathbb{Z}\}$.

## 46.6 Geometry

Many mathematical fallacies in geometry arise from using in an additive equality involving oriented quantities (such adding vectors along a given line or adding oriented angles in the plane) a valid identity, but which fixes only the absolute value of (one of) these quantities. This quantity is then incorporated into the equation with the wrong orientation, so as to produce an absurd conclusion. This wrong orientation is usually suggested implicitly by supplying an imprecise diagram of the situation, where relative positions of points or lines are chosen in a way that is actually impossible under the hypotheses of the argument, but non-obviously so. Such a fallacy is easy to expose by drawing a precise picture of the situation, in which some relative positions will be different form those in the provided diagram. In order to avoid such fallacies, a correct geometric argument using addition or subtraction of distances or angles should always prove that quantities are being incorporated with their correct orientation.

### 46.6.1   Fallacy of the isosceles triangle

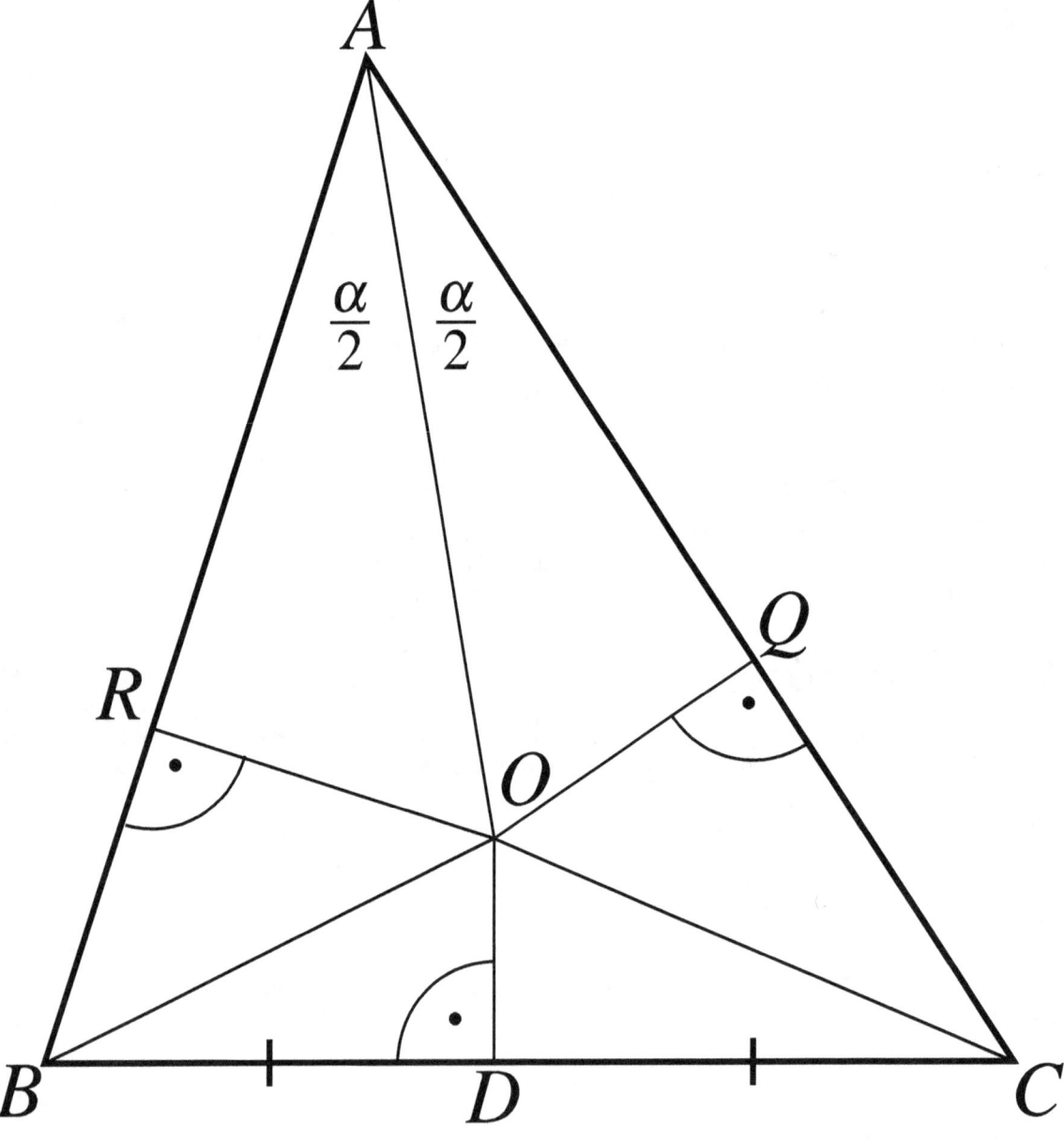

The fallacy of the isosceles triangle, from (Maxwell 1959, Chapter II, § 1), purports to show that every triangle is isosceles, meaning that two sides of the triangle are congruent. This fallacy has been attributed to Lewis Carroll.[12]

Given a triangle $\triangle ABC$, prove that AB = AC:

1. Draw a line bisecting $\angle A$

2. Draw the perpendicular bisector of segment BC, which bisects BC at a point D

3. Let these two lines meet at a point O.

4. Draw line OR perpendicular to AB, line OQ perpendicular to AC

5. Draw lines OB and OC

6. By AAS, $\triangle RAO \cong \triangle QAO$ ($\angle ORA = \angle OQA = 90$; $\angle RAO = \angle QAO$; AO=AO (COMMON SIDE))

7. By RHS,[13] $\triangle ROB \cong \triangle QOC$

8. Thus, AR = AQ, RB = QC, and AB = AR + RB = AQ + QC = AC

*Q.E.D.*

As a corollary, one can show that all triangles are equilateral, by showing that AB = BC and AC = BC in the same way.

The error in the proof is the assumption in the diagram that the point O is *inside* the triangle. In fact, O always lies at the circumcircle of the $\triangle ABC$ (except for isosceles and equilateral triangles where AO and OD coincides . Furthermore, it can be shown that, if AB is longer than AC, then R will lie *within* AB, while Q will lie *outside* of AC (and vice versa). (Any diagram drawn with sufficiently accurate instruments will verify the above two facts.) Because of this, AB is still AR + RB, but AC is actually AQ − QC; and thus the lengths are not necessarily the same.

## 46.7   Proof by induction

There exist several fallacious proofs by induction in which one of the components, basis case or inductive step, is incorrect. Intuituvely, proofs by induction work by arguing that, if a statement is true in one case, it is true in the next case, and hence by repeatedly applying this it can be shown to be true for all cases. This "proof" shows that all horses are the same colour.

1. Let us say that any group of N horses is all of the same colour.

2. If we remove a horse from the group, we have a group of N - 1 horses of the same colour. If we add another horse, we have another group of N horses. By our previous assumption, all the horses are of the same colour in this new group, since it is a group of N horses.

3. Thus we have constructed two groups of N horses all of the same colour, with N - 1 horses in common. Since these two groups have some horses in common, the two groups must be of the same colour as each other.

4. Therefore, combining all the horses used, we have a group of N + 1 horses of the same colour.

5. Thus if any N horses are all the same colour, any N + 1 horses are the same colour.

6. This is clearly true for N = 1 (i.e. one horse is a group where all the horses are the same colour). Thus, by induction, N horses are the same colour for any positive integer N. i.e. all horses are the same colour.

The fallacy in this proof arises in line 3. For N = 1, the two groups of horses have N − 1 = 0 horses in common, and thus are not necessarily the same colour as each other, so the group of N + 1 = 2 horses is not necessarily all of the same colour. The implication "Every N horses are of the same color, then N+1 horses are of the same color" works for any N greater than one, but fails to be true when N=1. The basis case is correct, but the induction step has a fundamental flaw.

## 46.8   See also

- List of incomplete proofs

- Paradox

- Proof by intimidation

## 46.9   Notes

[1] Maxwell 1959, p. 9

[2] Maxwell 1959

[3] Heath & Helberg 1908, Chapter II, §I

[4] Maxwell 1959

[5] Heuser, Harro (1989), *Lehrbuch der Analysis - Teil 1* (6th ed.), Teubner, p. 51, ISBN 978-3-8351-0131-9

[6] Barbeau, Ed (1990), "Fallacies, Flaws and Flimflam #19: Dolt's Theorem", *The College Mathematics Journal* **21** (3): 216–218

[7] Frohlichstein, Jack (1967). *Mathematical Fun, Games and Puzzles* (illustrated ed.). Courier Corporation. p. 207. ISBN 0-486-20789-7., Extract of page 207

[8] Maxwell 1959, Chapter VI, §I.2

[9] Maxwell 1959, Chapter VI, §I.1

[10] Maxwell 1959, Chapter VI, §II

[11] In general, the expression $\sqrt[n]{1}$ evaluates to $n$ complex numbers, called the $n$th roots of unity.

[12] Robin Wilson (2008), *Lewis Carroll in Numberland*, Penguin Books, pp. 169–170, ISBN 978-0-14-101610-8

[13] Hypotenuse-leg congruence

## 46.10   References

- Barbeau, Edward J. (2000), *Mathematical fallacies, flaws, and flimflam*, MAA Spectrum, Mathematical Association of America, ISBN 978-0-88385-529-4, MR 1725831.

- Bunch, Bryan (1997), *Mathematical fallacies and paradoxes*, New York: Dover Publications, ISBN 978-0-486-29664-7, MR 1461270.

- Heath, Sir Thomas Little; Heiberg, Johan Ludvig (1908), *The thirteen books of Euclid's Elements, Volume 1*, The University Press.

- Maxwell, E. A. (1959), *Fallacies in mathematics*, Cambridge University Press, ISBN 0-521-05700-0, MR 0099907.

## 46.11   External links

- Invalid proofs at Cut-the-knot (including literature references)

- Classic fallacies with some discussion

- More invalid proofs from AhaJokes.com

- Math jokes including an invalid proof

# Chapter 47

# Metalanguage

Not to be confused with Metalinguistics.
For the programming language, see ML (programming language).

Broadly, any **metalanguage** is language or symbols used when language itself is being discussed or examined.[1] In logic and linguistics, a metalanguage is a language used to make statements about statements in another language (the object language). Expressions in a metalanguage are often distinguished from those in an object language by the use of italics, quotation marks, or writing on a separate line.

## 47.1 Types of metalanguage

There is a variety of recognized metalanguages, including *embedded*, *ordered*, and *nested* (or, *hierarchical*).

### 47.1.1 Embedded metalanguage

An **embedded metalanguage** is a language formally, naturally and firmly fixed in an object language. This idea is found in Douglas Hofstadter's book, *Gödel, Escher, Bach*, in a discussion of the relationship between formal languages and number theory: "... it is in the nature of any formalization of number theory that its metalanguage is embedded within it.".[2] It occurs in natural, or informal, languages, as well—such as in English, where words such as *noun, verb,* or even *word* describe features and concepts pertaining to the English language itself.

### 47.1.2 Ordered metalanguage

An **ordered metalanguage** is analogous to ordered logic. An example of an ordered metalanguage is the construction of one metalanguage to discuss an object language, followed by the creation of another metalanguage to discuss the first, etc.

### 47.1.3 Nested metalanguage

A **nested** (or, *hierarchical*) **metalanguage** is similar to an ordered metalanguage in that each level represents a greater degree of abstraction. However, a nested metalanguage differs from an ordered one in that each level includes the one below. The paradigmatic example of a nested metalanguage comes from the Linnean taxonomic system in biology. Each level in the system incorporates the one below it. The language used to discuss genus is also used to discuss species; the one used to discuss orders is also used to discuss genera, etc., up to kingdoms.

## 47.2   Metalanguages in natural language

Natural language combines nested and ordered metalanguages. In a natural language there is an infinite regress of meta-languages, each with more specialized vocabulary and simpler syntax. Designating the language now as $L_0$, the grammar of the language is a discourse in the metalanguage $L_1$, which is a sublanguage[3] nested within $L_0$. The grammar of $L_1$, which has the form of a factual description, is a discourse in the metametalanguage $L_2$, which is also a sublanguage of $L_0$. The grammar of $L_2$, which has the form of a theory describing the syntactic structure of such factual descriptions, is stated in the metametametalanguage $L_3$, which likewise is a sublanguage of $L_0$. The grammar of $L_3$ has the form of a metatheory describing the syntactic structure of theories stated in $L_2$. $L_4$ and succeeding metalanguages have the same grammar as $L_3$, differing only in reference. Since all of these metalanguages are sublanguages of $L_0$, $L_1$ is a nested metalanguage, but $L_2$ and sequel are ordered metalanguages.[4] Since all these metalanguages are sublanguages of $L_0$ they are all embedded languages with respect to the language as a whole.

Metalanguages of formal systems all resolve ultimately to natural language, the 'common parlance' in which mathematicians and logicians converse to define their terms and operations and 'read out' their formulae.[5]

## 47.3   Types of expressions in a metalanguage

There are several entities commonly expressed in a metalanguage. In logic usually the object language that the metalanguage is discussing is a formal language, and very often the metalanguage as well.

### 47.3.1   Deductive systems

Main article: Deductive system

A **deductive system** (or, *deductive apparatus*) of a formal system) consists of the axioms (or axiom schemata) and rules of inference that can be used to derive the theorems of the system.[6]

### 47.3.2   Metavariables

Main article: Metavariable (logic)

A **metavariable** (or, *metalinguistic variable*) is a symbol or set of symbols in a metalanguage which stands for a symbol or set of symbols in some object language. For instance, in the sentence:

> Let $A$ and $B$ be arbitrary formula of a formal language $\mathcal{L}$ .

The symbols $A$ and $B$ are not symbols of the object language $\mathcal{L}$ , they are metavariables in the metalanguage (in this case, English) that is discussing the object language $\mathcal{L}$ .

### 47.3.3   Metatheories and metatheorems

Main articles: Metatheory and Metatheorem

A *metatheory* is a theory whose subject matter is some other theory (a theory about a theory). Statements made in the metatheory about the theory are called metatheorems. A **metatheorem** is a true statement about a formal system expressed in a metalanguage. Unlike theorems proved within a given formal system, a metatheorem is proved within a metatheory, and may reference concepts that are present in the metatheory but not the object theory.[7]

### 47.3.4 Interpretations

Main article: Interpretation (logic)

An **interpretation** is an assignment of meanings to the symbols and words of a language.

## 47.4 Role in metaphor

Michael J. Reddy (1979) discovered and has demonstrated that much of the language we use to talk about language is conceptualized and structured by what he refers to as the conduit metaphor.[8] This paradigm operates through two distinct, related frameworks.

The *major framework* views language as a sealed pipeline between people:
**1.** Language transfers people's thoughts and feelings (mental content) to others

*ex:* Try to get your thoughts across better.

**2.** Speakers and writers insert their mental content into words

*ex:* You have to put each concept into words more carefully.

**3.** Words are containers

*ex:* That sentence was filled with emotion.

**4.** Listeners and writers extract mental content from words

*ex:* Let me know if you find any new sensations in the poem.

The *minor framework* views language as an open pipe spilling mental content into the void:
**1.** Speakers and writers eject mental content into an external space

*ex:* Get those ideas out where they can do some good.

**2.** Mental content is reified (viewed as concrete) in this space

*ex:* That concept has been floating around for decades.

**3.** Listeners and writers extract mental content from this space

*ex:* Let me know if you find any good concepts in the essay.

## 47.5 Metaprogramming

Computers follow programs, sets of instructions in a formal language. The development of a programming language involves the use of a metalanguage. The act of working with metalanguages in programming is known as *metaprogramming*. Backus–Naur Form, developed in the 1960s by John Backus and Peter Naur, is one of the earliest metalanguages used in computing. Examples of modern-day programming languages which commonly find use in metaprogramming include Lisp, m4, and Yacc.

## 47.6 See also

- Category theory
- Conduit metaphor
- Jakobson's functions of language
- Language-oriented programming

- Metaethics

- Meta-communication

- Metafiction

- Metagraphy

- Metalinguistic abstraction

- Metalocutionary act

- Metaphilosophy

- Metaprogramming

- Natural Semantic Metalanguage

- Paralanguage

- Self reference

- Use–mention distinction

## 47.7   Dictionaries

- Audi, R. 1996. *The Cambridge Dictionary of Philosophy*. Cambridge: Cambridge University Press.

- Baldick, C. 1996. *Oxford Concise Dictionary of Literary Terms*. Oxford: Oxford University Press.

- Cuddon, J. A. 1999. *The Penguin Dictionary of Literary Terms and Literary Theory*. London: Penguin Books.

- Honderich, T. 1995. *The Oxford Companion to Philosophy*. Oxford: Oxford University Press.

- Matthews, P. H. 1997. *The Concise Oxford Dictionary of Linguistics*. Oxford: Oxford University Press. ISBN 978-0-19-280008-4

- McArthur, T. 1996. *The Concise Oxford Companion to the English Language*. Oxford: Oxford University Press.

## 47.8   References

[1] 2010. *Cambridge Advanced Learner's Dictionary*. Cambridge: Cambridge University Press. Dictionary online. Available from http://dictionary.cambridge.org/dictionary/british/metalanguage Internet. Retrieved 20 November 2010

[2] Hofstadter, Douglas. 1980. *Gödel, Escher, Bach: An Eternal Golden Braid*. New York: Vintage Books ISBN 0-14-017997-6

[3] Harris, Zellig S. (1991). *A theory of language and information: A mathematical approach*. Oxford: Clarendon Press. pp. 272–318. ISBN 0-19-824224-7.

[4] *Ibid*. p. 277.

[5] Borel, Félix Édouard Justin Émile (1928). *Leçons sur la theorie des fonctions* (in French) (3 ed.). Paris: Gauthier-Villars & Cie. p. 160.

[6] Hunter, Geoffrey. 1971. *Metalogic: An Introduction to the Metatheory of Standard First-Order Logic*. Berkeley:University of California Press ISBN 978-0-520-01822-8

[7] Ritzer, George. 1991. *Metatheorizing in Sociology*. New York: Simon Schuster ISBN 0-669-25008-2

[8] Reddy, Michael J. 1979. The conduit metaphor: A case of frame conflict in our language about language. In Andrew Ortony (ed.), *Metaphor and Thought*. Cambridge: Cambridge University Press

## 47.9 External links

- Metalanguage, *Principia Cybernetica*

- Willard McCarty (submitted 2006) Problematic Metaphors, *Humanist Discussion Group*, Vol. 20, No. 92.

# Chapter 48

# Natural deduction

In logic and proof theory, **natural deduction** is a kind of proof calculus in which logical reasoning is expressed by inference rules closely related to the "natural" way of reasoning. This contrasts with the axiomatic systems which instead use axioms as much as possible to express the logical laws of deductive reasoning.

## 48.1 Motivation

Natural deduction grew out of a context of dissatisfaction with the axiomatizations of deductive reasoning common to the systems of Hilbert, Frege, and Russell (see, e.g., Hilbert system). Such axiomatizations were most famously used by Russell and Whitehead in their mathematical treatise *Principia Mathematica*. Spurred on by a series of seminars in Poland in 1926 by Łukasiewicz that advocated a more natural treatment of logic, Jaśkowski made the earliest attempts at defining a more natural deduction, first in 1929 using a diagrammatic notation, and later updating his proposal in a sequence of papers in 1934 and 1935.[1] His proposals led to different notations such as Fitch-style calculus (or Fitch's diagrams) or Suppes' method of which e.g. Lemmon gave a variant called system L.

Natural deduction in its modern form was independently proposed by the German mathematician Gentzen in 1934, in a dissertation delivered to the faculty of mathematical sciences of the University of Göttingen.[2] The term *natural deduction* (or rather, its German equivalent *natürliches Schließen*) was coined in that paper:

> Ich wollte nun zunächst einmal einen Formalismus aufstellen, der dem wirklichen Schließen möglichst nahe kommt. So ergab sich ein "Kalkül des natürlichen Schließens".[3]

> (First I wished to construct a formalism that comes as close as possible to actual reasoning. Thus arose a "calculus of natural deduction".)

Gentzen was motivated by a desire to establish the consistency of number theory. He was unable to prove the main result required for the consistency result, the cut elimination theorem — the Hauptsatz — directly for Natural Deduction. For this reason he introduced his alternative system, the sequent calculus, for which he proved the Hauptsatz both for classical and intuitionistic logic. In a series of seminars in 1961 and 1962 Prawitz gave a comprehensive summary of natural deduction calculi, and transported much of Gentzen's work with sequent calculi into the natural deduction framework. His 1965 monograph *Natural deduction: a proof-theoretical study*[4] was to become a reference work on natural deduction, and included applications for modal and second-order logic.

In natural deduction, a proposition is deduced from a collection of premises by applying inference rules repeatedly. The system presented in this article is a minor variation of Gentzen's or Prawitz's formulation, but with a closer adherence to Martin-Löf's description of logical judgments and connectives.[5]

## 48.2 Judgments and propositions

A *judgment* is something that is knowable, that is, an object of knowledge. It is *evident* if one in fact knows it.[6] Thus "*it is raining*" is a judgment, which is evident for the one who knows that it is actually raining; in this case one may readily find evidence for the judgment by looking outside the window or stepping out of the house. In mathematical logic however, evidence is often not as directly observable, but rather deduced from more basic evident judgments. The process of deduction is what constitutes a *proof*; in other words, a judgment is evident if one has a proof for it.

The most important judgments in logic are of the form "*A is true*". The letter $A$ stands for any expression representing a *proposition*; the truth judgments thus require a more primitive judgment: "*A is a proposition*". Many other judgments have been studied; for example, "*A is false*" (see classical logic), "*A is true at time t*" (see temporal logic), "*A is necessarily true*" or "*A is possibly true*" (see modal logic), "*the program M has type τ*" (see programming languages and type theory), "*A is achievable from the available resources*" (see linear logic), and many others. To start with, we shall concern ourselves with the simplest two judgments "*A is a proposition*" and "*A is true*", abbreviated as "*A* prop" and "*A* true" respectively.

The judgment "*A* prop" defines the structure of valid proofs of $A$, which in turn defines the structure of propositions. For this reason, the inference rules for this judgment are sometimes known as *formation rules*. To illustrate, if we have two propositions $A$ and $B$ (that is, the judgments "*A* prop" and "*B* prop" are evident), then we form the compound proposition $A$ *and* $B$, written symbolically as " $A \wedge B$ ". We can write this in the form of an inference rule:

$$\frac{A \text{ prop} \quad B \text{ prop}}{(A \wedge B) \text{ prop}} \wedge_F$$

where the parentheses are omitted to make the inference rule more succinct:

$$\frac{A \text{ prop} \quad B \text{ prop}}{A \wedge B \text{ prop}} \wedge_F$$

This inference rule is *schematic*: $A$ and $B$ can be instantiated with any expression. The general form of an inference rule is:

$$\frac{J_1 \quad J_2 \quad \cdots \quad J_n}{J} \text{ name}$$

where each $J_i$ is a judgment and the inference rule is named "name". The judgments above the line are known as *premises*, and those below the line are *conclusions*. Other common logical propositions are disjunction ( $A \vee B$ ), negation ( $\neg A$ ), implication ( $A \supset B$ ), and the logical constants truth ( $\top$ ) and falsehood ( $\bot$ ). Their formation rules are below.

$$\frac{A \text{ prop} \quad B \text{ prop}}{A \vee B \text{ prop}} \vee_F \qquad \frac{A \text{ prop} \quad B \text{ prop}}{A \supset B \text{ prop}} \supset_F \qquad \frac{}{\top \text{ prop}} \top_F \qquad \frac{}{\bot \text{ prop}} \bot_F$$

$$\frac{A \text{ prop}}{\neg A \text{ prop}} \neg_F$$

## 48.3 Introduction and elimination

Now we discuss the "*A* true" judgment. Inference rules that introduce a logical connective in the conclusion are known as *introduction rules*. To introduce conjunctions, *i.e.*, to conclude "*A and B* true" for propositions $A$ and $B$, one requires evidence for "*A* true" and "*B* true". As an inference rule:

$$\frac{A \text{ true} \quad B \text{ true}}{(A \wedge B) \text{ true}} \wedge_I$$

It must be understood that in such rules the objects are propositions. That is, the above rule is really an abbreviation for:

$$\frac{A \text{ prop} \quad B \text{ prop} \quad A \text{ true} \quad B \text{ true}}{(A \wedge B) \text{ true}} \wedge_I$$

This can also be written:

$$\frac{A \wedge B \text{ prop} \quad A \text{ true} \quad B \text{ true}}{(A \wedge B) \text{ true}} \wedge_I$$

In this form, the first premise can be satisfied by the $\wedge_F$ formation rule, giving the first two premises of the previous form. In this article we shall elide the "prop" judgments where they are understood. In the nullary case, one can derive truth from no premises.

$$\frac{}{\top \text{ true}} \top_I$$

If the truth of a proposition can be established in more than one way, the corresponding connective has multiple introduction rules.

$$\frac{A \text{ true}}{A \vee B \text{ true}} \vee_{I1} \qquad \frac{B \text{ true}}{A \vee B \text{ true}} \vee_{I2}$$

Note that in the nullary case, *i.e.*, for falsehood, there are *no* introduction rules. Thus one can never infer falsehood from simpler judgments.

Dual to introduction rules are *elimination rules* to describe how to de-construct information about a compound proposition into information about its constituents. Thus, from "$A \wedge B$ true", we can conclude "$A$ true" and "$B$ true":

$$\frac{A \wedge B \text{ true}}{A \text{ true}} \wedge_{E1} \qquad \frac{A \wedge B \text{ true}}{B \text{ true}} \wedge_{E2}$$

As an example of the use of inference rules, consider commutativity of conjunction. If $A \wedge B$ is true, then $B \wedge A$ is true; This derivation can be drawn by composing inference rules in such a fashion that premises of a lower inference match the conclusion of the next higher inference.

$$\frac{\dfrac{A \wedge B \text{ true}}{B \text{ true}} \wedge_{E2} \qquad \dfrac{A \wedge B \text{ true}}{A \text{ true}} \wedge_{E1}}{B \wedge A \text{ true}} \wedge_{I}$$

The inference figures we have seen so far are not sufficient to state the rules of implication introduction or disjunction elimination; for these, we need a more general notion of *hypothetical derivation*.

## 48.4   Hypothetical derivations

A pervasive operation in mathematical logic is *reasoning from assumptions*. For example, consider the following derivation:

$$\frac{\dfrac{A \wedge (B \wedge C) \text{ true}}{B \wedge C \text{ true}} \wedge_{E_2}}{B \text{ true}} \wedge_{E_1}$$

This derivation does not establish the truth of $B$ as such; rather, it establishes the following fact:

> If $A \wedge (B \wedge C)$ is *true* then $B$ is *true*.

In logic, one says "*assuming $A \wedge (B \wedge C)$ is true, we show that $B$ is true*"; in other words, the judgment "$B$ *true*" depends on the assumed judgment "$A \wedge (B \wedge C)$ *true*". This is a *hypothetical derivation*, which we write as follows:

$$A \wedge (B \wedge C) \text{ true}$$
$$\vdots$$
$$B \textit{ true}$$

The interpretation is: "$B$ *true* is derivable from $A \wedge (B \wedge C)$ *true*". Of course, in this specific example we actually know the derivation of "$B$ *true*" from "$A \wedge (B \wedge C)$ *true*", but in general we may not *a-priori* know the derivation. The general form of a hypothetical derivation is:

$$D_1 \quad D_2 \cdots D_n$$
$$\vdots$$
$$J$$

Each hypothetical derivation has a collection of *antecedent* derivations (the $Di$) written on the top line, and a *succedent* judgment ($J$) written on the bottom line. Each of the premises may itself be a hypothetical derivation. (For simplicity, we treat a judgment as a premise-less derivation.)

The notion of hypothetical judgment is *internalised* as the connective of implication. The introduction and elimination rules are as follows.

$$\frac{\overline{A\ true}^{\,u}}{\vdots}$$

$$\frac{B\ true}{A \supset B\ true}\supset I^u \qquad \frac{A \supset B\ true \quad A\ true}{B\ true}\supset E$$

In the introduction rule, the antecedent named $u$ is *discharged* in the conclusion. This is a mechanism for delimiting the *scope* of the hypothesis: its sole reason for existence is to establish "$B$ *true*"; it cannot be used for any other purpose, and in particular, it cannot be used below the introduction. As an example, consider the derivation of "$A \supset (B \supset (A \wedge B))$ *true*":

$$\cfrac{\cfrac{\cfrac{\overline{A\ true}^{\,u} \quad \overline{B\ true}^{\,w}}{A \wedge B\ true}\wedge I}{B \supset (A \wedge B)\ true}\supset I^w}{A \supset (B \supset (A \wedge B))\ true}\supset I^u$$

This full derivation has no unsatisfied premises; however, sub-derivations *are* hypothetical. For instance, the derivation of "$B \supset (A \wedge B)$ *true*" is hypothetical with antecedent "$A$ *true*" (named $u$).

With hypothetical derivations, we can now write the elimination rule for disjunction:

$$\frac{A \vee B\ true \qquad \begin{array}{cc}\overline{A\ true}^{\,u} & \overline{B\ true}^{\,w} \\ \vdots & \vdots \\ C\ true & C\ true\end{array}}{C\ true}\vee E^{u,w}$$

In words, if $A \vee B$ is true, and we can derive $C$ *true* both from $A$ *true* and from $B$ *true*, then $C$ is indeed true. Note that this rule does not commit to either $A$ *true* or $B$ *true*. In the zero-ary case, *i.e.* for falsehood, we obtain the following elimination rule:

$$\frac{\bot\ true}{C\ true}\bot E$$

This is read as: if falsehood is true, then any proposition $C$ is true.

Negation is similar to implication.

$$\frac{\overline{A\ true}^{\,u}}{\vdots}$$

$$\frac{p\ true}{\neg A\ true}\neg I^{u,p} \qquad \frac{\neg A\ true \quad A\ true}{C\ true}\neg E$$

The introduction rule discharges both the name of the hypothesis $u$, and the succedent $p$, *i.e.*, the proposition $p$ must not occur in the conclusion $A$. Since these rules are schematic, the interpretation of the introduction rule is: if from "$A$ *true*" we can derive for every proposition $p$ that "$p$ *true*", then $A$ must be false, *i.e.*, "*not A true*". For the elimination, if both $A$ and *not A* are shown to be true, then there is a contradiction, in which case every proposition $C$ is true. Because the rules for implication and negation are so similar, it should be fairly easy to see that *not A* and $A \supset \boxed{?}$ are equivalent, i.e., each is derivable from the other.

## 48.5 Consistency, completeness, and normal forms

A theory is said to be consistent if falsehood is not provable (from no assumptions) and is complete if every theorem is provable using the inference rules of the logic. These are statements about the entire logic, and are usually tied to some notion of a model. However, there are local notions of consistency and completeness that are purely syntactic checks on the inference rules, and require no appeals to models. The first of these is local consistency, also known as local reducibility, which says that any derivation containing an introduction of a connective followed immediately by its elimination can be turned into an equivalent derivation without this detour. It is a check on the *strength* of elimination rules: they must not be so strong that they include knowledge not already contained in its premises. As an example, consider conjunctions.

Dually, local completeness says that the elimination rules are strong enough to decompose a connective into the forms suitable for its introduction rule. Again for conjunctions:

These notions correspond exactly to β-reduction (beta reduction) and η-conversion (eta conversion) in the lambda calculus, using the Curry–Howard isomorphism. By local completeness, we see that every derivation can be converted to an equivalent derivation where the principal connective is introduced. In fact, if the entire derivation obeys this ordering of eliminations followed by introductions, then it is said to be *normal*. In a normal derivation all eliminations happen above introductions. In most logics, every derivation has an equivalent normal derivation, called a *normal form*. The existence of normal forms is generally hard to prove using natural deduction alone, though such accounts do exist in the literature, most notably by Dag Prawitz in 1961.[7] It is much easier to show this indirectly by means of a cut-free sequent calculus presentation.

## 48.6    First and higher-order extensions

The logic of the earlier section is an example of a *single-sorted* logic, *i.e.*, a logic with a single kind of object: propositions. Many extensions of this simple framework have been proposed; in this section we will extend it with a second sort of *individuals* or *terms*. More precisely, we will add a new kind of judgment, "*t is a term*" (or "*t term*") where *t* is schematic. We shall fix a countable set *V* of *variables*, another countable set *F* of *function symbols*, and construct terms as follows:

For propositions, we consider a third countable set *P* of *predicates*, and define *atomic predicates over terms* with the following formation rule:

In addition, we add a pair of *quantified* propositions: universal (∀) and existential (∃):

These quantified propositions have the following introduction and elimination rules.

In these rules, the notation $[t/x] A$ stands for the substitution of *t* for every (visible) instance of *x* in *A*, avoiding capture; see the article on lambda calculus for more detail about this standard operation. As before the superscripts on the name stand for the components that are discharged: the term *a* cannot occur in the conclusion of ∀I (such terms are known as *eigenvariables* or *parameters*), and the hypotheses named *u* and *v* in ∃E are localised to the second premise in a hypothetical derivation. Although the propositional logic of earlier sections was decidable, adding the quantifiers makes the logic undecidable.

So far the quantified extensions are *first-order*: they distinguish propositions from the kinds of objects quantified over. Higher-order logic takes a different approach and has only a single sort of propositions. The quantifiers have as the domain of quantification the very same sort of propositions, as reflected in the formation rules:

A discussion of the introduction and elimination forms for higher-order logic is beyond the scope of this article. It is possible to be in between first-order and higher-order logics. For example, second-order logic has two kinds of propositions, one kind quantifying over terms, and the second kind quantifying over propositions of the first kind.

## 48.7    Different presentations of natural deduction

### 48.7.1    Tree-like presentations

Gentzen's discharging annotations used to internalise hypothetical judgments can be avoided by representing proofs as a tree of sequents $\Gamma \vdash A$ instead of a tree of *A true* judgments.

### 48.7.2    Sequential presentations

Jaśkowski's representations of natural deduction led to different notations such as Fitch-style calculus (or Fitch's diagrams) or Suppes' method, of which Lemmon gave a variant called system L. Such presentation systems, which are more

**contexts**

$\Sigma$     term variables

$\Gamma$     true hypotheses

**terms**

$$\frac{}{\Sigma, v \vdash v \ term} \ \text{var-}F \qquad \frac{f \in F \quad \Sigma \vdash t_1 \ term \quad \Sigma \vdash t_2 \ term \quad \cdots \quad \Sigma \vdash t_n \ term}{\Sigma \vdash f(t_1, t_2, \ldots, t_n) \ term} \ \text{app-}F$$

**propositions**

$$\frac{\varphi \in P \quad \Sigma \vdash t_1 \ term \quad \Sigma \vdash t_2 \ term \quad \cdots \quad \Sigma \vdash t_n \ term}{\Sigma \vdash \varphi(t_1, t_2, \ldots, t_n) \ prop} \ \text{pred-}F$$

$$\frac{\Sigma \vdash A \ prop \quad \Sigma \vdash B \ prop}{\Sigma \vdash A \wedge B \ prop} \ \wedge F \qquad \frac{}{\Sigma \vdash \top \ prop} \ \top F \qquad \frac{\Sigma \vdash A \ prop \quad \Sigma \vdash B \ prop}{\Sigma \vdash A \vee B \ prop} \ \vee F \qquad \frac{}{\Sigma \vdash \bot \ prop} \ \bot F$$

$$\frac{\Sigma \vdash A \ prop \quad \Sigma \vdash B \ prop}{\Sigma \vdash A \supset B \ prop} \ \supset F \qquad \frac{\Sigma, x \vdash A \ prop}{\Sigma \vdash \forall x. A \ prop} \ \forall F \qquad \frac{\Sigma, x \vdash A \ prop}{\Sigma \vdash \exists x. A \ prop} \ \exists F$$

**judgemental rules**

$$\frac{}{\Sigma : \Gamma, u : A \vdash u : A} \ \text{hyp}$$

**introduction rules**

$$\frac{\Sigma : \Gamma \vdash \pi_1 : A \quad \Sigma : \Gamma \vdash \pi_2 : B}{\Sigma : \Gamma \vdash (\pi_1, \pi_2) : A \wedge B} \ \wedge I \qquad \frac{}{\Sigma : \Gamma \vdash () : \top} \ \top I$$

$$\frac{\Sigma : \Gamma \vdash \pi : A}{\Sigma : \Gamma \vdash \mathbf{inl}\,\pi : A \vee B} \ \vee I_1 \qquad \frac{\Sigma : \Gamma \vdash \pi : B}{\Sigma : \Gamma \vdash \mathbf{inr}\,\pi : A \vee B} \ \vee I_2 \qquad \text{no} \ \bot I$$

$$\frac{\Sigma : \Gamma, u : A \vdash \pi : B}{\Sigma : \Gamma \vdash \lambda u. \pi : A \supset B} \ \supset I \qquad \frac{\Sigma, x : \Gamma \vdash \pi : A}{\Sigma : \Gamma \vdash \Lambda x. \pi : \forall x. A} \ \forall I \qquad \frac{\Sigma \vdash t \ term \quad \Sigma : \Gamma \vdash [t/x]\pi : [t/x]A}{\Sigma : \Gamma \vdash (t, \pi) : \exists x. A} \ \exists I$$

**elimination rules**

$$\frac{\Sigma : \Gamma \vdash \pi : A \wedge B}{\Sigma : \Gamma \vdash \mathbf{fst}\,\pi : A} \ \wedge E_1 \qquad \frac{\Sigma : \Gamma \vdash \pi : A \wedge B}{\Sigma : \Gamma \vdash \mathbf{snd}\,\pi : B} \ \wedge E_2 \qquad \text{no} \ \top E$$

$$\frac{\Sigma : \Gamma \vdash \pi : A \vee B \quad \Sigma : \Gamma, u : A \vdash \pi_1 : C \quad \Sigma : \Gamma, v : B \vdash \pi_2 : C}{\Sigma : \Gamma \vdash \mathbf{case}\,\pi\,\mathbf{of}\,\mathbf{inl}\,u \Rightarrow \pi_1 \mid \mathbf{inr}\,v \Rightarrow \pi_2 : C} \ \vee E \qquad \frac{\Sigma : \Gamma \vdash \pi : \bot}{\Sigma : \Gamma \vdash \mathbf{abort}\,\pi : C} \ \bot E$$

$$\frac{\Sigma : \Gamma \vdash \pi_1 : A \supset B \quad \Sigma : \Gamma \vdash \pi_2 : A}{\Sigma : \Gamma \vdash \pi_1 \ \pi_2 : B} \ \supset E \qquad \frac{\Sigma : \Gamma \vdash \pi : \forall x. A \quad \Sigma \vdash t \ term}{\Sigma : \Gamma \vdash \pi[t] : [t/x]A} \ \forall E$$

$$\frac{\Sigma : \Gamma \vdash \pi_1 : \exists x. A \quad \Sigma, x : \Gamma, u : A \vdash \pi_2 : C}{\Sigma : \Gamma \vdash \mathbf{let}\,(x, u) = \pi_1 \ \mathbf{in} \ \pi_2 : C} \ \exists E$$

*Summary of first-order system*

accurately described as tabular, include the following.

- 1940: In a textbook, Quine[8] indicated antecedent dependencies by line numbers in square brackets, anticipating

Suppes' 1957 line-number notation.

- 1950: In a textbook, Quine (1982, pp. 241–255) demonstrated a method of using one or more asterisks to the left of each line of proof to indicate dependencies. This is equivalent to Kleene's vertical bars. (It is not totally clear if Quine's asterisk notation appeared in the original 1950 edition or was added in a later edition.)

- 1957: An introduction to practical logic theorem proving in a textbook by Suppes (1999, pp. 25–150). This indicated dependencies (i.e. antecedent propositions) by line numbers at the left of each line.

- 1963: Stoll (1979, pp. 183–190, 215–219) uses sets of line numbers to indicate antecedent dependencies of the lines of sequential logical arguments based on natural deduction inference rules.

- 1965: The entire textbook by Lemmon (1965) is an introduction to logic proofs using a method based on that of Suppes.

- 1967: In a textbook, Kleene (2002, pp. 50–58, 128–130) briefly demonstrated two kinds of practical logic proofs, one system using explicit quotations of antecedent propositions on the left of each line, the other system using vertical bar-lines on the left to indicate dependencies.[9]

## 48.8   Proofs and type-theory

The presentation of natural deduction so far has concentrated on the nature of propositions without giving a formal definition of a *proof*. To formalise the notion of proof, we alter the presentation of hypothetical derivations slightly. We label the antecedents with *proof variables* (from some countable set $V$ of variables), and decorate the succedent with the actual proof. The antecedents or *hypotheses* are separated from the succedent by means of a *turnstile* ($\vdash$). This modification sometimes goes under the name of *localised hypotheses*. The following diagram summarises the change.

The collection of hypotheses will be written as $\Gamma$ when their exact composition is not relevant. To make proofs explicit, we move from the proof-less judgment "*A true*" to a judgment: "$\pi$ *is a proof of (A true)*", which is written symbolically as "$\pi : A$ *true*". Following the standard approach, proofs are specified with their own formation rules for the judgment "$\pi$ *proof*". The simplest possible proof is the use of a labelled hypothesis; in this case the evidence is the label itself.

For brevity, we shall leave off the judgmental label *true* in the rest of this article, *i.e.*, write "$\Gamma \vdash \pi : A$". Let us re-examine some of the connectives with explicit proofs. For conjunction, we look at the introduction rule $\wedge I$ to discover the form of proofs of conjunction: they must be a pair of proofs of the two conjuncts. Thus:

The elimination rules $\wedge E_1$ and $\wedge E_2$ select either the left or the right conjunct; thus the proofs are a pair of projections — first (**fst**) and second (**snd**).

For implication, the introduction form localises or *binds* the hypothesis, written using a $\lambda$; this corresponds to the discharged label. In the rule, "$\Gamma, u:A$" stands for the collection of hypotheses $\Gamma$, together with the additional hypothesis $u$.

With proofs available explicitly, one can manipulate and reason about proofs. The key operation on proofs is the substitution of one proof for an assumption used in another proof. This is commonly known as a *substitution theorem*, and can be proved by induction on the depth (or structure) of the second judgment.

**Substitution theorem**   *If $\Gamma \vdash \pi_1 : A$ and $\Gamma, u:A \vdash \pi_2 : B$, then $\Gamma \vdash [\pi_1/u]\,\pi_2 : B$.*

So far the judgment "$\Gamma \vdash \pi : A$" has had a purely logical interpretation. In type theory, the logical view is exchanged for a more computational view of objects. Propositions in the logical interpretation are now viewed as *types*, and proofs as programs in the lambda calculus. Thus the interpretation of "$\pi : A$" is "*the program $\pi$ has type $A$*". The logical connectives are also given a different reading: conjunction is viewed as product ($\times$), implication as the function arrow ($\rightarrow$), etc. The differences are only cosmetic, however. Type theory has a natural deduction presentation in terms of formation, introduction and elimination rules; in fact, the reader can easily reconstruct what is known as *simple type theory* from the previous sections.

The difference between logic and type theory is primarily a shift of focus from the types (propositions) to the programs (proofs). Type theory is chiefly interested in the convertibility or reducibility of programs. For every type, there are canonical programs of that type which are irreducible; these are known as *canonical forms* or *values*. If every program can be reduced to a canonical form, then the type theory is said to be *normalising* (or *weakly normalising*). If the canonical form is unique, then the theory is said to be *strongly normalising*. Normalisability is a rare feature of most non-trivial type theories, which is a big departure from the logical world. (Recall that almost every logical derivation has an equivalent normal derivation.) To sketch the reason: in type theories that admit recursive definitions, it is possible to write programs that never reduce to a value; such looping programs can generally be given any type. In particular, the looping program has type $\perp$, although there is no logical proof of "$\perp$ *true*". For this reason, the *propositions as types; proofs as programs* paradigm only works in one direction, if at all: interpreting a type theory as a logic generally gives an inconsistent logic.

Like logic, type theory has many extensions and variants, including first-order and higher-order versions. An interesting branch of type theory, known as dependent type theory, allows quantifiers to range over programs themselves. These quantified types are written as $\Pi$ and $\Sigma$ instead of $\forall$ and $\exists$, and have the following formation rules:

These types are generalisations of the arrow and product types, respectively, as witnessed by their introduction and elimination rules.

Dependent type theory in full generality is very powerful: it is able to express almost any conceivable property of programs directly in the types of the program. This generality comes at a steep price — either typechecking is undecidable (extensional type theory), or extensional reasoning is more difficult (intensional type theory). For this reason, some dependent type theories do not allow quantification over arbitrary programs, but rather restrict to programs of a given decidable *index domain*, for example integers, strings, or linear programs.

Since dependent type theories allow types to depend on programs, a natural question to ask is whether it is possible for programs to depend on types, or any other combination. There are many kinds of answers to such questions. A popular approach in type theory is to allow programs to be quantified over types, also known as *parametric polymorphism*; of this there are two main kinds: if types and programs are kept separate, then one obtains a somewhat more well-behaved system called *predicative polymorphism*; if the distinction between program and type is blurred, one obtains the type-theoretic analogue of higher-order logic, also known as *impredicative polymorphism*. Various combinations of dependency and polymorphism have been considered in the literature, the most famous being the lambda cube of Henk Barendregt.

The intersection of logic and type theory is a vast and active research area. New logics are usually formalised in a general type theoretic setting, known as a logical framework. Popular modern logical frameworks such as the calculus of constructions and LF are based on higher-order dependent type theory, with various trade-offs in terms of decidability and expressive power. These logical frameworks are themselves always specified as natural deduction systems, which is a testament to the versatility of the natural deduction approach.

## 48.9   Classical and modal logics

For simplicity, the logics presented so far have been intuitionistic. Classical logic extends intuitionistic logic with an additional axiom or principle of excluded middle:

> *For any proposition p, the proposition p $\vee$ $\neg p$ is true.*

This statement is not obviously either an introduction or an elimination; indeed, it involves two distinct connectives. Gentzen's original treatment of excluded middle prescribed one of the following three (equivalent) formulations, which were already present in analogous forms in the systems of Hilbert and Heyting:

($XM_3$ is merely $XM_2$ expressed in terms of E.) This treatment of excluded middle, in addition to being objectionable from a purist's standpoint, introduces additional complications in the definition of normal forms.

A comparatively more satisfactory treatment of classical natural deduction in terms of introduction and elimination rules alone was first proposed by Parigot in 1992 in the form of a classical lambda calculus called $\lambda\mu$. The key insight of his approach was to replace a truth-centric judgment $A$ *true* with a more classical notion, reminiscent of the sequent calculus: in localised form, instead of $\Gamma \vdash A$, he used $\Gamma \vdash \Delta$, with $\Delta$ a collection of propositions similar to $\Gamma$. $\Gamma$ was treated as a conjunction, and Δ as a disjunction. This structure is essentially lifted directly from classical sequent calculi, but the innovation in λμ was to give a computational meaning to classical natural deduction proofs in terms of a callcc or a throw/catch mechanism seen in LISP and its descendants. (See also: first class control.)

Another important extension was for modal and other logics that need more than just the basic judgment of truth. These were first described, for the alethic modal logics S4 and S5, in a natural deduction style by Prawitz in 1965,[4] and have since accumulated a large body of related work. To give a simple example, the modal logic S4 requires one new judgment, "*A valid*", that is categorical with respect to truth:

> *If "A true" under no assumptions of the form "B true", then "A valid".*

This categorical judgment is internalised as a unary connective □A (read "*necessarily A*") with the following introduction and elimination rules:

Note that the premise "*A valid*" has no defining rules; instead, the categorical definition of validity is used in its place. This mode becomes clearer in the localised form when the hypotheses are explicit. We write "$\Omega;\Gamma \vdash A\ true$" where $\Gamma$ contains the true hypotheses as before, and $\Omega$ contains valid hypotheses. On the right there is just a single judgment "*A true*"; validity is not needed here since "$\Omega \vdash A\ valid$" is by definition the same as "$\Omega;\cdot \vdash A\ true$". The introduction and elimination forms are then:

The modal hypotheses have their own version of the hypothesis rule and substitution theorem.

**Modal substitution theorem**   *If* $\Omega;\cdot \vdash \pi_1 : A\ true$ *and* $\Omega, u: (A\ valid)\ ;\ \Gamma \vdash \pi_2 : C\ true$, *then* $\Omega;\Gamma \vdash [\pi_1/u]\ \pi_2 : C\ true$.

This framework of separating judgments into distinct collections of hypotheses, also known as *multi-zoned* or *polyadic* contexts, is very powerful and extensible; it has been applied for many different modal logics, and also for linear and other substructural logics, to give a few examples. However, relatively few systems of modal logic can be formalised directly in natural deduction. To give proof-theoretic characterisations of these systems, extensions such as labelling or systems of deep inference.

The addition of labels to formulae permits much finer control of the conditions under which rules apply, allowing the more flexible techniques of analytic tableaux to be applied, as has been done in the case of labelled deduction. Labels also allow the naming of worlds in Kripke semantics; Simpson (1993) presents an influential technique for converting frame conditions of modal logics in Kripke semantics into inference rules in a natural deduction formalisation of hybrid logic. Stouppa (2004) surveys the application of many proof theories, such as Avron and Pottinger's hypersequents and Belnap's display logic to such modal logics as S5 and B.

## 48.10   Comparison with other foundational approaches

### 48.10.1   Sequent calculus

Main article: Sequent calculus

The sequent calculus is the chief alternative to natural deduction as a foundation of mathematical logic. In natural deduction the flow of information is bi-directional: elimination rules flow information downwards by deconstruction, and introduction rules flow information upwards by assembly. Thus, a natural deduction proof does not have a purely bottom-up or top-down reading, making it unsuitable for automation in proof search. To address this fact, Gentzen in 1935 proposed his sequent calculus, though he initially intended it as a technical device for clarifying the consistency of predicate logic. Kleene, in his seminal 1952 book *Introduction to Metamathematics*, gave the first formulation of the sequent calculus in the modern style.[10]

In the sequent calculus all inference rules have a purely bottom-up reading. Inference rules can apply to elements on both sides of the turnstile. (To differentiate from natural deduction, this article uses a double arrow ⇒ instead of the right tack ⊢ for sequents.) The introduction rules of natural deduction are viewed as *right rules* in the sequent calculus, and are structurally very similar. The elimination rules on the other hand turn into *left rules* in the sequent calculus. To give an example, consider disjunction; the right rules are familiar:

On the left:

Recall the ∨E rule of natural deduction in localised form:

The proposition $A \vee B$, which is the succedent of a premise in ∨E, turns into a hypothesis of the conclusion in the left rule ∨L. Thus, left rules can be seen as a sort of inverted elimination rule. This observation can be illustrated as follows:

In the sequent calculus, the left and right rules are performed in lock-step until one reaches the *initial sequent*, which corresponds to the meeting point of elimination and introduction rules in natural deduction. These initial rules are superficially similar to the hypothesis rule of natural deduction, but in the sequent calculus they describe a *transposition* or a *handshake* of a left and a right proposition:

The correspondence between the sequent calculus and natural deduction is a pair of soundness and completeness theorems, which are both provable by means of an inductive argument.

**Soundness of ⇒ wrt. ⊢** *If $\Gamma \Rightarrow A$, then $\Gamma \vdash A$.*

**Completeness of ⇒ wrt. ⊢** *If $\Gamma \vdash A$, then $\Gamma \Rightarrow A$.*

It is clear by these theorems that the sequent calculus does not change the notion of truth, because the same collection of propositions remain true. Thus, one can use the same proof objects as before in sequent calculus derivations. As an example, consider the conjunctions. The right rule is virtually identical to the introduction rule

The left rule, however, performs some additional substitutions that are not performed in the corresponding elimination rules.

The kinds of proofs generated in the sequent calculus are therefore rather different from those of natural deduction. The sequent calculus produces proofs in what is known as the *β-normal η-long* form, which corresponds to a canonical representation of the normal form of the natural deduction proof. If one attempts to describe these proofs using natural deduction itself, one obtains what is called the *intercalation calculus* (first described by John Byrnes), which can be used to formally define the notion of a *normal form* for natural deduction.

The substitution theorem of natural deduction takes the form of a structural rule or structural theorem known as *cut* in the sequent calculus.

**Cut (substitution)** *If $\Gamma \Rightarrow \pi_1 : A$ and $\Gamma, u{:}A \Rightarrow \pi_2 : C$, then $\Gamma \Rightarrow [\pi_1/u]\,\pi_2 : C$.*

In most well behaved logics, cut is unnecessary as an inference rule, though it remains provable as a meta-theorem; the superfluousness of the cut rule is usually presented as a computational process, known as *cut elimination*. This has an interesting application for natural deduction; usually it is extremely tedious to prove certain properties directly in natural deduction because of an unbounded number of cases. For example, consider showing that a given proposition is *not* provable in natural deduction. A simple inductive argument fails because of rules like ∨E or E which can introduce arbitrary propositions. However, we know that the sequent calculus is complete with respect to natural deduction, so it is enough to show this unprovability in the sequent calculus. Now, if cut is not available as an inference rule, then all sequent rules either introduce a connective on the right or the left, so the depth of a sequent derivation is fully bounded by the connectives in the final conclusion. Thus, showing unprovability is much easier, because there are only a finite number of cases to consider, and each case is composed entirely of sub-propositions of the conclusion. A simple instance of this is the *global consistency* theorem: "· ⊢ ⊥ *true*" is not provable. In the sequent calculus version, this is manifestly true because there is no rule that can have "· ⇒ ⊥" as a conclusion! Proof theorists often prefer to work on cut-free sequent calculus formulations because of such properties.

# 48.11 See also

- Mathematical logic

- Sequent calculus

- Gerhard Gentzen

- System L (tabular natural deduction)

## 48.12   Notes

[1] Jaśkowski 1934.

[2] Gentzen 1934, Gentzen 1935.

[3] Gentzen 1934, p. 176.

[4] Prawitz 1965, Prawitz 2006.

[5] Martin-Löf 1996.

[6] This is due to Bolzano, as cited by Martin-Löf 1996, p. 15.

[7] See also his book Prawitz 1965, Prawitz 2006.

[8] Quine (1981). See particularly pages 91–93 for Quine's line-number notation for antecedent dependencies.

[9] A particular advantage of Kleene's tabular natural deduction systems is that he proves the validity of the inference rules for both propositional calculus and predicate calculus. See Kleene 2002, pp. 44–45, 118–119.

[10] Kleene 2009, pp. 440–516. See also Kleene 1980.

## 48.13   References

- Barker-Plummer, Dave; Barwise, Jon; Etchemendy, John (2011). *Language Proof and Logic* (2nd ed.). CSLI Publications. ISBN 978-1575866321.

- Gallier, Jean (2005). "Constructive Logics. Part I: A Tutorial on Proof Systems and Typed λ-Calculi". Retrieved 12 June 2014.

- Gentzen, Gerhard Karl Erich (1934). "Untersuchungen über das logische Schließen. I". *Mathematische Zeitschrift* **39** (2): 176–210. doi:10.1007/BF01201353. (English translation *Investigations into Logical Deduction* in M. E. Szabo. The Collected Works of Gerhard Gentzen. North-Holland Publishing Company, 1969.)

- Gentzen, Gerhard Karl Erich (1935). "Untersuchungen über das logische Schließen. II". *Mathematische Zeitschrift* **39** (3): 405–431. doi:10.1007/bf01201363.

- Girard, Jean-Yves (1990). *Proofs and Types*. Cambridge Tracts in Theoretical Computer Science. Cambridge University Press, Cambridge, England. Translated and with appendices by Paul Taylor and Yves Lafont.

- Jaśkowski, Stanisław (1934). *On the rules of suppositions in formal logic*. Reprinted in *Polish logic 1920–39*, ed. Storrs McCall.

- Kleene, Stephen Cole (1980) [1952]. *Introduction to metamathematics* (Eleventh ed.). North-Holland. ISBN 978-0-7204-2103-3.

- Kleene, Stephen Cole (2009) [1952]. *Introduction to metamathematics*. Ishi Press International. ISBN 978-0-923891-57-2.

- Kleene, Stephen Cole (2002) [1967]. *Mathematical logic*. Mineola, New York: Dover Publications. ISBN 978-0-486-42533-7.

- Lemmon, Edward John (1965). *Beginning logic*. Thomas Nelson. ISBN 0-17-712040-1.

- Martin-Löf, Per (1996). "On the meanings of the logical constants and the justifications of the logical laws" (PDF). *Nordic Journal of Philosophical Logic* **1** (1): 11–60. Lecture notes to a short course at Università degli Studi di Siena, April 1983.

- Pfenning, Frank; Davies, Rowan (2001). "A judgmental reconstruction of modal logic" (PDF). *Mathematical Structures in Computer Science* **11** (4): 511–540. doi:10.1017/S0960129501003322.

- Prawitz, Dag (1965). *Natural deduction: A proof-theoretical study*. Acta Universitatis Stockholmiensis, Stockholm studies in philosophy 3. Stockholm, Göteborg, Uppsala: Almqvist & Wicksell.

- Prawitz, Dag (2006) [1965]. *Natural deduction: A proof-theoretical study*. Mineola, New York: Dover Publications. ISBN 978-0-486-44655-4.

- Quine, Willard Van Orman (1981) [1940]. *Mathematical logic* (Revised ed.). Cambridge, Massachusetts: Harvard University Press. ISBN 978-0-674-55451-1.

- Quine, Willard Van Orman (1982) [1950]. *Methods of logic* (Fourth ed.). Cambridge, Massachusetts: Harvard University Press. ISBN 978-0-674-57176-1.

- Simpson, Alex (1993). *The proof theory and semantics of intuitionistic modal logic*. University of Edinburgh. PhD thesis.

- Stoll, Robert Roth (1979) [1963]. *Set Theory and Logic*. Mineola, New York: Dover Publications. ISBN 978-0-486-63829-4.

- Stouppa, Phiniki (2004). *The Design of Modal Proof Theories: The Case of S5*. University of Dresden. MSc thesis.

- Suppes, Patrick Colonel (1999) [1957]. *Introduction to logic*. Mineola, New York: Dover Publications. ISBN 978-0-486-40687-9.

## 48.14 External links

- Clemente, Daniel, "Introduction to natural deduction."

- Domino On Acid. Natural deduction visualized as a game of dominoes.

- Pelletier, Jeff, "A History of Natural Deduction and Elementary Logic Textbooks."

- Levy, Michel, A Propositional Prover.

- A Propositional proving tool for iOS

# Chapter 49

# ω-consistent theory

In mathematical logic, an **ω-consistent** (or **omega-consistent**, also called **numerically segregative**[1]) **theory** is a theory (collection of sentences) that is not only (syntactically) consistent (that is, does not prove a contradiction), but also avoids proving certain infinite combinations of sentences that are intuitively contradictory. The name is due to Kurt Gödel, who introduced the concept in the course of proving the incompleteness theorem.[2]

## 49.1  Definition

A theory $T$ is said to interpret the language of arithmetic if there is a translation of formulas of arithmetic into the language of $T$ so that $T$ is able to prove the basic axioms of the natural numbers under this translation.

A $T$ that interprets arithmetic is **ω-inconsistent** if, for some property $P$ of natural numbers (defined by a formula in the language of $T$), $T$ proves $P(0)$, $P(1)$, $P(2)$, and so on (that is, for every standard natural number $n$, $T$ proves that $P(n)$ holds), but $T$ also proves that there is some (necessarily nonstandard) natural number $n$ such that $P(n)$ *fails*. This may not lead directly to an outright contradiction, because $T$ may not be able to prove for any *specific* value of $n$ that $P(n)$ fails, only that there *is* such an $n$.

$T$ is **ω-consistent** if it is *not* ω-inconsistent.

There is a weaker but closely related property of $\Sigma_1$-soundness. A theory $T$ is $\Sigma_1$-**sound** (or **1-consistent**, in another terminology) if every $\Sigma^0_1$-sentence[3] provable in $T$ is true in the standard model of arithmetic **N** (i.e., the structure of the usual natural numbers with addition and multiplication). If $T$ is strong enough to formalize a reasonable model of computation, $\Sigma_1$-soundness is equivalent to demanding that whenever $T$ proves that a computer program $C$ halts, then $C$ actually halts. Every ω-consistent theory is $\Sigma_1$-sound, but not vice versa.

More generally, we can define an analogous concept for higher levels of the arithmetical hierarchy. If $\Gamma$ is a set of arithmetical sentences (typically $\Sigma^0_n$ for some $n$), a theory $T$ is $\Gamma$-**sound** if every $\Gamma$-sentence provable in $T$ is true in the standard model. When $\Gamma$ is the set of all arithmetical formulas, $\Gamma$-soundness is called just (arithmetical) soundness. If the language of $T$ consists *only* of the language of arithmetic (as opposed to, for example, set theory), then a sound system is one whose model can be thought of as the set $\omega$, the usual set of mathematical natural numbers. The case of general $T$ is different, see ω-logic below.

$\Sigma n$-soundness has the following computational interpretation: if the theory proves that a program $C$ using a $\Sigma n_{-1}$-oracle halts, then $C$ actually halts.

## 49.2  Examples

### 49.2.1 Consistent, ω-inconsistent theories

Write PA for the theory Peano arithmetic, and Con(PA) for the statement of arithmetic that formalizes the claim "PA is consistent". Con(PA) could be of the form "For every natural number $n$, $n$ is not the Gödel number of a proof from PA that 0=1". (This formulation uses 0=1 instead of a direct contradiction; that gives the same result, because PA certainly proves ¬0=1, so if it proved 0=1 as well we would have a contradiction, and on the other hand, if PA proves a contradiction, then it proves anything, including 0=1.)

Now, assuming PA is really consistent, it follows that PA + ¬Con(PA) is also consistent, for if it were not, then PA would prove Con(PA) (since an inconsistent theory proves every sentence), contradicting Gödel's second incompleteness theorem. However, PA + ¬Con(PA) is *not* ω-consistent. This is because, for any particular natural number $n$, PA + ¬Con(PA) proves that $n$ is not the Gödel number of a proof that 0=1 (PA itself proves that fact; the extra assumption ¬Con(PA) is not needed). However, PA + ¬Con(PA) proves that, for *some* natural number $n$, $n$ *is* the Gödel number of such a proof (this is just a direct restatement of the claim ¬Con(PA) ).

In this example, the axiom ¬Con(PA) is $\Sigma_1$, hence the system PA + ¬Con(PA) is in fact $\Sigma_1$-unsound, not just ω-inconsistent.

### 49.2.2 Arithmetically sound, ω-inconsistent theories

Let $T$ be PA together with the axioms $c \neq n$ for each natural number $n$, where $c$ is a new constant added to the language. Then $T$ is arithmetically sound (as any nonstandard model of PA can be expanded to a model of $T$), but ω-inconsistent (as it proves $\exists x\, c = x$, and $c \neq n$ for every number $n$).

$\Sigma_1$-sound ω-inconsistent theories using only the language of arithmetic can be constructed as follows. Let $I\Sigma n$ be the subtheory of PA with the induction schema restricted to $\Sigma n$-formulas, for any $n > 0$. The theory $I\Sigma n_{+1}$ is finitely axiomatizable, let thus $A$ be its single axiom, and consider the theory $T = I\Sigma n + \neg A$. We can assume that $A$ is an instance of the induction schema, which has the form

$$\forall w \left[ B(0, w) \wedge \forall x \left( B(x, w) \to B(x + 1, w) \right) \to \forall x\, B(x, w) \right].$$

If we denote the formula

$$\forall w \left[ B(0, w) \wedge \forall x \left( B(x, w) \to B(x + 1, w) \right) \to B(n, w) \right]$$

by $P(n)$, then for every natural number $n$, the theory $T$ (actually, even the pure predicate calculus) proves $P(n)$. On the other hand, $T$ proves the formula $\exists x\, \neg P(x)$, because it is logically equivalent to the axiom $\neg A$. Therefore, $T$ is ω-inconsistent.

It is possible to show that $T$ is $\Pi n_{+3}$-sound. In fact, it is $\Pi n_{+3}$-conservative over the (obviously sound) theory $I\Sigma n$. The argument is more complicated (it relies on the provability of the $\Sigma n_{+2}$-reflection principle for $I\Sigma n$ in $I\Sigma n_{+1}$).

### 49.2.3 Arithmetically unsound, ω-consistent theories

Let ω-Con(PA) be the arithmetical sentence formalizing the statement "PA is ω-consistent". Then the theory PA + ¬ω-Con(PA) is unsound ($\Sigma_3$-unsound, to be precise), but ω-consistent. The argument is similar to the first example: a suitable version of the Hilbert-Bernays-Löb derivability conditions holds for the "provability predicate" ω-Prov($A$) = ¬ω-Con(PA + ¬$A$), hence it satisfies an analogue of Gödel's second incompleteness theorem.

## 49.3   ω-logic

Not to be confused with Ω-logic.

The concept of theories of arithmetic whose integers are the true mathematical integers is captured by **ω-logic**.[4] Let $T$ be a theory in a countable language which includes a unary predicate symbol $N$ intended to hold just of the natural numbers, as well as specified names 0, 1, 2, ..., one for each (standard) natural number (which may be separate constants, or constant terms such as 0, 1, 1+1, 1+1+1, ..., etc.). Note that $T$ itself could be referring to more general objects, such as real numbers or sets; thus in a model of $T$ the objects satisfying $N(x)$ are those that $T$ interprets as natural numbers, not all of which need be named by one of the specified names.

The system of ω-logic includes all axioms and rules of the usual first-order predicate logic, together with, for each $T$-formula $P(x)$ with a specified free variable $x$, an infinitary **ω-rule** of the form:

> From $P(0), P(1), P(2), \ldots$ infer $\forall x\,(N(x) \to P(x))$.

That is, if the theory asserts (i.e. proves) $P(n)$ separately for each natural number $n$ given by its specified name, then it also asserts $P$ collectively for all natural numbers at once via the evident finite universally quantified counterpart of the infinitely many antecedents of the rule. For a theory of arithmetic, meaning one with intended domain the natural numbers such as Peano arithmetic, the predicate $N$ is redundant and may be omitted from the language, with the consequent of the rule for each $P$ simplifying to $\forall x\, P(x)$.

An ω-model of $T$ is a model of $T$ whose domain includes the natural numbers and whose specified names and symbol $N$ are standardly interpreted, respectively as those numbers and the predicate having just those numbers as its domain (whence there are no nonstandard numbers). If $N$ is absent from the language then what would have been the domain of $N$ is required to be that of the model, i.e. the model contains only the natural numbers. (Other models of $T$ may interpret these symbols nonstandardly; the domain of $N$ need not even be countable, for example.) These requirements make the ω-rule sound in every ω-model. As a corollary to the omitting types theorem, the converse also holds: the theory $T$ has an ω-model if and only if it is consistent in ω-logic.

There is a close connection of ω-logic to ω-consistency. A theory consistent in ω-logic is also ω-consistent (and arithmetically sound). The converse is false, as consistency in ω-logic is a much stronger notion than ω-consistency. However, the following characterization holds: a theory is ω-consistent if and only if its closure under *unnested* applications of the ω-rule is consistent.

## 49.4   Relation to other consistency principles

If the theory $T$ is recursively axiomatizable, ω-consistency has the following characterization, due to C. Smoryński:[5]

> $T$ is ω-consistent if and only if $T + \mathrm{RFN}_T + \mathrm{Th}_{\Pi^0_2}(\mathbb{N})$ is consistent.

Here, $\mathrm{Th}_{\Pi^0_2}(\mathbb{N})$ is the set of all $\Pi^0_2$-sentences valid in the standard model of arithmetic, and $\mathrm{RFN}_T$ is the uniform reflection principle for $T$, which consists of the axioms

$$\forall x\,(\mathrm{Prov}_T(\ulcorner \varphi(\dot{x}) \urcorner) \to \varphi(x))$$

for every formula $\varphi$ with one free variable. In particular, a finitely axiomatizable theory $T$ in the language of arithmetic is ω-consistent if and only if $T + \mathrm{PA}$ is $\Sigma^0_2$-sound.

## 49.5   Notes

[1]  W.V.O. Quine, *Set Theory and its Logic*

[2] Smorynski, "The incompleteness theorems", *Handbook of Mathematical Logic*, 1977, p. 851.

[3] The definition of this symbolism can be found at arithmetical hierarchy.

[4] J. Barwise (ed.), *Handbook of Mathematical Logic*, North-Holland, Amsterdam, 1977.

[5] Smoryński, Craig (1985). *Self-reference and modal logic*. Berlin: Springer. ISBN 978-0-387-96209-2. Reviewed in Boolos, G.; Smorynski, C. (1988). "Self-Reference and Modal Logic". *The Journal of Symbolic Logic* **53**: 306. doi:10.2307/2274450. JSTOR 2274450.

## 49.6 Bibliography

- Kurt Gödel (1931). 'Über formal unentscheidbare Sätze der Principia Mathematica und verwandter Systeme I'. In *Monatshefte für Mathematik*. Translated into English as On Formally Undecidable Propositions of Principia Mathematica and Related Systems.

# Chapter 50

# Ordinal analysis

In proof theory, **ordinal analysis** assigns ordinals (often large countable ordinals) to mathematical theories as a measure of their strength. The field was formed when Gerhard Gentzen in 1934 used cut elimination to prove, in modern terms, that the **proof theoretic ordinal** of Peano arithmetic is $\varepsilon_0$.

## 50.1  Definition

Ordinal analysis concerns true, effective (recursive) theories that can interpret a sufficient portion of arithmetic to make statements about ordinal notations. The **proof theoretic ordinal** of such a theory $T$ is the smallest recursive ordinal that the theory cannot prove is well founded — the supremum of all ordinals $\alpha$ for which there exists a notation $o$ in Kleene's sense such that $T$ proves that $o$ is an ordinal notation. Equivalently, it is the supremum of all ordinals $\alpha$ such that there exists a recursive relation $R$ on $\omega$ (the set of natural numbers) that well-orders it with ordinal $\alpha$ and such that $T$ proves transfinite induction of arithmetical statements for $R$ .

The existence of any recursive ordinal that the theory fails to prove is well ordered follows from the $\Sigma_1^1$ bounding theorem, as the set of natural numbers that an effective theory proves to be ordinal notations is a $\Sigma_1^0$ set (see Hyperarithmetical theory). Thus the proof theoretic ordinal of a theory will always be a countable ordinal less than the Church-Kleene ordinal $\omega_1^{CK}$ .

In practice, the proof theoretic ordinal of a theory is a good measure of the strength of a theory. If theories have the same proof theoretic ordinal they are often equiconsistent, and if one theory has a larger proof theoretic ordinal than another it can often prove the consistency of the second theory.

## 50.2  Examples

### 50.2.1  Theories with proof theoretic ordinal $\omega^2$

- RFA, rudimentary function arithmetic.[1]

- $I\Delta_0$, arithmetic with induction on $\Delta_0$-predicates without any axiom asserting that exponentiation is total.

### 50.2.2  Theories with proof theoretic ordinal $\omega^3$

Friedman's grand conjecture suggests that much "ordinary" mathematics can be proved in weak systems having this as their proof-theoretic ordinal.

- EFA, elementary function arithmetic.

- $I\Delta_0 + \exp$, arithmetic with induction on $\Delta_0$-predicates augmented by an axiom asserting that exponentiation is total.

- RCA*

  0, a second order form of EFA sometimes used in reverse mathematics.

- WKL*

  0, a second order form of EFA sometimes used in reverse mathematics.

### 50.2.3 Theories with proof theoretic ordinal $\omega^n$

- $I\Delta_0$ or EFA augmented by an axiom ensuring that each element of the $n$-th level $\mathcal{E}^n$ of the Grzegorczyk hierarchy is total.

### 50.2.4 Theories with proof theoretic ordinal $\omega^\omega$

- $RCA_0$, recursive comprehension.

- $WKL_0$, weak König's lemma.

- PRA, primitive recursive arithmetic.

- $I\Sigma_1$, arithmetic with induction on $\Sigma_1$-predicates.

### 50.2.5 Theories with proof theoretic ordinal $\varepsilon_0$

- PA, Peano arithmetic (shown by Gentzen using cut elimination).

- $ACA_0$, arithmetical comprehension.

### 50.2.6 Theories with proof theoretic ordinal the Feferman-Schütte ordinal $\Gamma_0$

This ordinal is sometimes considered to be the upper limit for "predicative" theories.

- $ATR_0$, arithmetical transfinite recursion.

- Martin-Löf type theory with arbitrarily many finite level universes.

### 50.2.7 Theories with proof theoretic ordinal the Bachmann-Howard ordinal

- $ID_1$, the theory of inductive definitions.

- KP, Kripke-Platek set theory with the axiom of infinity.

- CZF, Aczel's constructive Zermelo-Fraenkel set theory.

- MLW, Martin-Löf Type Theory with indexed W-Types

- EON, a weak variant of the Feferman's explicit mathematics system $T_0$.

### 50.2.8 Theories with larger proof theoretic ordinals

- $\Pi_1^1$-$CA_0$ , $\Pi_1{}^1$ comprehension has a rather large proof theoretic ordinal, which was described by Takeuti in terms of "ordinal diagrams", and which is bounded by $\psi_0(\Omega\omega)$ in Buchholz's notation. It is also the ordinal of $ID_{<\omega}$ , the theory of finitely iterated inductive definitions.

- $T_0$, Feferman's constructive system of explicit mathematics has a larger proof-theoretic ordinal, which is also the proof-theoretic ordinal of the KPi, Kripke-Platek Set theory with iterated admissibles and $\Sigma_2^1$-AC + BI .

- KPM, an extension of Kripke-Platek set theory based on a Mahlo cardinal, has a very large proof theoretic ordinal $\vartheta$, which was described by Rathjen (1990).

- MLM, an extension of Martin-Löf type theory by one Mahlo-universe, has an even larger proof theoretic ordinal $\psi\Omega_1(\Omega M_+\omega)$.

Most theories capable of describing the power set of the natural numbers have proof theoretic ordinals that are so large that no explicit combinatorial description has yet (as of 2008) been given. This includes second order arithmetic and set theories with powersets. (The CZF and Kripke-Platek set theories mentioned above are weak set theories without powersets.)

## 50.3 See also

- Equiconsistency

- Large cardinal property

- Feferman–Schütte ordinal

- Bachmann–Howard ordinal

## 50.4 References

- Buchholz, W.; Feferman, S.; Pohlers, W.; Sieg, W. (1981), *Iterated inductive definitions and sub-systems of analysis*, Lecture Notes in Math. **897**, Berlin: Springer-Verlag, doi:10.1007/BFb0091894, ISBN 978-3-540-11170-2

- Pohlers, Wolfram (1989), *Proof theory*, Lecture Notes in Mathematics **1407**, Berlin: Springer-Verlag, ISBN 3-540-51842-8, MR 1026933

- Pohlers, Wolfram (1998), "Handbook of Proof Theory", *Handbook of Proof Theory*, Studies in Logic and the Foundations of Mathematics (Amsterdam: Elsevier Science B. V.) **137**: 210–335, ISBN 0-444-89840-9, MR 1640328 |chapter= ignored (help)

- Rathjen, Michael (1990), "Ordinal notations based on a weakly Mahlo cardinal.", *Arch. Math. Logic* **29** (4): 249–263, doi:10.1007/BF01651328, MR 1062729

- Rathjen, Michael (2006), "The art of ordinal analysis", *International Congress of Mathematicians* (PDF) **II**, Zürich,: Eur. Math. Soc., pp. 45–69, MR 2275588

- Rose, H.E. (1984), *Subrecursion. Functions and Hierarchies*, Oxford logic guides **9**, Oxford, New York: Clarendon Press, Oxford University Press

- Schütte, Kurt (1977), *Proof theory*, Grundlehren der Mathematischen Wissenschaften **225**, Berlin-New York: Springer-Verlag, pp. xii+299, ISBN 3-540-07911-4, MR 0505313

- Takeuti, Gaisi (1987), *Proof theory*, Studies in Logic and the Foundations of Mathematics **81** (Second ed.), Amsterdam: North-Holland Publishing Co., ISBN 0-444-87943-9, MR 0882549

[1] Krajicek, Jan (1995). *Bounded Arithmetic, Propositional Logic and Complexity Theory*. Cambridge University Press. pp. 18–20. ISBN 9780521452052. defines the rudimentary sets and rudimentary functions, and proves them equivalent to the $\Delta_0$-predicates on the naturals. An ordinal analysis of the system can be found in Rose, H. E. (1984). *Subrecursion: functions and hierarchies.* University of Michigan: Clarendon Press. ISBN 9780198531890.

# Chapter 51

# Ordinal notation

In mathematical logic and set theory, an **ordinal notation** is a finite sequence of symbols from a finite alphabet that names an ordinal number according to some scheme that gives meaning to the language. Given such a scheme, one should be able to define a recursive well-ordering of a subset of the natural numbers by associating a natural number with each finite sequence of symbols via a Gödel numbering.

There are many such schemes of ordinal notations, including schemes by Wilhelm Ackermann, Heinz Bachmann, Wilfried Buchholz, Georg Cantor, Solomon Feferman, Gerhard Jäger, Isles, Pfeiffer, Wolfram Pohlers, Kurt Schütte, Gaisi Takeuti (called **ordinal diagrams**), Oswald Veblen. Stephen Cole Kleene has a system of notations, called Kleene's O, which includes ordinal notations but it is not as well behaved as the other systems described here.

Usually one proceeds by defining several functions from ordinals to ordinals and representing each such function by a symbol. In many systems, such as Veblen's well known system, the functions are normal functions, that is, they are strictly increasing and continuous in at least one of their arguments, and increasing in other arguments. Another desirable property for such functions is that the value of the function is greater than each of its arguments, so that an ordinal is always being described in terms of smaller ordinals. There are several such desirable properties. Unfortunately, no one system can have all of them since they contradict each other.

## 51.1 A simplified example using a pairing function

As usual, we must start off with a constant symbol for zero, "0", which we may consider to be a function of arity zero. This is necessary because there are no smaller ordinals in terms of which zero can be described. The most obvious next step would be to define a unary function, "S", which takes an ordinal to the smallest ordinal greater than it; in other words, S is the successor function. In combination with zero, successor allows one to name any natural number.

The third function might be defined as one that maps each ordinal to the smallest ordinal that cannot yet be described with the above two functions and previous values of this function. This would map $\beta$ to $\omega \cdot \beta$ except when $\beta$ is a fixed point of that function plus a finite number in which case one uses $\omega \cdot (\beta+1)$.

The fourth function would map $\alpha$ to $\omega^\omega \cdot \alpha$ except when $\alpha$ is a fixed point of that plus a finite number in which case one uses $\omega^\omega \cdot (\alpha+1)$.

### 51.1.1 $\xi$-notation

One could continue in this way, but it would give us an infinite number of functions. So instead let us merge the unary functions together into a binary function. By transfinite recursion on $\alpha$, we can use transfinite recursion on $\beta$ to define $\xi(\alpha,\beta)$ = the smallest ordinal $\gamma$ such that $\alpha < \gamma$ and $\beta < \gamma$ and $\gamma$ is not the value of $\xi$ for any smaller $\alpha$ or for the same $\alpha$ with a smaller $\beta$.

Thus, define $\xi$-notations as follows:

- "0" is a ξ-notation for zero.

- If "A" and "B" are replaced by ξ-notations for α and β in "ξAB", then the result is a ξ-notation for ξ(α,β).

- There are no other ξ-notations.

ξ is defined for all pairs of ordinals and is one-to-one. It always gives values larger than its arguments and its range is all ordinals other than 0 and the epsilon numbers ($\varepsilon=\omega^\varepsilon$).

ξ(α,β)<ξ(γ,δ) if and only if either (α=γ and β<δ) or (α<γ and β<ξ(γ,δ)) or (α>γ and ξ(α,β)≤δ).

With this definition, the first few ξ-notations are:

"0" for 0. "ξ00" for 1. "ξ0ξ00" for ξ(0,1)=2. "ξξ000" for ξ(1,0)=ω. "ξ0ξ0ξ00" for 3. "ξ0ξξ000" for ω+1. "ξξ00ξ00" for ω·2. "ξξ0ξ000" for $\omega^\omega$. "ξξξ0000" for $\omega^{\omega^{\tilde\omega}}$.

In general, ξ(0,β) = β+1. While ξ(1+α,β) = $\omega^{\omega^\alpha}\cdot(\beta+k)$ for k = 0 or 1 or 2 depending on special situations:
k = 2 if α is an epsilon number and β is finite.
Otherwise, k = 1 if β is a multiple of $\omega^{\omega^{\alpha+1}}$ plus a finite number.
Otherwise, k = 0.

The ξ-notations can be used to name any ordinal less than $\varepsilon_0$ with an alphabet of only two symbols ("0" and "ξ"). If these notations are extended by adding functions that enumerate epsilon numbers, then they will be able to name any ordinal less than the first epsilon number that cannot be named by the added functions. This last property, adding symbols within an initial segment of the ordinals gives names within that segment, is called repleteness (after Solomon Feferman).

# 51.2  Systems of ordinal notation

There are many different systems for ordinal notation introduced by various authors. It is often quite hard to convert between the different systems.

## 51.2.1  Cantor

Main article: Cantor normal form

"Exponential polynomials" in 0 and ω gives a system of ordinal notation for ordinals less than epsilon zero. There are many equivalent ways to write these; instead of exponential polynomials, one can use rooted trees, or nested parentheses, or the system described above.

## 51.2.2  Veblen

Main article: Veblen function

The 2-variable Veblen functions (Veblen 1908) can be used to give a system of ordinal notation for ordinals less than the Feferman-Schutte ordinal. The Veblen functions in a finite or transfinite number of variables give systems of ordinal notations for ordinals less than the small and large Veblen ordinals.

## 51.2.3  Ackermann

Ackermann (1951) described a system of ordinal notation rather weaker than the system described earlier by Veblen. The limit of his system is sometimes called the Ackermann ordinal.

### 51.2.4   Bachmann

Bachmann (1950) introduced the key idea of using uncountable ordinals to produce new countable ordinals. His original system was rather cumbersome to use as it required choosing a special sequence converging to each ordinal. Later systems of notation introduced by Feferman and others avoided this complication.

### 51.2.5   Takeuti (ordinal diagrams)

Takeuti (1987) described a very powerful system of ordinal notation called "ordinal diagrams", which is hard to understand but was later simplified by Feferman.

### 51.2.6   Feferman's θ functions

Feferman introduced theta functions, described in Buchholz (1986) as follows. The function for an ordinal $\alpha$, $\theta\alpha$ is a function from ordinals to ordinals. Often $\theta\alpha(\beta)$ is written as $\theta\alpha\beta$. The set $C(\alpha,\beta)$ is defined by induction on $\alpha$ to be the set of ordinals that can be generated from 0, $\omega_1$, $\omega_2$, ..., $\omega\omega$, together with the ordinals less than $\beta$ by the operations of ordinal addition and the functions $\theta\xi$ for $\xi<\alpha$. And the function $\theta_\gamma$ is defined to be the function enumerating the ordinals $\delta$ with $\delta\notin C(\gamma,\delta)$.

### 51.2.7   Buchholz

Main article: Ordinal collapsing function

Buchholz (1986) described the following system of ordinal notation as a simplification of Feferman's theta functions. Define:

- $\Omega\xi = \omega\xi$ if $\xi > 0$, $\Omega_0 = 1$

The functions $\psi v(\alpha)$ for $\alpha$ an ordinal, $v$ an ordinal at most $\omega$, are defined by induction on $\alpha$ as follows:

- $\psi v(\alpha)$ is the smallest ordinal not in $Cv(\alpha)$

where $Cv(\alpha)$ is the smallest set such that

- $Cv(\alpha)$ contains all ordinals less than $\Omega v$
- $Cv(\alpha)$ is closed under ordinal addition
- $Cv(\alpha)$ is closed under the functions $\psi u$ (for $u\leq\omega$) applied to arguments less than $\alpha$.

This system has about the same strength as Fefermans system, as $\theta\epsilon_{\Omega_v+1}0 = \psi_0(\epsilon_{\Omega_v+1})$ for $v \leq \omega$.

### 51.2.8   Kleene's $\mathcal{O}$

Main article: Kleene's O

Kleene (1938) described a system of notation for all recursive ordinals (those less than the Church–Kleene ordinal). It uses a subset of the natural numbers instead of finite strings of symbols. Unfortunately, unlike the other systems described above there is in general no effective way to tell whether some natural number represents an ordinal, or whether two numbers represent the same ordinal. However, one can effectively find notations that represent the ordinal sum, product, and power (see ordinal arithmetic) of any two given notations in Kleene's $\mathcal{O}$ ; and given any notation for an ordinal, there is a recursively enumerable set of notations that contains one element for each smaller ordinal and is effectively ordered. Kleene's $\mathcal{O}$ denotes a canonical (and very non-computable) set of notations.

## 51.3  See also

- Large countable ordinals

- Ordinal arithmetic

- Ordinal analysis

## 51.4  References

- Ackermann, Wilhelm (1951), "Konstruktiver Aufbau eines Abschnitts der zweiten Cantorschen Zahlenklasse", *Math. Z.* **53** (5): 403–413, doi:10.1007/BF01175640, MR 0039669

- Buchholz, W. (1986), "A new system of proof-theoretic ordinal functions", *Ann. Pure Appl. Logic* **32** (3): 195–207, doi:10.1016/0168-0072(86)90052-7, MR 0865989

- "Constructive Ordinal Notation Systems" by Fredrick Gass

- Kleene, S. C. (1938), "On Notation for Ordinal Numbers", *The Journal of Symbolic Logic* (The Journal of Symbolic Logic, Vol. 3, No. 4) **3** (4): 150–155, doi:10.2307/2267778, JSTOR 2267778

- "Hyperarithmetical Index Sets In Recursion Theory" by Stephen Lempp

- Hilbert Levitz, *Transfinite Ordinals and Their Notations: For The Uninitiated*, expository article (8 pages, in PostScript)

- Miller, Larry W. (1976), "Normal Functions and Constructive Ordinal Notations", *The Journal of Symbolic Logic* (The Journal of Symbolic Logic, Vol. 41, No. 2) **41** (2): 439 to 459, doi:10.2307/2272243, JSTOR 2272243

- Pohlers, Wolfram (1989), *Proof theory*, Lecture Notes in Mathematics **1407**, Berlin: Springer-Verlag, ISBN 3-540-51842-8, MR 1026933

- Rogers, Hartley (1987) [1967], *The Theory of Recursive Functions and Effective Computability*, First MIT press paperback edition, ISBN 978-0-262-68052-3

- Schütte, Kurt (1977), *Proof theory*, Grundlehren der Mathematischen Wissenschaften **225**, Berlin-New York: Springer-Verlag, pp. xii+299, ISBN 3-540-07911-4, MR 0505313

- Takeuti, Gaisi (1987), *Proof theory*, Studies in Logic and the Foundations of Mathematics **81** (Second ed.), Amsterdam: North-Holland Publishing Co., ISBN 0-444-87943-9, MR 0882549

- Veblen, Oswald (1908), "Continuous Increasing Functions of Finite and Transfinite Ordinals", *Transactions of the American Mathematical Society* (Transactions of the American Mathematical Society, Vol. 9, No. 3) **9** (3): 280–292, doi:10.2307/1988605, JSTOR 1988605

# Chapter 52

# Paraconsistent mathematics

**Paraconsistent mathematics** (sometimes called **inconsistent mathematics**) represents an attempt to develop the classical infrastructure of mathematics (e.g. analysis) based on a foundation of paraconsistent logic instead of classical logic. A number of reformulations of analysis can be developed, for example functions which both do and do not have a given value simultaneously.

Chris Mortensen claims (see references):

> One could hardly ignore the examples of analysis and its special case, the calculus. There prove to be many places where there are distinctive inconsistent insights; see Mortensen (1995) for example. (1) Robinson's non-standard analysis was based on infinitesimals, quantities smaller than any real number, as well as their reciprocals, the infinite numbers. This has an inconsistent version, which has some advantages for calculation in being able to discard higher-order infinitesimals. The theory of differentiation turned out to have these advantages, while the theory of integration did not. (2)

## 52.1 References

- *Inconsistent Mathematics*, by Chris Mortensen, Dordrecht, Kluwer Academic Publishers, 1995 Kluwer *Mathematics and Its Applications Series*, Vol 312 ISBN 0-7923-3186-9

## 52.2 External links

- Entry in the *Internet Encyclopedia of Philosophy*

- Entry in the *Stanford Encyclopedia of Philosophy*

- Lectures by Manuel Bremer of the University of Düsseldorf

# Chapter 53

# Peano-Russell notation

**Peano-Russell notation** was Bertrand Russell's application of Peano's logical notation to the logical notions of Frege and was used in the writing of *Principia Mathematica* in collaboration with Alfred North Whitehead:[1]

> "The notation adopted in the present work is based upon that of Peano, and the following explanations are to some extent modelled on those which he prefixes to his *Formulario Mathematico.*" (Chapter I: Preliminary Explanations of Ideas and Notations, page 4)

## 53.1  Variables

In the notation, variables are ambiguous in denotation, preserve a recognizable identity appearing in various places in logical statements within a given context, and have a range of possible determination between any two variables which is the same or different. When the possible determination is the same for both variables, then one implies the other; otherwise, the possible determination of one given to the other produces a meaningless phrase. The alphabetic symbol set for variables includes the lower and upper case Roman letters as well as many from the Greek alphabet.

## 53.2  Fundamental functions of propositions

The four fundamental functions are the *contradictory function*, the *logical sum*, the *logical product*, and the *implicative function*.[2]

### 53.2.1  Contradictory function

The contradictory function applied to a proposition returns its negation.

$\sim p$

### 53.2.2  Logical sum

The logical sum applied to two propositions returns their disjunction.

$p \vee q$

### 53.2.3  Logical product

The logical product applied to two propositions returns the truth-value of both propositions being simultaneously true.

$p \cdot q$

### 53.2.4  Implicative function

The implicative function applied to two ordered propositions returns the truth value of the first implying the second proposition.

$p \supset q$

## 53.3  More complex functions of propositions

*Equivalence* is written as $p \equiv q$ , standing for $p \supset q \cdot q \supset p$ .[3]

*Assertion* is same as the making of a statement between two full stops.

$\vdash p$

An asserted proposition is either true or an error on the part of the writer.[4]

*Inference* is equivalent to the rule *modus ponens*, where $p \cdot p \supset q. \supset q$ [5]

In addition to the logical product, *dots* are also used to show groupings of functions of propositions. In the above example, the dot before the final implication function symbol groups all of the previous functions on that line together as the antecedent to the final consequent.

The notation includes *definitions* as complex functions of propositions, using the equals sign "=" to separate the defined term from its symbolic definition, ending with the letters "def."[6]

## 53.4  Notes

[1] Russell, p. 4

[2] Russell, p. 6

[3] Russell, p. 7

[4] Russell, p. 8

[5] Russell, pp. 8-9

[6] Russell, p. 11

## 53.5  References

Russell, Bertrand and Alfred North Whitehead (1910). *Principia Mathematica* Cambridge, England: The University Press.

## 53.6 External links

- The Notation in *Principia Mathematica* entry by Bernard Linsky in the *Stanford Encyclopedia of Philosophy*

# Chapter 54

# Presburger arithmetic

**Presburger arithmetic** is the first-order theory of the natural numbers with addition, named in honor of Mojżesz Presburger, who introduced it in 1929. The signature of Presburger arithmetic contains only the addition operation and equality, omitting the multiplication operation entirely. The axioms include a schema of induction.

Presburger arithmetic is much weaker than Peano arithmetic, which includes both addition and multiplication operations. Unlike Peano arithmetic, Presburger arithmetic is a decidable theory. This means it is possible to algorithmically determine, for any sentence in the language of Presburger arithmetic, whether that sentence is provable from the axioms of Presburger arithmetic. The asymptotic running-time computational complexity of this decision problem is doubly exponential, however, as shown by Fischer & Rabin (1974).

## 54.1 Overview

The language of Presburger arithmetic contains constants 0 and 1 and a binary function +, interpreted as addition. In this language, the axioms of Presburger arithmetic are the universal closures of the following:

1. $\neg(0 = x + 1)$

2. $x + 1 = y + 1 \rightarrow x = y$

3. $x + 0 = x$

4. $x + (y + 1) = (x + y) + 1$

5. Let $P(x)$ be a first-order formula in the language of Presburger arithmetic with a free variable $x$ (and possibly other free variables). Then the following formula is an axiom:

$$(P(0) \wedge \forall x (P(x) \rightarrow P(x + 1))) \rightarrow \forall y\, P(y).$$

(5) is an axiom schema of induction, representing infinitely many axioms. Since the axioms in the schema in (5) cannot be replaced by any finite number of axioms, Presburger arithmetic is not finitely axiomatizable in first-order logic.

Presburger arithmetic cannot formalize concepts such as divisibility or prime number. Generally, any number concept leading to multiplication cannot be defined in Presburger arithmetic, since that leads to incompleteness and undecidability. However, it can formulate individual instances of divisibility; for example, it proves "for all $x$, there exists $y$ : $(y + y = x)$ $\vee$ $(y + y + 1 = x)$". This states that every number is either even or odd.

## 54.2 Properties

Mojżesz Presburger proved Presburger arithmetic to be:

- consistent: There is no statement in Presburger arithmetic which can be deduced from the axioms such that its negation can also be deduced.

- complete: For each statement in the language of Presburger arithmetic, either it is possible to deduce it from the axioms or it is possible to deduce its negation.

- decidable: There exists an algorithm which decides whether any given statement in Presburger arithmetic is a theorem or a nontheorem.

The decidability of Presburger arithmetic can be shown using quantifier elimination, supplemented by reasoning about arithmetical congruence (Enderton 2001, p. 188).

Peano arithmetic, which is Presburger arithmetic augmented with multiplication, is not decidable, as a consequence of the negative answer to the Entscheidungsproblem. By Gödel's incompleteness theorem, Peano arithmetic is incomplete and its consistency is not internally provable (but see Gentzen's consistency proof).

The decision problem for Presburger arithmetic is an interesting example in computational complexity theory and computation. Let $n$ be the length of a statement in Presburger arithmetic. Then Fischer and Rabin (1974) proved that any decision algorithm for Presburger arithmetic has a worst-case runtime of at least $2^{2^{cn}}$, for some constant $c>0$. Hence, the decision problem for Presburger arithmetic is an example of a decision problem that has been proved to require more than exponential run time. Fischer and Rabin also proved that for any reasonable axiomatization (defined precisely in their paper), there exist theorems of length $n$ which have doubly exponential length proofs. Intuitively, this means there are computational limits on what can be proven by computer programs. Fischer and Rabin's work also implies that Presburger arithmetic can be used to define formulas which correctly calculate any algorithm as long as the inputs are less than relatively large bounds. The bounds can be increased, but only by using new formulas. On the other hand, a triply exponential upper bound on a decision procedure for Presburger Arithmetic was proved by Oppen (1978). A more tight complexity bound was shown using alternating complexity classes by Berman (1980).

## 54.3 Applications

Because Presburger arithmetic is decidable, automatic theorem provers for Presburger arithmetic exist. For example, the Coq proof assistant system features the tactic omega for Presburger arithmetic and the Isabelle (proof assistant) contains a verified quantifier elimination procedure by Nipkow (2010). The double exponential complexity of the theory makes it infeasible to use the theorem provers on complicated formulas, but this behavior occurs only in the presence of nested quantifiers: Oppen and Nelson (1980) describe an automatic theorem prover which uses the simplex algorithm on an extended Presburger arithmetic without nested quantifiers to prove some of the instances of quantifier-free Presburger arithmetic formulas. More recent Satisfiability Modulo Theories solvers use complete integer programming techniques to handle quantifier-free fragment of Presburger arithmetic theory (King, Barrett, Tinelli 2014).

Presburger arithmetic can be extended to include multiplication by constants, since multiplication is repeated addition. Most array subscript calculations then fall within the region of decidable problems. This approach is the basis of at least five proof-of-correctness systems for computer programs, beginning with the Stanford Pascal Verifier in the late 1970s and continuing through to Microsoft's Spec# system of 2005.

## 54.4 See also

- Robinson arithmetic

## 54.5   References

- Cooper, D. C., 1972, "Theorem Proving in Arithmetic without Multiplication" in B. Meltzer and D. Michie, eds., *Machine Intelligence Vol. 7.* Edinburgh University Press: 91–99.

- Enderton, Herbert (2001). *A mathematical introduction to logic* (2nd ed.). Boston, MA: Academic Press. ISBN 978-0-12-238452-3.

- Ferrante, Jeanne, and Charles W. Rackoff, 1979. *The Computational Complexity of Logical Theories.* Lecture Notes in Mathematics 718. Springer-Verlag.

- Fischer, Michael J.; Rabin, Michael O. (1974). "Super-Exponential Complexity of Presburger Arithmetic". *Proceedings of the SIAM-AMS Symposium in Applied Mathematics* **7**: 27–41.

- G. Nelson and D. C. Oppen (Apr 1978). "A simplifier based on efficient decision algorithms". *Proc. 5th ACM SIGACT-SIGPLAN symposium on Principles of programming languages*: 141–150. doi:10.1145/512760.512775.

- Mojżesz Presburger, 1929, "Über die Vollständigkeit eines gewissen Systems der Arithmetik ganzer Zahlen, in welchem die Addition als einzige Operation hervortritt" in *Comptes Rendus du I congrès de Mathématiciens des Pays Slaves.* Warszawa: 92–101. — see Stansifer (1984)for an English translation

- Ryan Stansifer (Sep 1984). Presburger's Article on Integer Arithmetic: Remarks and Translation (PDF) (Technical Report). TR84-639. Ithaca/NY: Dept. of Computer Science, Cornell University.

- William Pugh, 1991, "The Omega test: a fast and practical integer programming algorithm for dependence analysis,".

- Reddy, C. R., and D. W. Loveland, 1978, "Presburger Arithmetic with Bounded Quantifier Alternation." *ACM Symposium on Theory of Computing*: 320–325.

- Young, P., 1985, "Gödel theorems, exponential difficulty and undecidability of arithmetic theories: an exposition" in A. Nerode and R. Shore, Recursion Theory, American Mathematical Society: 503-522.

- Oppen, Derek C. (1978). "A $2^{2^{2^{pn}}}$ Upper Bound on the Complexity of Presburger Arithmetic" (PDF). *J. Comput. Syst. Sci.* **16** (3): 323–332. doi:10.1016/0022-0000(78)90021-1.

- Berman, L. (1980). "The Complexity of Logical Theories".*Theoretical Computer Science***11**(1): 71–77. doi:10.10 3975(80)90037-7.

- Nipkow, T (2010). "Linear Quantifier Elimination".*Journal of Automated Reasoning***45**(2): 189–212. doi:10.10 010-9183-0.

- King, Tim; Barrett, Clark W.; Tinelli, Cesare (2014). "Leveraging linear and mixed integer programming for SMT". *FMCAD* **2014**: 139–146. doi:10.1109/FMCAD.2014.6987606.

## 54.6   External links

- online prover A Java applet proves or disproves arbitrary formulas of Presburger arithmetic (In German)

- A complete Theorem Prover for Presburger Arithmetic by Philipp Rümmer

# Chapter 55

# Primitive recursive functional

In mathematical logic, the **primitive recursive functionals** are a generalization of primitive recursive functions into higher type theory. They consist of a collection of functions in all pure finite types.

The primitive recursive functionals are important in proof theory and constructive mathematics They are a central part of the Dialectica interpretation of intuitionistic arithmetic developed by Kurt Gödel.

In recursion theory, the primitive recursive functionals are an example of higher-type computability, as primitive recursive functions are examples of Turing computability.

## 55.1  Background

Every primitive recursive functional has a type, which tells what kind of inputs it takes and what kind of output it produces. An object of type 0 is simply a natural number; it can also be viewed as a constant function that takes no input and returns an output in the set **N** of natural numbers.

For any two types $\sigma$ and $\tau$, the type $\sigma \to \tau$ represents a function that takes an input of type $\sigma$ and returns an output of type $\tau$. Thus the function $f(n) = n+1$ is of type $0 \to 0$. The types $(0 \to 0) \to 0$ and $0 \to (0 \to 0)$ are different; by convention, the notation $0 \to 0 \to 0$ refers to $0 \to (0 \to 0)$. In the jargon of type theory, objects of type $0 \to 0$ are called *functions* and objects that take inputs of type other than 0 are called *functionals*.

For any two types $\sigma$ and $\tau$, the type $\sigma \times \tau$ represents an ordered pair, the first element of which has type $\sigma$ and the second element of which has type $\tau$. For example, consider the functional $A$ takes as inputs a function $f$ from **N** to **N**, and a natural number $n$, and returns $f(n)$. Then $A$ has type $(0 \times (0 \to 0)) \to 0$. This type can also be written as $0 \to (0 \to 0) \to 0$, by Currying.

The set of (pure) *finite types* is the smallest collection of types that includes 0 and is closed under the operations of $\times$ and $\to$. A superscript is used to indicate that a variable $x^\tau$ is assumed to have a certain type $\tau$; the superscript may be omitted when the type is clear from context.

## 55.2  Definition

The primitive recursive functionals are the smallest collection of objects of finite type such that:

- The constant function $f(n) = 0$ is a primitive recursive functional

- The successor function $g(n) = n + 1$ is a primitive recursive functional

- For any type $\sigma \times \tau$, the functional $K(x^\sigma, y^\tau) = x$ is a primitive recursive functional

- For any types $\rho$, $\sigma$, $\tau$, the functional

$$S(r^{\rho \to \sigma \to \tau}, s^{\rho \to \sigma}, t^{\rho}) = (r(t))(s(t))$$

  is a primitive recursive functional

- For any type $\tau$, and $f$ of type $\tau$, and any $g$ of type $0 \to \tau \to \tau$, the functional $R(f,g)^{0 \to \tau}$ defined recursively as

$$R(f,g)(0) = f,$$
$$R(f,g)(n+1) = g(n, R(f,g)(n))$$

  is a primitive recursive functional

## 55.3   References

- Jeremy Avigad and Solomon Feferman (1999). *Gödel's functional ("Dialectica") interpretation* (PDF). in S. Buss ed., The Handbook of Proof Theory, North-Holland. pp. 337–405.

# Chapter 56

# Proof (truth)

For other uses, see Proof.

A **proof** is sufficient evidence or an argument for the truth of a proposition.[1][2][3][4]

The concept is applied in a variety of disciplines, with both the nature of the evidence or justification and the criteria for sufficiency being area-dependent. In the area of oral and written communication such as conversation, dialog, rhetoric, etc., a proof is a persuasive perlocutionary speech act, which demonstrates the truth of a proposition.[5] In any area of mathematics defined by its assumptions or axioms, a proof is an argument establishing a theorem of that area via accepted rules of inference starting from those axioms and other previously established theorems.[6] The subject of logic, in particular proof theory, formalizes and studies the notion of formal proof.[7] In the areas of epistemology and theology, the notion of justification plays approximately the role of proof,[8] while in jurisprudence the corresponding term is evidence,[9] with burden of proof as a concept common to both philosophy and law.

## 56.1 On proof

In most disciplines, evidence is required to prove something. Evidence is drawn from experience of the world around us, with science obtaining its evidence from nature,[10] law obtaining its evidence from witnesses and forensic investigation,[11] and so on. A notable exception is mathematics, whose proofs are drawn from a mathematical world begun with axioms and further developed and enriched by theorems proved earlier.

Exactly what evidence is sufficient to prove something is also strongly area-dependent, usually with no absolute threshold of sufficiency at which evidence becomes proof.[12][13][14] In law, the same evidence that may convince one jury may not persuade another. Formal proof provides the main exception, where the criteria for proofhood are ironclad and it is impermissible to defend any step in the reasoning as "obvious";[15] for a well-formed formula to qualify as part of a formal proof, it must be the result of applying a rule of the deductive apparatus of some formal system to the previous well-formed formulae in the proof sequence.[16]

Proofs have been presented since antiquity. Aristotle used the observation that patterns of nature never display the machine-like uniformity of determinism as proof that chance is an inherent part of nature.[17] On the other hand, Thomas Aquinas used the observation of the existence of rich patterns in nature as proof that nature is *not* ruled by chance.[18]

Proofs need not be verbal. Before Galileo, people took the apparent motion of the Sun across the sky as proof that the Sun went round the Earth.[19] Suitably incriminating evidence left at the scene of a crime may serve as proof of the identity of the perpetrator. Conversely, a verbal entity need not assert a proposition to constitute a proof of that proposition. For example, a signature constitutes direct proof of authorship; less directly, handwriting analysis may be submitted as proof of authorship of a document.[20] Privileged information in a document can serve as proof that the document's author had access to that information; such access might in turn establish the location of the author at certain time, which might then provide the author with an alibi.

## 56.2 See also

- Mathematical proof

- Proof theory

- Proof of concept

- Provability logic

- Evidence, information which tends to determine or demonstrate the truth of a proposition

- Proof procedure

- Proof complexity

- Standard of proof

## 56.3 References

[1] *Proof and other dilemmas: mathematics and philosophy* by Bonnie Gold, Roger A. Simons 2008 ISBN 0883855674 pages 12–20

[2] *Philosophical Papers, Volume 2* by Imre Lakatos, John Worrall, Gregory Currie, ISBN Philosophical Papers, Volume 2 by Imre Lakatos, John Worrall, Gregory Currie 1980 ISBN 0521280303 pages 60–63

[3] *Evidence, proof, and facts: a book of sources* by Peter Murphy 2003 ISBN 0199261954 pages 1–2

[4] *Logic in Theology – And Other Essays* by Isaac Taylor 2010 ISBN 1445530139 pages 5–15

[5] John Langshaw Austin: *How to Do Things With Words.* Cambridge (Mass.) 1962 – Paperback: Harvard University Press, 2nd edition, 2005, ISBN 0-674-41152-8.

[6] Cupillari, Antonella. The Nuts and Bolts of Proofs. Academic Press, 2001. Page 3.

[7] Alfred Tarski, Introduction to Logic and to the Methodology of the Deductive Sciences (ed. Jan Tarski). 4th Edition. Oxford Logic Guides, No. 24. New York and Oxford: Oxford University Press, 1994, xxiv + 229 pp. ISBN 0-19-504472-X

[8] http://plato.stanford.edu/entries/justep-foundational/

[9] http://dictionary.reference.com/browse/proof

[10] Reference Manual on Scientific Evidence, 2nd Ed. (2000), p. 71. Accessed May 13, 2007.

[11] John Henry Wigmore, *A Treatise on the System of Evidence in Trials at Common Law,* 2nd ed., Little, Brown, and Co., Boston, 1915

[12] Simon, Rita James, and Mahan, Linda. (1971). "Quantifying Burdens of Proof—A View from the Bench, the Jury, and the Classroom". *Law and Society Review* **5** (3): 319–330. doi:10.2307/3052837. JSTOR 3052837.

[13] Katie Evans, David Osthus, Ryan G. Spurrier. "Distributions of Interest for Quantifying Reasonable Doubt and Their Applications" (PDF). Retrieved 2007-01-14.

[14] The Principle of Sufficient Reason: A Reassessment by Alexander R. Pruss

[15] A. S. Troelstra, H. Schwichtenberg (1996). *Basic Proof Theory.* In series *Cambridge Tracts in Theoretical Computer Science*, Cambridge University Press, ISBN 0-521-77911-1.

[16] Hunter, Geoffrey, Metalogic: An Introduction to the Metatheory of Standard First-Order Logic, University of California Pres, 1971

[17] *Aristotle's Physics: a Guided Study*, Joe Sachs, 1995 ISBN 0813521920 p. 70

[18] *The treatise on the divine nature: Summa theologiae I*, 1–13, by Saint Thomas Aquinas, Brian J. Shanley, 2006 ISBN 0872208052 p. 198

[19] Thomas S. Kuhn, The Copernican Revolution, pp. 5–20

[20] *Trial tactics by Stephen A. Saltzburg, 2007 ISBN 159031767X page 47*

# Chapter 57

# Proof calculus

In mathematical logic, a **proof calculus** corresponds to a family of formal systems that use a common style of formal inference for its inference rules. The specific inference rules of a member of such a family characterize the theory of a logic.

Usually a given proof calculus encompasses more than a single particular formal system, since many proof calculi are under-determining and can be used for radically different logics. For example, a paradigmatic case is the sequent calculus, which can be used to express the consequence relations of both intuitionistic logic and relevance logic. Thus, loosely speaking, a proof calculus is a template or design pattern, characterized by a certain style of formal inference, that may be specialized to produce specific formal systems, namely by specifying the actual inference rules for such a system. There is no consensus among logicians on how best to define the term.

## 57.1   Examples of proof calculi

The most widely known proof calculi are those classical calculi that are still in widespread use:

- The class of Hilbert systems, of which the most famous example is the 1928 Hilbert-Ackermann system of first-order logic;

- Gerhard Gentzen's calculus of natural deduction, which is the first formalism of structural proof theory, and which is the cornerstone of the formulae-as-types correspondence relating logic to functional programming;

- Gentzen's sequent calculus, which is the most studied formalism of structural proof theory.

Many other proof calculi were, or might have been, seminal, but are not widely used today.

- Aristotle's syllogistic calculus, presented in the *Organon*, readily admits formalisation. There is still some modern interest in syllogistic, carried out under the aegis of term logic.

- Gottlob Frege's two-dimensional notation of the *Begriffsschrift* is usually regarded as introducing the modern concept of quantifier to logic.

- C.S. Peirce's existential graph might easily have been seminal, had history worked out differently.

Modern research in logic teems with rival proof calculi:

- Several systems have been proposed which replace the usual textual syntax with some graphical syntax. Proof nets and cirquent calculus are among such systems.

- Recently, many logicians interested in structural proof theory have proposed calculi with deep inference, for instance display logic, hypersequents, the calculus of structures, and bunched implication.

## 57.2  See also

- propositional proof system
- Proof nets

# Chapter 58

# Proof compression

In proof theory, an area of mathematical logic, **proof compression** is the problem of algorithmically compressing formal proofs. The developed algorithms can be used to improve the proofs generated by automated theorem proving tools such as sat-solvers, SMT-solvers, first-order theorem provers and proof assistants.

## 58.1 Problem Representation

In propositional logic a resolution proof of a clause $\kappa$ from a set of clauses C is a directed acyclic graph (DAG): the input nodes are axiom inferences (without premises) whose conclusions are elements of C, the resolvent nodes are resolution inferences, and the proof has a node with conclusion $\kappa$ .[1]

The DAG contains an edge from a node $\eta_1$ to a node $\eta_2$ if and only if a premise of $\eta_1$ is the conclusion of $\eta_2$ . In this case, $\eta_1$ is a child of $\eta_2$ , and $\eta_2$ is a parent of $\eta_1$ . A node with no children is a root.

A proof compression algorithm will try to create a new DAG with fewer nodes that represents a valid proof of $\kappa$ or, in some cases, a valid proof of a subset of $\kappa$ .

### 58.1.1 A simple example

Let's take a resolution proof for the clause $\{a, b, c\}$ from the set of clauses

$$\{\eta_1 : \{a, b, p\}, \eta_2 : \{c, \neg p\}\} \qquad \frac{\eta_1 : a, b, p \qquad \eta_2 : c, \neg p}{\eta_3 : a, b, c} p$$

Here we can see:

- $\eta_1$ and $\eta_2$ are input nodes.

- The node $\eta_3$ has a pivot $p$ ,

  - left resolved literal $p$

  - right resolved literal $\neg p$

- $\eta_3$ conclusion is the clause $\{a, b, c\}$

- $\eta_3$ premises are the conclusion of nodes $\eta_1$ and $\eta_2$ (its parents)

- The DAG would be

$$\eta_1 \qquad \eta_2$$
$$\searrow \nearrow$$
$$\eta_3$$

- $\eta_1$ and $\eta_2$ are parents of $\eta_3$

- $\eta_3$ is a child of $\eta_1$ and $\eta_2$

- $\eta_3$ is a root of the proof

A (resolution) refutation of C is a resolution proof of $\perp$ from C. It is a common that given a node $\eta$, to refer to the clause $\eta$ or $\eta$'s clause meaning the conclusion clause of $\eta$, and (sub)proof $\eta$ meaning the (sub)proof having $\eta$ as its only root.

In some works it can be found an algebraic representation of a resolution inference. The resolvent of $\kappa_1$ and $\kappa_2$ with pivot $p$ can be denoted as $\kappa_1 \odot_p \kappa_2$. When the pivot is uniquely defined or irrelevant, we omit it and write simply $\kappa_1 \odot \kappa_2$. In this way, the set of clauses can be seen as an algebra with a commutative operator; and terms in the corresponding term algebra denote resolution proofs in a notation style that is more compact and more convenient for describing resolution proofs than the usual graph notation.

In our last example the notation of the DAG would be $\{a, b, p\} \odot_p \{c, \neg p\}$ or simply $\{a, b, p\} \odot \{c, \neg p\}$.

We can identify $\underbrace{\overbrace{\{a, b, p\}}^{\eta_1} \odot \overbrace{\{c, \neg p\}}^{\eta_2}}_{\eta_3}$

## 58.2 Compression algorithms

Algorithms for compression of sequent calculus proofs include Cut-introduction and Cut-elimination.

Algorithms for compression of propositional resolution proofs include RecycleUnits,[2] RecyclePivots,[3]Recycle Pivots, With Intersection LowerUnits, [5] LowerUnivalents, [6] Split, [7] Reduce&Reconstruct,[8] and Subsumption.

## 58.3 Notes

[1] Fontaine, Pascal; Merz, Stephan; Woltzenlogel Paleo, Bruno. *Compression of Propositional Resolution Proofs via Partial Regularization.* 23rd International Conference on Automated Deduction, 2011.

[2] Bar-Ilan, O.; Fuhrmann, O.; Hoory, S. ; Shacham, O. ; Strichman, O. *Linear-time Reductions of Resolution Proofs.* Hardware and Software: Verification and Testing, p. 114–128, Springer, 2011.

[3] Bar-Ilan, O.; Fuhrmann, O.; Hoory, S. ; Shacham, O. ; Strichman, O. *Linear-time Reductions of Resolution Proofs.* Hardware and Software: Verification and Testing, p. 114–128, Springer, 2011.

[4] Fontaine, Pascal; Merz, Stephan; Woltzenlogel Paleo, Bruno. *Compression of Propositional Resolution Proofs via Partial Regularization.* 23rd International Conference on Automated Deduction, 2011.

[5] Fontaine, Pascal; Merz, Stephan; Woltzenlogel Paleo, Bruno. *Compression of Propositional Resolution Proofs via Partial Regularization.* 23rd International Conference on Automated Deduction, 2011.

[6] https://github.com/Paradoxika/Skeptik/tree/develop/doc/papers/LUniv

[7] Cotton, Scott. "Two Techniques for Minimizing Resolution Proofs". 13th International Conference on Theory and Applications of Satisfiability Testing, 2010.

[8] Simone, S.F. ; Brutomesso, R. ; Sharygina, N. "An Efficient and Flexible Approach to Resolution Proof Reduction". 6th Haifa Verification Conference, 2010.

# Chapter 59

# Proof mining

In proof theory, a branch of mathematical logic, **proof mining** (or **unwinding**) is a research program that analyzes formalized proofs, especially in analysis, to obtain explicit bounds or rates of convergence from proofs that, when expressed in natural language, appear to be nonconstructive.[1] This research has led to improved results in analysis obtained from the analysis of classical proofs.

## 59.1 References

[1] Ulrich Kohlenbach (2008). *Applied Proof Theory: Proof Interpretations and Their Use in Mathematics*. Springer Verlag, Berlin. pp. 1–536.

- Ulrich Kohlenbach and Paulo Oliva, "Proof Mining: A systematic way of analysing proofs in mathematics", *Proc. Steklov Inst. Math*, 242:136–164, 2003

- Paulo Oliva, "Proof Mining in Subsystems of Analysis", BRICS PhD thesis citeseer

# Chapter 60

# Proof net

In proof theory, **proof nets** are a geometrical method of representing proofs that eliminates two forms of *bureaucracy* that differentiates proofs: (A) irrelevant syntactical features of regular proof calculi such as the natural deduction calculus and the sequent calculus, and (B) the order of rules applied in a derivation. In this way, the formal properties of proof identity correspond more closely to the intuitively desirable properties. Proof nets were introduced by Jean-Yves Girard.

For instance, these two linear logic proofs are "morally" identical:

And their corresponding nets will be the same.

## 60.1 Correctness criteria

Several correctness criteria are known to check if a sequential proof structure (i.e. something which seems to be a proof net) is actually a concrete proof structure (i.e. something which encodes a valid derivation in linear logic). The first such criterion is the long-trip criterion[1] which was described by Jean-Yves Girard.

## 60.2 See also

- Linear logic

- Ludics

- Geometry of interaction

- Coherent space

- Deep inference

- Interaction nets

## 60.3 References

[1] Girard, Jean-Yves. *Linear logic*, Theoretical Computer Science, Vol 50, no 1, pp. 1–102, 1987

## 60.4 Sources

- *Proofs and Types*. Girard J-Y, Lafont Y, and Taylor P. Cambridge Press, 1989.

- Roberto Di Cosmo and Vincent Danos, The Linear Logic Primer

- Sean A. Fulop, A survey of proof nets and matrices for substructural logics

# Chapter 61

# Proof procedure

In logic, and in particular proof theory, a **proof procedure** for a given logic is a systematic method for producing proofs in some proof calculus of (provable) statements.

## 61.1 Types of proof calculi used

There are several types of proof calculi. The most popular are natural deduction, sequent calculi (i.e., Gentzen type systems), Hilbert systems, and semantic tableaux or trees. A given proof procedure will target a specific proof calculus, but can often be reformulated so as to produce proofs in other proof styles.

## 61.2 Completeness

A proof procedure for a logic is *complete* if it produces a proof for each provable statement. The theorems of logical systems are typically recursively enumerable, which implies the existence of a complete but extremely inefficient proof procedure; however, a proof procedure is only of interest if it is reasonably efficient.

Faced with an unprovable statement, a complete proof procedure may sometimes succeed in detecting and signalling its unprovability. In the general case, where provability is a semidecidable property, this is not possible, and instead the procedure will diverge (not terminate).

## 61.3 See also

- Automated theorem proving
- Proof complexity
- Proof tableaux
- Deductive system
- Proof (truth)

## 61.4 References

- W. Quine 1982 (1950). *Methods of Logic*. Harvard Univ. Press.

# Chapter 62

# Proof-theoretic semantics

**Proof-theoretic semantics** is an approach to the semantics of logic that attempts to locate the meaning of propositions and logical connectives not in terms of interpretations, as in Tarskian approaches to semantics, but in the role that the proposition or logical connective plays within the system of inference.

Gerhard Gentzen is the founder of proof-theoretic semantics, providing the formal basis for it in his account of cut-elimination for the sequent calculus, and some provocative philosophical remarks about locating the meaning of logical connectives in their introduction rules within natural deduction. The history of proof-theoretic semantics since then has been devoted to exploring the consequences of these ideas.

Dag Prawitz extended Gentzen's notion of analytic proof to natural deduction, and suggested that the value of a proof in natural deduction may be understood as its normal form. This idea lies at the basis of the Curry–Howard isomorphism, and of intuitionistic type theory. His inversion principle lies at the heart of most modern accounts of proof-theoretic semantics.

Michael Dummett introduced the very fundamental idea of logical harmony, building on a suggestion of Nuel Belnap. In brief, a language, which is understood to be associated with certain patterns of inference, has logical harmony if it is always possible to recover analytic proofs from arbitrary demonstrations, as can be shown for the sequent calculus by means of cut-elimination theorems and for natural deduction by means of normalisation theorems. A language that lacks logical harmony will suffer from the existence of incoherent forms of inference: it will likely be inconsistent.

## 62.1 References

- Proof-Theoretic Semantics, at the Stanford Encyclopedia of Philosophy

- Logical Consequence, Deductive-Theoretic Conceptions, at the Internet Encyclopedia of Philosophy.

- Nissim Francez, "On a Distinction of Two Facets of Meaning and its Role in Proof-theoretic Semantics", *Logica Universalis* 9, 2015. doi:10.1007/s11787-015-0118-8

## 62.2 See also

- Inferential role semantics

- Truth-conditional semantics

## 62.3 External links

- Arché Bibliography on Proof-Theoretic Semantics.

# Chapter 63

# Provability logic

**Provability logic** is a modal logic, in which the box (or "necessity") operator is interpreted as 'it is provable that'. The point is to capture the notion of a proof predicate of a reasonably rich formal theory, such as Peano arithmetic.

## 63.1 Examples

There are a number of provability logics, some of which are covered in the literature mentioned in the References section. The basic system is generally referred to as GL (for Gödel-Löb) or L or K4W. It can be obtained by adding the modal version of Löb's theorem to the logic K (or K4).

Namely, the **axioms** of GL are all tautologies of classical propositional logic plus all formulas of one of the following forms:

- **Distribution Axiom**: $\Box(p \to q) \to (\Box p \to \Box q)$;

- **Löb's Axiom**: $\Box(\Box p \to p) \to \Box p$.

And the **rules of inference** are:

- **Modus Ponens**: From $p \to q$ and $p$ conclude $q$;

- **Necessitation**: From $p$ conclude $\Box p$.

## 63.2 History

The GL model was pioneered by Robert M. Solovay in 1976. Since then until his death in 1996 the prime inspirer of the field was George Boolos. Significant contributions to the field have been made by Sergei N. Artemov, Lev Beklemishev, Giorgi Japaridze, Dick de Jongh, Franco Montagna, Vladimir Shavrukov, Albert Visser and others.

## 63.3 Generalizations

Interpretability logics and Japaridze's Polymodal Logic present natural extensions of provability logic.

## 63.4 See also

- Hilbert–Bernays provability conditions

- Interpretability logic

- Kripke semantics

- Japaridze's Polymodal Logic

## 63.5 References

- George Boolos, **The Logic of Provability**. Cambridge University Press, 1993.

- Giorgi Japaridze and Dick de Jongh, *The logic of provability*. In: **Handbook of Proof Theory**, S. Buss, ed. Elsevier, 1998, pp. 475-546.

- Sergei N. Artemov and Lev Beklemishev, *Provability logic*. In: **Handbook of Philosophical Logic**, D. Gabbay and F. Guenthner, eds., vol. 13, 2nd ed., pp. 189-360. Springer, 2005.

- Per Lindström, *Provability logic - a short introduction*. Theoria 62 (1996), pp. 19-61.

- Craig Smoryński, **Self-reference and modal logic**. Springer, Berlin, 1985.

- Robert M. Solovay, ``Provability Interpretations of Modal Logic``, Israel Journal of Mathematics, Vol. 25 (1976): 287-304.

- Provability logic, from the Stanford Encyclopedia of Philosophy.

# Chapter 64

# $\Psi_0(\Omega\omega)$

The correct title of this article is $\Psi_0(\Omega\omega)$. It appears incorrectly here because of technical restrictions.

In mathematics, $\Psi_0(\Omega\omega)$ is a large countable ordinal that is used to measure the proof-theoretic strength of some mathematical systems. In particular, it is the proof theoretic ordinal of the subsystem $\Pi_1^1$-$CA_0$ of second-order arithmetic; this is one of the "big five" subsystems studied in reverse mathematics (Simpson 1999).

## 64.1 Definition

Main article: Ordinal collapsing function

- $\Omega_0 = 0$, and $\Omega_n = \aleph_n$ for $n > 0$.

- $C_i(\alpha)$ is the smallest set of ordinals that contains $\Omega_n$ for $n$ finite, and contains all ordinals less than $\Omega_i$, and is closed under ordinal addition and exponentiation, and contains $\Psi_j(\xi)$ if $j \geq i$ and $\xi \in C_i(\alpha)$ and $\xi < \alpha$.

- $\Psi_i(\alpha)$ is the smallest ordinal not in $C_i(\alpha)$

## 64.2 References

- G. Takeuti, *Proof theory*, 2nd edition 1987 ISBN 0-444-10492-5

- K. Schütte, *Proof theory*, Springer 1977 ISBN 0-387-07911-4

- Simpson, Stephen G. (2009), *Subsystems of second order arithmetic*, Perspectives in Logic (2nd ed.), Cambridge University Press, ISBN 978-0-521-88439-6, MR 2517689

# Chapter 65

# Pure type system

In the branches of mathematical logic known as proof theory and type theory, a **pure type system** (**PTS**), previously known as a **generalized type system** (**GTS**), is a form of typed lambda calculus that allows an arbitrary number of sorts and dependencies between any of these. The framework can be seen as a generalisation of Barendregt's lambda cube, in the sense that all corners of the cube can be represented as instances of a PTS with just two sorts.[1][2] In fact Barendregt (1991) framed his cube in this setting.[3] Pure type systems may obscure the distinction between *types* and *terms* and collapse the type hierarchy, as is the case with the calculus of constructions, but this is not generally the case, e.g. the simply typed lambda calculus allows only terms to depend on types.

Pure type systems were independently introduced by Stefano Berardi (1988) and Jan Terlouw (1989).[1][2] Barendregt discussed them at length in his subsequent papers.[4] In his PhD thesis,[5] Berardi defined a cube of constructive logics akin to the lambda cube (these specifications are non-dependent). A modification of this cube was later called the L-cube by Geuvers, who in his PhD thesis extended the Curry–Howard correspondence to this setting.[6] Based on these ideas, Barthe and others defined **classical pure type systems** (**CPTS**) by adding a double negation operator.[7] Similarly, in 1998, Tijn Borghuis introduced **modal pure type systems** (**MPTS**).[8] Roorda has discussed the application of pure type systems to functional programming; and Roorda and Jeuring have proposed a programming language based on pure type systems.[9]

The systems from the lambda cube are all known to be strongly normalizing. Pure type systems in general need not be, for example U from **Girard's paradox** is not. (Roughly speaking, Girard found pure systems in which one can express the sentence "the types form a type".) Furthermore, all known examples of pure type systems that are not strongly normalizing are not even (weakly) normalizing: they contain expressions that do not have normal forms, just like the untyped lambda calculus. It is a major open problem in the field whether this is always the case, i.e. whether a (weakly) normalizing PTS always has the strong normalization property. This is known as the **Barendregt–Geuvers–Klop conjecture**[10] (named after Henk Barendregt, Herman Geuvers, and Jan Willem Klop).

## 65.1 Implementations

The following programming languages have pure type systems:

- SAGE

- Yarrow

- Henk 2000

## 65.2 See also

- Lambda-mu calculus uses a different approach to control than CPTS.

## 65.3 Notes

[1] Pierce, Benjamin (2002). *Types and Programming Languages*. MIT Press. p. 466. ISBN 0-262-16209-1.

[2] Kamareddine, Fairouz D.; Laan, Twan; Nederpelt, Rob P. (2004). "Section 4c: Pure type systems". *A modern perspective on type theory: from its origins until today*. Springer. p. 116. ISBN 1-4020-2334-0.

[3] Barendregt, H. P. (1991). Introduction to generalized type systems "Introduction to generalized type systems" (PDF). *Journal of Functional Programming* **1** (2): 125–154.

[4] H. Barendregt (1992). "Lambda calculi with types". In S. Abramsky, D. Gabbay and T. Maibaum. *Handbook of Logic in Computer Science*. Oxford Science Publications.

[5] Berardi, S. (1990). *Type dependence and Constructive Mathematics* (PhD thesis). University of Torino.

[6] Geuvers, H. (1993). *Logics and Type Systems* (PhD thesis). University of Nijmegen. CiteSeerX: 10.1.1.56.7045.

[7] G. Barthe; J. Hatcliff; M. H. Sørensen (1997). "A Notion of Classical Pure Type System". *Electronic Notes in Theoretical Computer Science* **6**: 4–59. doi:10.1016/S1571-0661(05)80170-7. CiteSeerX: 10.1.1.32.1371.

[8] Borghuis, Tijn (1998). "Modal Pure Type Systems".*Journal of Logic, Language and Information***7**(3): 265–296. doi:10.1023/

[9] Jan-Willem Roorda; Johan Jeuring. "Pure Type Systems for Functional Programming". Roorda's masters' thesis (linked from the cited page) also contains a general introduction to pure type systems.

[10] Sørensen, Morten Heine; Urzyczyn, Paweł (2006). "Pure type systems and the lambda cube". *Lectures on the Curry–Howard isomorphism*. Elsevier. p. 358. ISBN 0-444-52077-5.

## 65.4 References

- Morten Heine Sørensen, Paweł Urzyczyn, *Lectures on the Curry–Howard isomorphism*, Elsevier, 2006, ISBN 0-444-52077-5, chapter 14, "Pure type systems and the lambda cube."

- Berardi, Stefano. *Towards a mathematical analysis of the Coquand–Huet calculus of constructions and the other systems in Barendregt's cube*. Technical report, Department of Computer Science, CMU, and Dipartimento Matematica, Universita di Torino, 1988.

- Terlouw, J. (in Dutch) *Een nadere bewijstheoretische analyse van GSTTs*. Manuscript, University of Nijmegen, Netherlands, 1989.

## 65.5 Further reading

- David A. Schmidt, *The structure of typed programming languages*, MIT Press, 1994, ISBN 0-262-19349-3, section 8.3, "Generalized Type Systems"

## 65.6 External links

- Pure type system in *nLab*
- Pure Type Systems overview by Roger Bishop Jones

# Chapter 66

# Realizability

In mathematical logic, **realizability** is a collection of methods in proof theory used to study constructive proofs and extract additional information from them.[1] Formulas from a formal theory are "realized" by objects, known as "realizers", in a way that knowledge of the realizer gives knowledge about the truth of the formula. There are many variations of realizability; exactly which class of formulas is studied and which objects are realizers differ from one variation to another.

Realizability can be seen as a formalization of the BHK interpretation of intuitionistic logic; in realizability the notion of "proof" (which is left undefined in the BHK interpretation) is replaced with a formal notion of "realizer". Most variants of realizability begin with a theorem that any statement that is provable in the formal system being studied is realizable. The realizer, however, usually gives more information about the formula than a formal proof would directly provide.

Beyond giving insight into intuitionistic provability, realizability can be applied to prove the disjunction and existence properties for intuitionistic theories and to extract programs from proofs, as in proof mining. It is also related to topos theory via the realizability topos.

## 66.1 Example: realizability by numbers

Kleene's original version of realizability uses natural numbers as realizers for formulas in Heyting arithmetic. The following clauses are used to define a relation "$n$ realizes $A$" between natural numbers $n$ and formulas $A$ in the language of Heyting arithmetic. A few pieces of notation are required: first, an ordered pair $(n,m)$ is treated as a single number using a fixed effective pairing function; second, for each natural number $n$, $\varphi n$ is the computable function with index $n$.

- A number $n$ realizes an atomic formula $s=t$ if and only if $s=t$ is true. Thus every number realizes a true equation, and no number realizes a false equation.

- A pair $(n,m)$ realizes a formula $A \wedge B$ if and only if $n$ realizes $A$ and $m$ realizes $B$. Thus a realizer for a conjunction is a pair of realizers for the conjuncts.

- A pair $(n,m)$ realizes a formula $A \vee B$ if and only if the following hold: $n$ is 0 or 1; and if $n$ is 0 then $m$ realizes $A$; and if $n$ is 1 then $m$ realizes $B$. Thus a realizer for a disjunction explicitly picks one of the disjuncts (with $n$) and provides a realizer for it (with $m$).

- A number $n$ realizes a formula $A \rightarrow B$ if and only if, for every $m$ that realizes $A$, $\varphi n(m)$ realizes $B$. Thus a realizer for an implication is a computable function that takes a realizer for the hypothesis and produces a realizer for the conclusion.

- A pair $(n,m)$ realizes a formula $(\exists x)A(x)$ if and only if $m$ is a realizer for $A(n)$. Thus a realizer for an existential formula produces an explicit witness for the quantifier along with a realizer for the formula instantiated with that witness.

- A number $n$ realizes a formula $(\forall x)A(x)$ if and only if, for all $m$, $\varphi n(m)$ is defined and realizes $A(m)$. Thus a realizer for a universal statement is a computable function that produces, for each $m$, a witness for the formula instantiated with $m$.

With this definition, the following theorem is obtained:[2]

Let $A$ be a sentence of Heyting arithmetic (HA). If HA proves $A$ then there is an $n$ such that $n$ realizes $A$.

On the other hand, there are formulas that are realized but which are not provable in HA, a fact first established by Rose.[3] Further analysis of the method can be used to prove that HA has the "disjunction and existence properties":[4]

- If HA proves a sentence $(\exists x)A(x)$, then there is an $n$ such that HA proves $A(n)$

- If HA proves a sentence $A \lor B$, then HA proves $A$ or HA proves $B$.

## 66.2   Later developments

Kreisel introduced **modified realizability**, which uses typed lambda calculus as the language of realizers. Modified realizability is one way to show that Markov's principle is not derivable in intuitionistic logic. On the contrary, it allows to constructively justify the principle of independence of premiss:

$$(A \to \exists x\, P(x)) \to \exists x\, (A \to P(x))$$

Relative realizability[5] is an intuitionist analysis of recursive or recursively enumerable elements of data structures that are not necessarily computable, such as computable operations on all real numbers when reals can be only approximated on digital computer systems.

## 66.3   Applications

Realizability is one of the methods used in proof mining to extract concrete "programs" from seemingly nonconstructive mathematical proof. Program extraction using realizability is implemented in some proof assistants such as Coq.

## 66.4   See also

- Curry–Howard correspondence

- Dialectica interpretation

## 66.5   Notes

[1]  van Oosten 2000

[2]  van Oosten 2000, p. 7

[3]  Rose 1953

[4]  van Oosten 2000, p. 6

[5]  Birkedal 2000

## 66.6 References

- Birkedal, Lars; Jaap van Oosten (2000). *Relative and modified relative realizability.*

- Kreisel G. (1959). "Interpretation of Analysis by Means of Constructive Functionals of Finite Types", in: Constructivity in Mathematics, edited by A. Heyting, North-Holland, pp. 101–128.

- Kleene, S. C. (1945). "On the interpretation of intuitionistic number theory". *Journal of Symbolic Logic* **10** (4): 109–124. doi:10.2307/2269016. JSTOR 2269016.

- Kleene, S. C. (1973). "Realizability: a retrospective survey" from Mathias, A. R. D.; Hartley Rogers (1973). *Cambridge Summer School in Mathematical Logic : held in Cambridge/England, August 1–21, 1971.* Berlin: Springer. ISBN 3-540-05569-X., pp. 95–112.

- van Oosten, Jaap (2000). *Realizability: An Historical Essay.*

- Rose, G. F. (1953). "Propositional calculus and realizability". *Transactions of the American Mathematical Society* **75** (1): 1–19. doi:10.2307/1990776. JSTOR 1990776.

## 66.7 External links

- Realizability Collection of links to recent papers on realizability and related topics.

# Chapter 67

# Redundant proof

In mathematical logic, a **redundant proof** is a proof that has a subset that is a shorter proof of the same result. That is, a proof $\psi$ of $\kappa$ is considered redundant if there exists another proof $\psi'$ of $\kappa'$ such that $\kappa' \subseteq \kappa$ (i.e. $\kappa'$ subsumes $\kappa$ ) and $|\psi'| < |\psi|$ where $|\varphi|$ is the number of nodes in $\varphi$ .[1]

## 67.1   Local redundancy

A proof containing a subproof of the shapes (here omitted pivots indicate that the resolvents must be uniquely defined)

$$(\eta \odot \eta_1) \odot (\eta \odot \eta_2) \text{ or } \eta \odot (\eta_1 \odot (\eta \odot \eta_2))$$

is locally redundant.

Indeed, both of these subproofs can be equivalently replaced by the shorter subproof $\eta \odot (\eta_1 \odot \eta_2)$ . In the case of local redundancy, the pairs of redundant inferences having the same pivot occur close to each other in the proof. However, redundant inferences can also occur far apart in the proof.

The following definition generalizes local redundancy by considering inferences with the same pivot that occur within different contexts. We write $\psi[\eta]$ to denote a proof-context $\psi[-]$ with a single placeholder replaced by the subproof $\eta$ .

## 67.2   Global redundancy

A proof

$$\psi[\psi_1[\eta \odot_p \eta_1] \odot \psi_2[\eta \odot_p \eta_2]] \text{ or } \psi[\psi_1[\eta \odot_p (\eta_1 \odot \psi_2[\eta \odot_p \eta_2])]]$$

is potentially (globally) redundant. Furthermore, it is (globally) redundant if it can be rewritten to one of the following shorter proofs:

$$\psi[\eta \odot_p (\psi_1[\eta_1] \odot \psi_2[\eta_2])] \text{ or } \eta \odot_p \psi[\psi_1[\eta_1] \odot \psi_2[\eta_2]] \text{ or } \psi[\psi_1[\eta_1] \odot \psi_2[\eta_2]].$$

### 67.2.1   Example

The proof

$$\frac{\dfrac{\eta : p, q \quad \eta_1 : \neg p, r}{q, r}p \qquad \eta_3 : \neg q}{\dfrac{r}{\psi : s}}q \qquad \frac{\dfrac{\eta \qquad \eta_2 : \neg p, s, \neg r}{q, s, \neg r}p \qquad \eta_3}{s, \neg r}q \qquad r$$

is locally redundant as it is an instance of the first pattern in the definition $((\eta \odot_p \eta_1) \odot \eta_3) \odot ((\eta \odot_p \eta_2) \odot \eta_3)$.

- The pattern is $\psi[\psi_1[\eta \odot_p \eta_1] \odot \psi_2[\eta \odot_p \eta_2]]$

- $\psi_1[-] = \psi_2[-] = \_ \odot \eta_3$ and $\psi[-] = \_$

But it is not globally redundant because the replacement terms according to the definition contain $\psi_1[\eta_1] \odot \psi_2[\eta_2]$ in all the cases and $\psi_1[\eta_1] \odot \psi_2[\eta_2] = (\eta_1 \odot \eta_3) \odot (\eta_2 \odot \eta_3)$ does not correspond to a proof. In particular, neither $\eta_1$ nor $\eta_2$ can be resolved with $\eta_3$, as they do not contain the literal $q$.

The second pattern of potentially globally redundant proofs appearing in global redundancy definition is related to the well-known notion of regularity. [This link to "regularity" is (obviously) a link to a disambiguation page.] Informally, a proof is irregular if there is a path from a node to the root of the proof such that a literal is used more than once as a pivot in this path.

## 67.3 Notes

[1] Fontaine, Pascal; Merz, Stephan; Woltzenlogel Paleo, Bruno. *Compression of Propositional Resolution Proofs via Partial Regularization*. 23rd International Conference on Automated Deduction, 2011.

# Chapter 68

# Resolution inference

In propositional logic, a resolution inference is an instance of the following rule:[1]

$$\frac{\Gamma_1 \cup \{\ell\} \quad \Gamma_2 \cup \{\overline{\ell}\}}{\Gamma_1 \cup \Gamma_2} |\ell|$$

We call:

- The clauses $\Gamma_1 \cup \{\ell\}$ and $\Gamma_2 \cup \{\overline{\ell}\}$ are the inference's premises

- $\Gamma_1 \cup \Gamma_2$ (the resolvent of the premises) is its conclusion.

- The literal $\ell$ is the left resolved literal,

- The literal $\overline{\ell}$ is the right resolved literal,

- $|\ell|$ is the resolved atom or pivot.

This rule can be generalized to first-order logic to:[2]

$$\frac{\Gamma_1 \cup \{L_1\} \quad \Gamma_2 \cup \{L_2\}}{(\Gamma_1 \cup \Gamma_2)\phi} \phi$$

where $\phi$ is a most general unifier of $L_1$ and $\overline{L_2}$ and $\Gamma_1$ and $\Gamma_2$ have no common variables.

## 68.1 Example

The clauses $P(x), Q(x)$ and $\neg P(b)$ can apply this rule with $[b/x]$ as unifier.

Here x is a variable and b is a constant.

$$\frac{P(x), Q(x) \quad \neg P(b)}{Q(B)} [b/x]$$

Here we see that

- The clauses $P(x), Q(x)$ and $\neg P(x)$ are the inference's premises

- $Q(b)$ (the resolvent of the premises) is its conclusion.

- The literal $P(x)$ is the left resolved literal,

- The literal $\neg P(b)$ is the right resolved literal,

- $P$ is the resolved atom or pivot.

- $[b/x]$ is the most general unifier of the resolved literals.

## 68.2 Notes

[1] Fontaine, Pascal; Merz, Stephan; Woltzenlogel Paleo, Bruno. *Compression of Propositional Resolution Proofs via Partial Regularization.* 23rd International Conference on Automated Deduction, 2011.

[2] Enrique P. Arís, Juan L. González y Fernando M. Rubio, Lógica Computacional, Thomson, (2005).

# Chapter 69

# Resolution proof compression by splitting

In mathematical logic, **proof compression by splitting** is an algorithm that operates as a post-process on resolution proofs. It was proposed by Scott Cotton in his paper "Two Techniques for Minimizing Resolution Proof".[1]

The Splitting algorithm is based on the following observation:

Given a proof of unsatisfiability $\pi$ and a variable $x$, it is easy to re-arrange (split) the proof in a proof of $x$ and a proof of $\neg x$ and the recombination of these two proofs (by an additional resolution step) may result in a proof smaller than the original.

Note that applying Splitting in a proof $\pi$ using a variable $x$ does not invalidates a latter application of the algorithm using a differente variable $y$. Actually, the method proposed by Cotton[1] generates a sequence of proofs $\pi_1 \pi_2 \ldots$, where each proof $\pi_{i+1}$ is the result of applying Splitting to $\pi_i$. During the construction of the sequence, if a proof $\pi_j$ happens to be too large, $\pi_{j+1}$ is set to be the smallest proof in $\{\pi_1, \pi_2, \ldots, \pi_j\}$.

For achieving a better compression/time ratio, a heuristic for variable selection is desirable. For this purpose, Cotton[1] defines the "additivity" of a resolution step (with antecedents $p$ and $n$ and resolvent $r$):

$$\mathrm{add}(r) := \max(|r| - \max(|p|, |n|), 0)$$

Then, for each variable $v$, a score is calculated summing the additivity of all the resolution steps in $\pi$ with pivot $v$ together with the number of these resolution steps. Denoting each score calculated this way by $\mathrm{add}(v, \pi)$, each variable is selected with a probability proportional to its score:

$$p(v) = \frac{\mathrm{add}(v, \pi_i)}{\sum_x \mathrm{add}(x, \pi_i)}$$

To split a proof of unsatisfiability $\pi$ in a proof $\pi_x$ of $x$ and a proof $\pi_{\neg x}$ of $\neg x$, Cotton [1] proposes the following:

Let $l$ denote a literal and $p \oplus_x n$ denote the resolvent of clauses $p$ and $n$ where $x \in p$ and $\neg x \in n$. Then, define the map $\pi_l$ on nodes in the resolution dag of $\pi$:

$$\pi_l(c) := \begin{cases} c, & \text{if } c \text{ an input is} \\ \pi_l(p), & \text{if } c = p \oplus_x n \text{ and } (l = x \text{ or } x \notin \pi_l(p)) \\ \pi_l(n), & \text{if } c = p \oplus_x n \text{ and } (l = \neg x \text{ or } \neg x \notin \pi_l(n)) \\ \pi_l(p) \oplus_x \pi_l(p), & \text{if } x \in \pi_l(p) \text{ and } \neg x \in \pi_l(n) \end{cases}$$

Also, let $o$ be the empty clause in $\pi$. Then, $\pi_x$ and $\pi_{\neg x}$ are obtained by computing $\pi_x(o)$ and $\pi_{\neg x}(o)$, respectively.

216

# 69.1 Notes

[1] Cotton, Scott. "Two Techniques for Minimizing Resolution Proofs". 13th International Conference on Theory and Applications of Satisfiability Testing, 2010.

# Chapter 70

# Resolution proof reduction via local context rewriting

In proof theory, an area of mathematical logic, **resolution proof reduction via local context rewriting** is a technique for resolution proof reduction via local context rewriting.[1] This proof compression method was presented as an algorithm named *ReduceAndReconstruct*, that operates as a post-processing of resolution proofs.

ReduceAndReconstruct is based on a set of local proof rewriting rules that transform a subproof into an equivalent or stronger one.[1] Each rule is defined to match a specific context.

A context[1] involves two pivots ( $p$ and $q$ ) and five clauses ( $\alpha$ , $\beta$ , $\gamma$ , $\delta$ and $\eta$ ). The structure of a context is shown in (**1**). Note that this implies that $p$ is contained in $\beta$ and $\gamma$ (with opposite polarity) and $q$ is contained in $\delta$ and $\alpha$ (also with opposite polarity).

The table below shows the rewriting rules proposed by Simone *et al.*.[1] The idea of the algorithm is to reduce proof size by opportunistically applying these rules.

The first five rules were introduced in an earlier paper.[2] In addition:

- Rule A2 does not perform any reduction on its own. However, it is still useful, because of its "shuffling" effect that can create new opportunities for applying the other rules;

- Rule A1 is not used in practice, because it may increase proof size;

- Rules B1, B2, B2' and B3 are directly responsible for the reduction, as they produce a transformed root clause stronger than the original one;

- The application of a B rule may lead to an illegal proof (see the example below), as some literals missing in the transformed root clause may be involved in another resolution step along the path to the proof root. Therefore, the algorithm also has to "reconstruct" a legal proof when this happen.

The following example[1] shows a situation where the proof becomes illegal after the application of B2' rule:

Applying rule B2' to the highlighted context:

The proof is now illegal because the literal $o$ is missing from the transformed root clause. To reconstruct the proof, one can remove $o$ together with the last resolution step (that is now redundant). The final result is the following legal (and stronger) proof:

A further reduction of this proof by applying rule A2 to create a new opportunity to apply rule B2'.[1]

There are usually a huge number of contexts where rule A2 may be applied, so an exhaustive approach is not feasible in general. One proposal[1] is to execute ReduceAndReconstruct as a loop with two termination criteria: number of iterations and a timeout (what is reached first). The pseudocode[1] below shows this.

1 **function** ReduceAndReconstruct( $\pi$ /* *a proof* */, *timelimit*, *maxIterations*): 2 **for** i = 1 to *maxIterations* **do** 3 Reduce-AndReconstructLoop(); 4 **if** *time* > *timelimit* **then** // *timeout* 5 **break**; 6 **end for** 7 **end function**

ReduceAndReconstruct uses the function ReduceAndReconstructLoop, which is specified below. The first part of the algorithm does a topological ordering of the resolution graph (considering that edges goes from antecedentes to resolvents). This is done to ensure that each node is visited after its antecedents (this way, broken resolution steps are always found and fixed).[1]

1 **function** ReduceAndReconstructLoop( $\pi$ /* *a proof* */): 2 *TS* = TopologicalSorting( $\pi$ ); 3 **for each** node $n$ **in** *TS* 4 **if** $n$ is not a leaf 5 **if** $n_{piv} \in n_{clause}^{left}$ *and* $\overline{n_{piv}} \in n_{clause}^{right}$ **then** 6 $n_{\text{clause}}$ = Resolution( $n_{\text{clause}}^{\text{left}}$ , $n_{\text{clause}}^{\text{right}}$ ); 7 Determine left context of $n$ , if any; 8 Determine right context of $n$ , if any; 9 Heuristically choose one context (if any) and apply the corresponding rule; 10 **else if** $n_{piv} \notin n_{clause}^{left}$ *and* $\overline{n_{piv}} \in n_{clause}^{right}$ **then** 11 Substitute $n$ with $n^{\text{left}}$ ; 12 **else if** $n_{piv} \in n_{clause}^{left}$ *and* $\overline{n_{piv}} \notin n_{clause}^{right}$ **then** 13 Substitute $n$ with $n^{\text{right}}$ ; 14 **else if** $n_{piv} \notin n_{clause}^{left}$ *and* $\overline{n_{piv}} \notin n_{clause}^{right}$ **then** 15 Heuristically choose an antecedent $n^{\text{left}}$ or $n^{\text{right}}$ ; 16 Substitute $n$ with $n^{\text{left}}$ or $n^{\text{right}}$ ; 17 **end for** 18 **end function**

If the input proof is not a tree (in general, resolution graphs are directed acyclic graphs), then the clause $\delta$ of a context may be involved in more than one resolution step. In this case, to ensure that an application of a rewriting rule is not going to interfere with other resolution steps, a safe solution is to create a copy of the node represented by clause $\delta$ .[1] This solution increases proof size and some caution is needed when doing this.

The heuristic for rule selection is important to achieve a good compression performance. Simone *et al.* [1] use the following order of preference for the rules (if applicable to the given context): B2 > B3 > { B2', B1 } > A1' > A2 (X > Y means that X is preferred over Y).

Experiments have shown that ReduceAndReconstruct alone has a worse compression/time ratio than the algorithm RecyclePivots.[3] However, while RecyclePivots can be applied only once to a proof, ReduceAndReconstruct may be applied multiple times to produce a better compression. An attempt to combine ReduceAndReconstruct and RecyclePivots algorithms has led to good results.[1]

## 70.1  Notes

[1] Simone, S.F. ; Brutomesso, R. ; Sharygina, N. "An Efficient and Flexible Approach to Resolution Proof Reduction". 6th Haifa Verification Conference, 2010.

[2] Bruttomesso, R. ; Rollini, S. ; Sharygina, N.; Tsitovich, A. "Flexible Interpolation with Local Proof Transformations". The International Conference on Computer-Aided Design, 2010.

[3] Bar-Ilan, O. ; Fuhrmann, O. ; Hoory, S. ; Shacham, O. ; Strichman, O. "Linear-Time Reductions of Resolution Proofs". HVC, 2008.

# Chapter 71

# Reverse mathematics

**Reverse mathematics** is a program in mathematical logic that seeks to determine which axioms are required to prove theorems of mathematics. Its defining method can briefly be described as "going backwards from the theorems to the axioms", in contrast to the ordinary mathematical practice of deriving theorems from axioms. The reverse mathematics program was foreshadowed by results in set theory such as the classical theorem that the axiom of choice and Zorn's lemma are equivalent over ZF set theory. The goal of reverse mathematics, however, is to study possible axioms of ordinary theorems of mathematics rather than possible axioms for set theory.

Reverse mathematics is usually carried out using subsystems of second-order arithmetic, where many of its definitions and methods are inspired by previous work in constructive analysis and proof theory. The use of second-order arithmetic also allows many techniques from recursion theory to be employed; many results in reverse mathematics have corresponding results in computable analysis.

The program was founded by Harvey Friedman (1975, 1976). A standard reference for the subject is (Simpson 2009).

## 71.1 General principles

In reverse mathematics, one starts with a framework language and a base theory—a core axiom system—that is too weak to prove most of the theorems one might be interested in, but still powerful enough to develop the definitions necessary to state these theorems. For example, to study the theorem "Every bounded sequence of real numbers has a supremum" it is necessary to use a base system which can speak of real numbers and sequences of real numbers.

For each theorem that can be stated in the base system but is not provable in the base system, the goal is to determine the particular axiom system (stronger than the base system) that is necessary to prove that theorem. To show that a system $S$ is required to prove a theorem $T$, two proofs are required. The first proof shows $T$ is provable from $S$; this is an ordinary mathematical proof along with a justification that it can be carried out in the system $S$. The second proof, known as a **reversal**, shows that $T$ itself implies $S$; this proof is carried out in the base system. The reversal establishes that no axiom system $S'$ that extends the base system can be weaker than $S$ while still proving $T$.

### 71.1.1 Use of second-order arithmetic

Most reverse mathematics research focuses on subsystems of second-order arithmetic. The body of research in reverse mathematics has established that weak subsystems of second-order arithmetic suffice to formalize almost all undergraduate-level mathematics. In second-order arithmetic, all objects can be represented as either natural numbers or sets of natural numbers. For example, in order to prove theorems about real numbers, the real numbers can be represented as Cauchy sequences of rational numbers, each of which can be represented as a set of natural numbers.‹The template *Elucidate* is being considered for deletion.›

The axiom systems most often considered in reverse mathematics are defined using axiom schemes called **comprehension**

**schemes**. Such a scheme states that any set of natural numbers definable by a formula of a given complexity exists. In this context, the complexity of formulas is measured using the arithmetical hierarchy and analytical hierarchy.

The reason that reverse mathematics is not carried out using set theory as a base system is that the language of set theory is too expressive. Extremely complex sets of natural numbers can be defined by simple formulas in the language of set theory (which can quantify over arbitrary sets). In the context of second-order arithmetic, results such as Post's theorem establish a close link between the complexity of a formula and the (non)computability of the set it defines.

Another effect of using second-order arithmetic is the need to restrict general mathematical theorems to forms that can be expressed within arithmetic. For example, second-order arithmetic can express the principle "Every countable vector space has a basis" but it cannot express the principle "Every vector space has a basis". In practical terms, this means that theorems of algebra and combinatorics are restricted to countable structures, while theorems of analysis and topology are restricted to separable spaces. Many principles that imply the axiom of choice in their general form (such as "Every vector space has a basis") become provable in weak subsystems of second-order arithmetic when they are restricted. For example, "every field has an algebraic closure" is not provable in ZF set theory, but the restricted form "every countable field has an algebraic closure" is provable in $RCA_0$, the weakest system typically employed in reverse mathematics.

## 71.2 The big five subsystems of second order arithmetic

Second order arithmetic is a formal theory of the natural numbers and sets of natural numbers. Many mathematical objects, such as countable rings, groups, and fields, as well as points in effective Polish spaces, can be represented as sets of natural numbers, and modulo this representation can be studied in second order arithmetic.

Reverse mathematics makes use of several subsystems of second order arithmetic. A typical reverse mathematics theorem shows that a particular mathematical theorem $T$ is equivalent to a particular subsystem $S$ of second order arithmetic over a weaker subsystem $B$. This weaker system $B$ is known as the **base system** for the result; in order for the reverse mathematics result to have meaning, this system must not itself be able to prove the mathematical theorem $T$.

Simpson (2009) describes five particular subsystems of second order arithmetic, which he calls the **Big Five**, that occur frequently in reverse mathematics. In order of increasing strength, these systems are named by the initialisms $RCA_0$, $WKL_0$, $ACA_0$, $ATR_0$, and $\Pi^1_1$-$CA_0$.

The following table summarizes the "big five" systems Simpson (2009, p.42)

The subscript $_0$ in these names means that the induction scheme has been restricted from the full second-order induction scheme (Simpson 2009, p. 6). For example, $ACA_0$ includes the induction axiom $(0 \in X \wedge \forall n(n \in X \rightarrow n+1 \in X)) \rightarrow \forall n$ $n \in X$. This together with the full comprehension axiom of second order arithmetic implies the full second-order induction scheme given by the universal closure of $(\varphi(0) \wedge \forall n(\varphi(n) \rightarrow \varphi(n+1))) \rightarrow \forall n \varphi(n)$ for any second order formula $\varphi$. However $ACA_0$ does not have the full comprehension axiom, and the subscript $_0$ is a reminder that it does not have the full second-order induction scheme either. This restriction is important: systems with restricted induction have significantly lower proof-theoretical ordinals than systems with the full second-order induction scheme.

### 71.2.1 The base system $RCA_0$

$RCA_0$ is the fragment of second-order arithmetic whose axioms are the axioms of Robinson arithmetic, induction for $\Sigma0$ 1 formulas, and comprehension for $\Delta0$ 1 formulas.

The subsystem $RCA_0$ is the one most commonly used as a base system for reverse mathematics. The initials "RCA" stand for "recursive comprehension axiom", where "recursive" means "computable", as in recursive function. This name is used because $RCA_0$ corresponds informally to "computable mathematics". In particular, any set of natural numbers that can be proven to exist in $RCA_0$ is computable, and thus any theorem which implies that noncomputable sets exist is not provable in $RCA_0$. To this extent, $RCA_0$ is a constructive system, although it does not meet the requirements of the program of constructivism because it is a theory in classical logic including the excluded middle.

Despite its seeming weakness (of not proving any noncomputable sets exist), $RCA_0$ is sufficient to prove a number of classical theorems which, therefore, require only minimal logical strength. These theorems are, in a sense, below the reach of the reverse mathematics enterprise because they are already provable in the base system. The classical theorems provable in $RCA_0$ include:

- Basic properties of the natural numbers, integers, and rational numbers (for example, that the latter form an ordered field).

- Basic properties of the real numbers (the real numbers are an Archimedean ordered field; any nested sequence of closed intervals whose lengths tend to zero has a single point in its intersection; the real numbers are not countable).

- The Baire category theorem for a complete separable metric space (the separability condition is necessary to even state the theorem in the language of second-order arithmetic).

- The intermediate value theorem on continuous real functions.

- The Banach–Steinhaus theorem for a sequence of continuous linear operators on separable Banach spaces.

- A weak version of Gödel's completeness theorem (for a set of sentences, in a countable language, that is already closed under consequence).

- The existence of an algebraic closure for a countable field (but not its uniqueness).

- The existence and uniqueness of the real closure of a countable ordered field.

The first-order part of $RCA_0$ (the theorems of the system that do not involve any set variables) is the set of theorems of first-order Peano arithmetic with induction limited to $\Sigma^0_1$ formulas. It is provably consistent, as is $RCA_0$, in full first-order Peano arithmetic.

### 71.2.2   Weak König's lemma $WKL_0$

The subsystem $WKL_0$ consists of $RCA_0$ plus a weak form of König's lemma, namely the statement that every infinite subtree of the full binary tree (the tree of all finite sequences of 0's and 1's) has an infinite path. This proposition, which is known as *weak König's lemma*, is easy to state in the language of second-order arithmetic. $WKL_0$ can also be defined as the principle of $\Sigma^0_1$ separation (given two $\Sigma^0_1$ formulas of a free variable $n$ which are exclusive, there is a class containing all $n$ satisfying the one and no $n$ satisfying the other).

The following remark on terminology is in order. The term "weak König's lemma" refers to the sentence which says that any infinite subtree of the binary tree has an infinite path. When this axiom is added to $RCA_0$, the resulting subsystem is called $WKL_0$. A similar distinction between particular axioms, on the one hand, and subsystems including the basic axioms and induction, on the other hand, is made for the stronger subsystems described below.

In a sense, weak König's lemma is a form of the axiom of choice (although, as stated, it can be proven in classical Zermelo–Fraenkel set theory without the axiom of choice). It is not constructively valid in some senses of the word constructive.

To show that $WKL_0$ is actually stronger than (not provable in) $RCA_0$, it is sufficient to exhibit a theorem of $WKL_0$ which implies that noncomputable sets exist. This is not difficult; $WKL_0$ implies the existence of separating sets for effectively inseparable recursively enumerable sets.

It turns out that $RCA_0$ and $WKL_0$ have the same first-order part, meaning that they prove the same first-order sentences. $WKL_0$ can prove a good number of classical mathematical results which do not follow from $RCA_0$, however. These results are not expressible as first order statements but can be expressed as second-order statements.

The following results are equivalent to weak König's lemma and thus to $WKL_0$ over $RCA_0$:

- The Heine–Borel theorem for the closed unit real interval, in the following sense: every covering by a sequence of open intervals has a finite subcovering.

- The Heine–Borel theorem for complete totally bounded separable metric spaces (where covering is by a sequence of open balls).

- A continuous real function on the closed unit interval (or on any compact separable metric space, as above) is bounded (or: bounded and reaches its bounds).

- A continuous real function on the closed unit interval can be uniformly approximated by polynomials (with rational coefficients).

- A continuous real function on the closed unit interval is uniformly continuous.

- A continuous real function on the closed unit interval is Riemann integrable.

- The Brouwer fixed point theorem (for continuous functions on a finite product of copies of the closed unit interval).

- The separable Hahn–Banach theorem in the form: a bounded linear form on a subspace of a separable Banach space extends to a bounded linear form on the whole space.

- The Jordan curve theorem

- Gödel's completeness theorem (for a countable language).

- Determinacy for open (or even clopen) games on $\{0,1\}$ of length $\omega$.

- Every countable commutative ring has a prime ideal.

- Every countable formally real field is orderable.

- Uniqueness of algebraic closure (for a countable field).

### 71.2.3 Arithmetical comprehension $ACA_0$

$ACA_0$ is $RCA_0$ plus the comprehension scheme for arithmetical formulas (which is sometimes called the "arithmetical comprehension axiom"). That is, $ACA_0$ allows us to form the set of natural numbers satisfying an arbitrary arithmetical formula (one with no bound set variables, although possibly containing set parameters). Actually, it suffices to add to $RCA_0$ the comprehension scheme for $\Sigma_1$ formulas in order to obtain full arithmetical comprehension.

The first-order part of $ACA_0$ is exactly first-order Peano arithmetic; $ACA_0$ is a *conservative* extension of first-order Peano arithmetic. The two systems are provably (in a weak system) equiconsistent. $ACA_0$ can be thought of as a framework of predicative mathematics, although there are predicatively provable theorems that are not provable in $ACA_0$. Most of the fundamental results about the natural numbers, and many other mathematical theorems, can be proven in this system.

One way of seeing that $ACA_0$ is stronger than $WKL_0$ is to exhibit a model of $WKL_0$ that doesn't contain all arithmetical sets. In fact, it is possible to build a model of $WKL_0$ consisting entirely of low sets using the low basis theorem, since low sets relative to low sets are low.

The following assertions are equivalent to $ACA_0$ over $RCA_0$:

- The sequential completeness of the real numbers (every bounded increasing sequence of real numbers has a limit).

- The Bolzano–Weierstrass theorem.

- Ascoli's theorem: every bounded equicontinuous sequence of real functions on the unit interval has a uniformly convergent subsequence.

- Every countable commutative ring has a maximal ideal.

- Every countable vector space over the rationals (or over any countable field) has a basis.

- Every countable field has a transcendence basis.

- König's lemma (for arbitrary finitely branching trees, as opposed to the weak version described above).

- Various theorems in combinatorics, such as certain forms of Ramsey's theorem.

### 71.2.4   Arithmetical transfinite recursion ATR$_0$

The system ATR$_0$ adds to ACA$_0$ an axiom which states, informally, that any arithmetical functional (meaning any arithmetical formula with a free number variable $n$ and a free class variable $X$, seen as the operator taking $X$ to the set of $n$ satisfying the formula) can be iterated transfinitely along any countable well ordering starting with any set. ATR$_0$ is equivalent over ACA$_0$ to the principle of $\Sigma^1_1$ separation. ATR$_0$ is impredicative, and has the proof-theoretic ordinal $\Gamma_0$ , the supremum of that of predicative systems.

ATR$_0$ proves the consistency of ACA$_0$, and thus by Gödel's theorem it is strictly stronger.

The following assertions are equivalent to ATR$_0$ over RCA$_0$:

- Any two countable well orderings are comparable. That is, they are isomorphic or one is isomorphic to a proper initial segment of the other.

- Ulm's theorem for countable reduced Abelian groups.

- The perfect set theorem, which states that every uncountable closed subset of a complete separable metric space contains a perfect closed set.

- Lusin's separation theorem (essentially $\Sigma^1_1$ separation).

- Determinacy for open sets in the Baire space.

### 71.2.5   $\Pi^1_1$ comprehension $\Pi^1_1$-CA$_0$

$\Pi^1_1$-CA$_0$ is stronger than arithmetical transfinite recursion and is fully impredicative. It consists of RCA$_0$ plus the comprehension scheme for $\Pi^1_1$ formulas.

In a sense, $\Pi^1_1$-CA$_0$ comprehension is to arithmetical transfinite recursion ($\Sigma^1_1$ separation) as ACA$_0$ is to weak König's lemma ($\Sigma^0_1$ separation). It is equivalent to several statements of descriptive set theory whose proofs make use of strongly impredicative arguments; this equivalence shows that these impredicative arguments cannot be removed.

The following theorems are equivalent to $\Pi^1_1$-CA$_0$ over RCA$_0$:

- The Cantor–Bendixson theorem (every closed set of reals is the union of a perfect set and a countable set).

- Every countable abelian group is the direct sum of a divisible group and a reduced group.

## 71.3   Additional systems

- Weaker systems than recursive comprehension can be defined. The weak system RCA*
  0 consists of elementary function arithmetic EFA (the basic axioms plus $\Delta^0_0$ induction in the enriched language with an exponential operation) plus $\Delta^0_1$ comprehension. Over RCA*
  0, recursive comprehension as defined earlier (that is, with $\Sigma^0_1$ induction) is equivalent to the statement that a polynomial (over a countable field) has only finitely many roots and to the classification theorem for finitely generated Abelian groups. The system RCA*
  0 has the same proof theoretic ordinal $\omega^3$ as EFA and is conservative over EFA for $\Pi0$
  2 sentences.

- Weak Weak König's Lemma is the statement that a subtree of the infinite binary tree having no infinite paths has an asymptotically vanishing proportion of the leaves at length $n$ (with a uniform estimate as to how many leaves of length $n$ exist). An equivalent formulation is that any subset of Cantor space that has positive measure is nonempty (this is not provable in RCA$_0$). WWKL$_0$ is obtained by adjoining this axiom to RCA$_0$. It is equivalent to the statement that if the unit real interval is covered by a sequence of intervals then the sum of their lengths is at least one. The model theory of WWKL$_0$ is closely connected to the theory of algorithmically random sequences. In particular, an ω-model of RCA$_0$ satisfies weak weak König's lemma if and only if for every set $X$ there is a set $Y$ which is 1-random relative to $X$.

- DNR (short for "diagonally non-recursive") adds to RCA$_0$ an axiom asserting the existence of a diagonally non-recursive function relative to every set. That is, DNR states that, for any set $A$, there exists a total function $f$ such that for all $e$ the $e$th partial recursive function with oracle $A$ is not equal to $f$. DNR is strictly weaker than WWKL (Lempp *et al.*, 2004).

- $\Delta^1_1$-comprehension is in certain ways analogous to arithmetical transfinite recursion as recursive comprehension is to weak König's lemma. It has the hyperarithmetical sets as minimal ω-model. Arithmetical transfinite recursion proves $\Delta^1_1$-comprehension but not the other way around.

- $\Sigma^1_1$-choice is the statement that if $\eta(n,X)$ is a $\Sigma^1_1$ formula such that for each $n$ there exists an $X$ satisfying $\eta$ then there is a sequence of sets $Xn$ such that $\eta(n,Xn)$ holds for each $n$. $\Sigma^1_1$-choice also has the hyperarithmetical sets as minimal ω-model. Arithmetical transfinite recursion proves $\Sigma^1_1$-choice but not the other way around.

# 71.4  ω-models and β-models

The ω in ω-model stands for the set of non-negative integers (or finite ordinals). An ω-model is a model for a fragment of second-order arithmetic whose first-order part is the standard model of Peano arithmetic, but whose second-order part may be non-standard. More precisely, an ω-model is given by a choice $S \subseteq 2^\omega$ of subsets of ω. The first order variables are interpreted in the usual way as elements of ω, and +, × have their usual meanings, while second order variables are interpreted as elements of $S$. There is a standard ω model where one just takes $S$ to consist of all subsets of the integers. However there are also other ω-models; for example, RCA$_0$ has a minimal ω-model where $S$ consists of the recursive subsets of ω.

A β model is an ω model that is equivalent to the standard ω-model for Π1
1 and Σ1
1 sentences (with parameters).

Non-ω models are also useful, especially in the proofs of conservation theorems.

# 71.5  References

- Ambos-Spies, K.; Kjos-Hanssen, B.; Lempp, S.; Slaman, T.A. (2004), "Comparing DNR and WWKL", *Journal of Symbolic Logic* **69** (4): 1089, doi:10.2178/jsl/1102022212.

- Friedman, Harvey (1975), "Some systems of second order arithmetic and their use", *Proceedings of the International Congress of Mathematicians (Vancouver, B. C., 1974), Vol. 1*, Canad. Math. Congress, Montreal, Que., pp. 235–242, MR 0429508

- Friedman, Harvey; Martin, D. A.; Soare, R. I.; Tait, W. W. (1976), "Meeting of the Association for Symbolic Logic: Systems of second order arithmetic with restricted induction, I, II", *The Journal of Symbolic Logic* (Association for Symbolic Logic) **41** (2): 557–559, doi:10.2307/2272259

- Simpson, Stephen G. (2009), *Subsystems of second order arithmetic*, Perspectives in Logic (2nd ed.), Cambridge University Press, ISBN 978-0-521-88439-6, MR 2517689

- Solomon, Reed (1999), "Ordered groups: a case study in reverse mathematics", *The Bulletin of Symbolic Logic* **5** (1): 45–58, doi:10.2307/421140, ISSN 1079-8986, JSTOR 421140, MR 1681895

## 71.6   External links

- Harvey Friedman's home page

- Stephen G. Simpson's home page

# Chapter 72

# Self-verifying theories

**Self-verifying theories** are consistent first-order systems of arithmetic much weaker than Peano arithmetic that are capable of proving their own consistency. Dan Willard was the first to investigate their properties, and he has described a family of such systems. According to Gödel's incompleteness theorem, these systems cannot contain the theory of Peano arithmetic, and in fact, not even the weak fragment of Robinson arithmetic; nonetheless, they can contain strong theorems.

In outline, the key to Willard's construction of his system is to formalise enough of the Gödel machinery to talk about provability internally without being able to formalise diagonalisation. Diagonalisation depends upon being able to prove that multiplication is a total function (and in the earlier versions of the result, addition also). Addition and multiplication are not function symbols of Willard's language; instead, subtraction and division are, with the addition and multiplication predicates being defined in terms of these. Here, one cannot prove the $\Pi_2^0$ sentence expressing totality of multiplication:

$$(\forall x, y) \, (\exists z) \, \text{multiply}(x, y, z).$$

where multiply is the three-place predicate which stands for $z/y = x$. When the operations are expressed in this way, provability of a given sentence can be encoded as an arithmetic sentence describing termination of an analytic tableau. Provability of consistency can then simply be added as an axiom. The resulting system can be proven consistent by means of a relative consistency argument with respect to ordinary arithmetic.

We can add any true $\Pi_1^0$ sentence of arithmetic to the theory and still remain consistent.

## 72.1 References

- Solovay, R., 1989. "Injecting Inconsistencies into Models of PA". Annals of Pure and Applied Logic 44(1-2): 101—132.

- Willard, D., 2001. "Self Verifying Axiom Systems, the Incompleteness Theorem and the Tangibility Reflection Principle". Journal of Symbolic Logic 66:536—596.

- Willard, D., 2002. "How to Extend the Semantic Tableaux and Cut-Free Versions of the Second Incompleteness Theorem to Robinson's Arithmetic Q". Journal of Symbolic Logic 67:465—496.

## 72.2 External links

- Dan Willard's home page.

# Chapter 73

# Sequent

For other uses, see Sequent (disambiguation).

In mathematical logic, a **sequent** is a very general kind of conditional assertion.

$$A_1, \ldots, A_m \vdash B_1, \ldots, B_n.$$

A sequent may have any number $m$ of condition formulas $Ai$ (called "antecedents") and any number $n$ of asserted formulas $Bj$ (called "succedents" or "consequents"). A sequent is understood to mean that if all of the antecedent conditions are true, then at least one of the consequent formulas is true. This style of conditional assertion is almost always associated with the conceptual framework of sequent calculus.

## 73.1 Introduction

### 73.1.1 The form and semantics of sequents

Sequents are best understood in the context of general logical assertions, which may be classified into the following three cases.

1. **Unconditional assertion**. No antecedent formulas.

   - Example: $\vdash B$
   - Meaning: $B$ is true.

2. **Conditional assertion**. Any number of antecedent formulas.

   (a) **Simple conditional assertion**. Single consequent formula.
   - Example: $A_1, A_2, A_3 \vdash B$
   - Meaning: IF $A_1$ AND $A_2$ AND $A_3$ are true, THEN $B$ is true.
   (b) **Sequent**. Any number of consequent formulas.
   - Example: $A_1, A_2, A_3 \vdash B_1, B_2, B_3, B_4$
   - Meaning: IF $A_1$ AND $A_2$ AND $A_3$ are true, THEN $B_1$ OR $B_2$ OR $B_3$ OR $B_4$ is true.

Thus sequents are a generalization of simple conditional assertions, which are a generalization of unconditional assertions.

The word "OR" here is the inclusive OR.[1] The motivation for disjunctive semantics on the right side of a sequent comes from three main benefits.

228

1. The symmetry of the classical inference rules for sequents with such semantics.

2. The ease and simplicity of converting such classical rules to intuitionistic rules.

3. The ability to prove completeness for predicate calculus when it is expressed in this way.

All three of these benefits were identified in the founding paper by Gentzen (1934, p. 194).

Not all authors have adhered to Gentzen's original meaning for the word "sequent". For example, Lemmon (1965) used the word "sequent" strictly for simple conditional assertions with one and only one consequent formula.[2] The same single-consequent definition for a sequent is given by Huth & Ryan 2004, p. 5.

## 73.1.2  Syntax details

In a general sequent of the form

$$\Gamma \vdash \Sigma$$

both $\Gamma$ and $\Sigma$ are sequences of logical formulas, not sets. Therefore both the number and order of occurrences of formulas are significant. In particular, the same formula may appear twice in the same sequence. The full set of sequent calculus inference rules contains rules to swap adjacent formulas on the left and on the right of the assertion symbol (and thereby arbitrarily permute the left and right sequences), and also to insert arbitrary formulas and remove duplicate copies within the left and the right sequences. (However, Smullyan (1995, pp. 107–108), uses *sets* of formulas in sequents instead of sequences of formulas. Consequently the three pairs of *structural rules* called "thinning", "contraction" and "interchange" are not required.)

The symbol ' $\vdash$ ' is often referred to as the "turnstile", "right tack", "tee", "assertion sign" or "assertion symbol". It is often read, suggestively, as "yields", "proves" or "entails".

## 73.1.3  Properties

### Effects of inserting and removing propositions

Since every formula in the antecedent (the left side) must be true to conclude the truth of at least one formula in the succedent (the right side), adding formulas to either side results in a weaker sequent, while removing them from either side gives a stronger one. This is one of the symmetry advantages which follows from the use of disjunctive semantics on the right hand side of the assertion symbol, whereas conjunctive semantics is adhered to on the left hand side.

### Consequences of empty lists of formulas

In the extreme case where the list of *antecedent* formulas of a sequent is empty, the consequent is unconditional. This differs from the simple unconditional assertion because the number of consequents is arbitrary, not necessarily a single consequent. Thus for example, ' $\vdash B_1, B_2$ ' means that either $B_1$, or $B_2$, or both must be true. An empty antecedent formula list is equivalent to the "always true" proposition, called the "verum", denoted "$\top$". (See Tee (symbol).)

In the extreme case where the list of *consequent* formulas of a sequent is empty, the rule is still that at least one term on the right be true, which is clearly impossible. This is signified by the 'always false' proposition, called the "falsum", denoted "$\bot$". Since the consequence is false, at least one of the antecedents must be false. Thus for example, ' $A_1, A_2 \vdash$ ' means that at least one of the antecedents $A_1$ and $A_2$ must be false.

One sees here again a symmetry because of the disjunctive semantics on the right hand side. If the left side is empty, then one or more right-side propositions must be true. If the right side is empty, then one or more of the left-side propositions must be false.

The doubly extreme case ' ⊢ ', where both the antecedent and consequent lists of formulas are empty is "not satisfiable".[3] In this case, the meaning of the sequent is effectively ' ⊤ ⊢ ⊥ '. This is equivalent to the sequent ' ⊢ ⊥ ', which clearly cannot be valid.

### 73.1.4   Examples

A sequent of the form ' ⊢ α, β ', for logical formulas α and β, means that either α is true or β is true. But it does not mean that either α is a tautology or β is a tautology. To clarify this, consider the example ' ⊢ B ∨ A, C ∨ ¬A '. This is a valid sequent because either B ∨ A is true or C ∨ ¬A is true. But neither of these expressions is a tautology in isolation. It is the *disjunction* of these two expressions which is a tautology.

Similarly, a sequent of the form ' α, β ⊢ ', for logical formulas α and β, means that either α is false or β is false. But it does not mean that either α is a contradiction or β is a contradiction. To clarify this, consider the example ' B ∧ A, C ∧ ¬A ⊢ '. This is a valid sequent because either B ∧ A is false or C ∧ ¬A is false. But neither of these expressions is a contradiction in isolation. It is the *conjunction* of these two expressions which is a contradiction.

### 73.1.5   Rules

Most proof systems provide ways to deduce one sequent from another. These inference rules are written with a list of sequents above and below a line. This rule indicates that if everything above the line is true, so is everything under the line.

A typical rule is:

$$\frac{\Gamma, \alpha \vdash \Sigma \qquad \Gamma \vdash \alpha}{\Gamma \vdash \Sigma}$$

This indicates that if we can deduce that $\Gamma, \alpha$ yields $\Sigma$, and that $\Gamma$ yields $\alpha$, then we can also deduce that $\Gamma$ yields $\Sigma$. (See also the full set of sequent calculus inference rules.)

## 73.2   Interpretation

### 73.2.1   History of the meaning of sequent assertions

The assertion symbol in sequents originally meant exactly the same as the implication operator. But over time, its meaning has changed to signify provability within a theory rather than semantic truth in all models.

In 1934, Gentzen did not define the assertion symbol ' ⊢ ' in a sequent to signify provability. He defined it to mean exactly the same as the implication operator ' ⇒ '. He wrote: "The sequent $A_1, ..., A\mu \rightarrow B_1, ..., B\nu$ signifies, as regards content, exactly the same as the formula $(A_1 \& ... \& A\mu) \supset (B_1 \vee ... \vee B\nu)$".[4] (Gentzen employed the right-arrow symbol between the antecedents and consequents of sequents. He employed the symbol ' ⊃ ' for the logical implication operator.)

In 1939, Hilbert and Bernays stated likewise that a sequent has the same meaning as the corresponding implication formula.[5]

In 1944, Alonzo Church emphasized that Gentzen's sequent assertions did not signify provability.

> "Employment of the deduction theorem as primitive or derived rule must not, however, be confused with the use of *Sequenzen* by Gentzen. For Gentzen's arrow, →, is not comparable to our syntactical notation, ⊢, but belongs to his object language (as is clear from the fact that expressions containing it appear as premises and conclusions in applications of his rules of inference)."[6]

Numerous publications after this time have stated that the assertion symbol in sequents does signify provability within the theory where the sequents are formulated. Curry in 1963,[7] Lemmon in 1965,[2] and Huth and Ryan in 2004[8] all state that the sequent assertion symbol signifies provability. However, Ben-Ari (2012, p. 69) states that the assertion symbol in Gentzen-system sequents, which he denotes as ' $\Rightarrow$ ', is part of the object language, not the metalanguage.[9]

According to Prawitz (1965): "The calculi of sequents can be understood as meta-calculi for the deducibility relation in the corresponding systems of natural deduction."[10] And furthermore: "A proof in a calculus of sequents can be looked upon as an instruction on how to construct a corresponding natural deduction."[11] In other words, the assertion symbol is part of the object language for the sequent calculus, which is a kind of meta-calculus, but simultaneously signifies deducibility in an underlying natural deduction system.

### 73.2.2  Intuitive meaning

A sequent is a formalized statement of provability that is frequently used when specifying calculi for deduction. In the sequent calculus, the name *sequent* is used for the construct, which can be regarded as a specific kind of judgment, characteristic to this deduction system.

The intuitive meaning of the sequent $\Gamma \vdash \Sigma$ is that under the assumption of $\Gamma$ the conclusion of $\Sigma$ is provable. Classically, the formulae on the left of the turnstile can be interpreted conjunctively while the formulae on the right can be considered as a disjunction. This means that, when all formulae in $\Gamma$ hold, then at least one formula in $\Sigma$ also has to be true. If the succedent is empty, this is interpreted as falsity, i.e. $\Gamma \vdash$ means that $\Gamma$ proves falsity and is thus inconsistent. On the other hand an empty antecedent is assumed to be true, i.e., $\vdash \Sigma$ means that $\Sigma$ follows without any assumptions, i.e., it is always true (as a disjunction). A sequent of this form, with $\Gamma$ empty, is known as a logical assertion.

Of course, other intuitive explanations are possible, which are classically equivalent. For example, $\Gamma \vdash \Sigma$ can be read as asserting that it cannot be the case that every formula in $\Gamma$ is true and every formula in $\Sigma$ is false (this is related to the double-negation interpretations of classical intuitionistic logic, such as Glivenko's theorem).

In any case, these intuitive readings are only pedagogical. Since formal proofs in proof theory are purely syntactic, the meaning of (the derivation of) a sequent is only given by the properties of the calculus that provides the actual rules of inference.

Barring any contradictions in the technically precise definition above we can describe sequents in their introductory logical form. $\Gamma$ represents a set of assumptions that we begin our logical process with, for example "Socrates is a man" and "All men are mortal". The $\Sigma$ represents a logical conclusion that follows under these premises. For example "Socrates is mortal" follows from a reasonable formalization of the above points and we could expect to see it on the $\Sigma$ side of the *turnstile*. In this sense, $\vdash$ means the process of reasoning, or "therefore" in English.

## 73.3  Variations

The general notion of sequent introduced here can be specialized in various ways. A sequent is said to be an **intuitionistic sequent** if there is at most one formula in the succedent (although multi-succedent calculi for intuitionistic logic are also possible). More precisely, the restriction of the general sequent calculus to single-succedent-formula sequents, *with the same inference rules* as for general sequents, constitutes an intuitionistic sequent calculus. (This restricted sequent calculus is denoted LJ.)

Similarly, one can obtain calculi for dual-intuitionistic logic (a type of paraconsistent logic) by requiring that sequents be singular in the antecedent.

In many cases, sequents are also assumed to consist of multisets or sets instead of sequences. Thus one disregards the order or even the numbers of occurrences of the formulae. For classical propositional logic this does not yield a problem, since the conclusions that one can draw from a collection of premises do not depend on these data. In substructural logic, however, this may become quite important.

Natural deduction systems use single-consequence conditional assertions, but they typically do not use the same sets of inference rules as Gentzen introduced in 1934. In particular, tabular natural deduction systems, which are very convenient for practical theorem-proving in propositional calculus and predicate calculus, were applied by Suppes (1957) and Lemmon (1965) for teaching introductory logic in textbooks.

## 73.4   Etymology

Historically, sequents have been introduced by Gerhard Gentzen in order to specify his famous sequent calculus.[12] In his German publication he used the word "Sequenz". However, in English, the word "sequence" is already used as a translation to the German "Folge" and appears quite frequently in mathematics. The term "sequent" then has been created in search for an alternative translation of the German expression.

Kleene[13] makes the following comment on the translation into English: "Gentzen says 'Sequenz', which we translate as 'sequent', because we have already used 'sequence' for any succession of objects, where the German is 'Folge'."

## 73.5   See also

- Intuitionistic logic

- Gerhard Gentzen

- Sequent calculus

- Natural deduction

## 73.6   Notes

[1] The disjunctive semantics for the right side of a sequent is stated and explained by Curry 1977, pp. 189–190, Kleene 2002, pp. 290, 297, Kleene 2009, p. 441, Hilbert & Bernays 1970, p. 385, Smullyan 1995, pp. 104–105, Takeuti 2013, p. 9, and Gentzen 1934, p. 180.

[2] Lemmon 1965, p. 12, wrote: "Thus a sequent is an argument-frame containing a set of assumptions and a conclusion which is claimed to follow from them. [...] The propositions to the left of '⊢' become assumptions of the argument, and the proposition to the right becomes a conclusion validly drawn from those assumptions."

[3] Smullyan 1995, p. 105.

[4] Gentzen 1934, p. 180.

> 2.4. Die Sequenz $A_1, ..., A\mu \rightarrow B_1, ..., B\nu$ bedeutet inhaltlich genau dasselbe wie die Formel
>
> $$(A_1 \, \& \, ... \, \& \, A\mu) \supset (B_1 \vee ... \vee B\nu).$$

[5] Hilbert & Bernays 1970, p. 385.

> Für die inhaltliche Deutung ist eine Sequenz
>
> $$A_1, ..., A_r \rightarrow B_1, ..., B_s,$$
>
> worin die Anzahlen r und s von 0 verschieden sind, gleichbedeutend mit der Implikation
>
> $$(A_1 \, \& \, ... \, \& \, A_r) \rightarrow (B_1 \vee ... \vee B_s)$$

[6] Church 1996, p. 165.

[7] Curry 1977, p. 184

[8] Huth & Ryan (2004, p. 5)

[9] Ben-Ari 2012, p. 69, defines sequents to have the form $U \Rightarrow V$ for (possibly non-empty) sets of formulas $U$ and $V$. Then he writes:

> "Intuitively, a sequent represents 'provable from' in the sense that the formulas in $U$ are assumptions for the set of formulas $V$ that are to be proved. The symbol $\Rightarrow$ is similar to the symbol ⊢ in Hilbert systems, except that $\Rightarrow$ is part of the object language of the deductive system being formalized, while ⊢ is a metalanguage notation used to reason about deductive systems."

[10] Prawitz 2006, p. 90.

[11] See Prawitz 2006, p. 91, for this and further details of interpretation.

[12] Gentzen 1934, Gentzen 1935.

[13] Kleene 2002, p. 441

## 73.7 References

- Ben-Ari, Mordechai (2012) [1993]. *Mathematical logic for computer science*. London: Springer. ISBN 978-1-4471-4128-0.

- Church, Alonzo (1996) [1944]. *Introduction to mathematical logic*. Princeton, New Jersey: Princeton University Press. ISBN 978-0-691-02906-1.

- Curry, Haskell Brooks (1977) [1963]. *Foundations of mathematical logic*. New York: Dover Publications Inc. ISBN 978-0-486-63462-3.

- Gentzen, Gerhard (1934). "Untersuchungen über das logische Schließen. I". *Mathematische Zeitschrift*. 39 (2): 176–210. doi:10.1007/bf01201353.

- Gentzen, Gerhard (1935). "Untersuchungen über das logische Schließen. II". *Mathematische Zeitschrift*. 39 (3): 405–431. doi:10.1007/bf01201363.

- Hilbert, David; Bernays, Paul (1970) [1939]. *Grundlagen der Mathematik II* (Second ed.). Berlin, New York: Springer-Verlag. ISBN 978-3-642-86897-9.

- Huth, Michael; Ryan, Mark (2004). *Logic in Computer Science* (Second ed.). Cambridge, United Kingdom: Cambridge University Press. ISBN 978-0-521-54310-1.

- Kleene, Stephen Cole (2009) [1952]. *Introduction to metamathematics*. Ishi Press International. ISBN 978-0-923891-57-2.

- Kleene, Stephen Cole (2002) [1967]. *Mathematical logic*. Mineola, New York: Dover Publications. ISBN 978-0-486-42533-7.

- Lemmon, Edward John (1965). *Beginning logic*. Thomas Nelson. ISBN 0-17-712040-1.

- Prawitz, Dag (2006) [1965]. *Natural deduction: A proof-theoretical study*. Mineola, New York: Dover Publications. ISBN 978-0-486-44655-4.

- Smullyan, Raymond Merrill (1995) [1968]. *First-order logic*. New York: Dover Publications. ISBN 978-0-486-68370-6.

- Suppes, Patrick Colonel (1999) [1957]. *Introduction to logic*. Mineola, New York: Dover Publications. ISBN 978-0-486-40687-9.

- Takeuti, Gaisi (2013) [1975]. *Proof theory* (Second ed.). Mineola, New York: Dover Publications. ISBN 978-0-486-49073-1.

## 73.8 External links

- Hazewinkel, Michiel, ed. (2001), "Sequent (in logic)", *Encyclopedia of Mathematics*, Springer, ISBN 978-1-55608-010-4

# Chapter 74

# Sequent calculus

**Sequent calculus** is, in essence, a style of formal logical argumentation where every line of a proof is a conditional tautology (called a sequent by Gerhard Gentzen) instead of an unconditional tautology. Each conditional tautology is inferred from other conditional tautologies on earlier lines in a formal argument according to rules and procedures of inference, giving a better approximation to the style of natural deduction used by mathematicians than David Hilbert's earlier style of formal logic where every line was an unconditional tautology. (This is the essence of the idea, but there are several over-simplifications here. For example, there may be non-logical axioms upon which all propositions are implicitly dependent. Then sequents signify conditional theorems in a first-order language rather than conditional tautologies.)

Sequent calculus is one of several extant styles of proof calculus for expressing line-by-line logical arguments.

- Hilbert style. Every line is an unconditional tautology (or theorem).

- Gentzen style. Every line is a conditional tautology (or theorem) with zero or more conditions on the left.

  - Natural deduction. Every (conditional) line has exactly one asserted proposition on the right.

  - Sequent calculus. Every (conditional) line has zero or more asserted propositions on the right.

In other words, natural deduction and sequent calculus systems are particular distinct kinds of Gentzen-style systems. Hilbert-style systems typically have a very small number of inference rules, relying more on sets of axioms. Gentzen-style systems typically have very few axioms, if any, relying more on sets of rules.

Gentzen-style systems have significant practical and theoretical advantages compared to Hilbert-style systems. For example, both natural deduction and sequent calculus systems facilitate the elimination and introduction of universal and existential quantifiers so that unquantified logical expressions can be manipulated according to the much simpler rules of propositional calculus. In a typical argument, quantifiers are eliminated, then propositional calculus is applied to unquantified expressions (which typically contain free variables), and then the quantifiers are reintroduced. This very much parallels the way in which mathematical proofs are carried out in practice by mathematicians. Predicate calculus proofs are generally much easier to discover with this approach, and are often shorter. Natural deduction systems are more suited to practical theorem-proving. Sequent calculus systems are more suited to theoretical analysis.

## 74.1 Introduction

In proof theory and mathematical logic, **sequent calculus** is a family of formal systems sharing a certain style of inference and certain formal properties. The first sequent calculi, systems **LK** and **LJ**, were introduced in 1934/1935 by Gerhard Gentzen[1] as a tool for studying natural deduction in first-order logic (in classical and intuitionistic versions, respectively). Gentzen's so-called "Main Theorem" (*Hauptsatz*) about LK and LJ was the cut-elimination theorem,[2][3] a result with far-reaching meta-theoretic consequences, including consistency. Gentzen further demonstrated the power and flexibility of this technique a few years later, applying a cut-elimination argument to give a (transfinite) proof of the consistency of Peano arithmetic, in surprising response to Gödel's incompleteness theorems. Since this early work, sequent calculi, also called **Gentzen systems**,[4][5][6][7] and the general concepts relating to them, have been widely applied in the fields of proof theory, mathematical logic, and automated deduction.

### 74.1.1 Hilbert-style deduction systems

One way to classify different styles of deduction systems is to look at the form of *judgments* in the system, *i.e.*, which things may appear as the conclusion of a (sub)proof. The simplest judgment form is used in Hilbert-style deduction systems, where a judgment has the form

$$B$$

where $B$ is any formula of first-order-logic (or whatever logic the deduction system applies to, *e.g.*, propositional calculus or a higher-order logic or a modal logic). The theorems are those formulae that appear as the concluding judgment in a valid proof. A Hilbert-style system needs no distinction between formulae and judgments; we make one here solely for comparison with the cases that follow.

The price paid for the simple syntax of a Hilbert-style system is that complete formal proofs tend to get extremely long. Concrete arguments about proofs in such a system almost always appeal to the deduction theorem. This leads to the idea of including the deduction theorem as a formal rule in the system, which happens in natural deduction.

### 74.1.2 Natural deduction systems

In natural deduction, judgments have the shape

$$A_1, A_2, \ldots, A_n \vdash B$$

where the $A_i$'s and $B$ are again formulae and $n \geq 0$. Permutations of the $A_i$'s are immaterial. In other words, a judgment consists of a list (possibly empty) of formulae on the left-hand side of a turnstile symbol " $\vdash$ ", with a single formula on the right-hand side.[8][9][10] The theorems are those formulae $B$ such that $\vdash B$ (with an empty left-hand side) is the conclusion of a valid proof. (In some presentations of natural deduction, the $A_i$ s and the turnstile are not written down explicitly; instead a two-dimensional notation from which they can be inferred is used.)

The standard semantics of a judgment in natural deduction is that it asserts that whenever[11] $A_1$, $A_2$, etc., are all true, $B$ will also be true. The judgments

$$A_1, \ldots, A_n \vdash B$$

and

$$\vdash (A_1 \wedge \cdots \wedge A_n) \to B$$

are equivalent in the strong sense that a proof of either one may be extended to a proof of the other.

### 74.1.3 Sequent calculus systems

Finally, *sequent calculus* generalizes the form of a natural deduction judgment to

$$A_1, \ldots, A_n \vdash B_1, \ldots, B_k,$$

a syntactic object called a sequent. The formulas on left-hand side of the turnstile are called the *antecedent*, and the formulas on right-hand side are called the *succedent* or *consequent*; together they are called *cedents* or *sequents*.[12] Again, $A_i$ and $B_i$ are formulae, and $n$ and $k$ are nonnegative integers, that is, the left-hand-side or the right-hand-side (or neither or both) may be empty. As in natural deduction, theorems are those $B$ where $\vdash B$ is the conclusion of a valid proof. The empty sequent, having both cedents empty, is defined to be false.[13]

The standard semantics of a sequent is an assertion that whenever *every* $A_i$ is true, *at least one* $B_i$ will also be true.[14] One way to express this is that a comma to the left of the turnstile should be thought of as an "and", and a comma to the right of the turnstile should be thought of as an (inclusive) "or". The sequents

$$A_1, \ldots, A_n \vdash B_1, \ldots, B_k$$

and

$$\vdash (A_1 \wedge \cdots \wedge A_n) \rightarrow (B_1 \vee \cdots \vee B_k)$$

are equivalent in the strong sense that a proof of either one may be extended to a proof of the other.

At first sight, this extension of the judgment form may appear to be a strange complication — it is not motivated by an obvious shortcoming of natural deduction, and it is initially confusing that the comma seems to mean entirely different things on the two sides of the turnstile. However, in a classical context the semantics of the sequent can also (by propositional tautology) be expressed either as

$$\vdash \neg A_1 \vee \neg A_2 \vee \cdots \vee \neg A_n \vee B_1 \vee B_2 \vee \cdots \vee B_k$$

(at least one of the As is false, or one of the Bs is true) or as

$$\vdash \neg(A_1 \wedge A_2 \wedge \cdots \wedge A_n \wedge \neg B_1 \wedge \neg B_2 \wedge \cdots \wedge \neg B_k)$$

(it cannot be the case that all of the As are true and all of the Bs are false). In these formulations, the only difference between formulae on either side of the turnstile is that one side is negated. Thus, swapping left for right in a sequent corresponds to negating all of the constituent formulae. This means that a symmetry such as De Morgan's laws, which manifests itself as logical negation on the semantic level, translates directly into a left-right symmetry of sequents — and indeed, the inference rules in sequent calculus for dealing with conjunction ($\wedge$) are mirror images of those dealing with disjunction ($\vee$).

Many logicians feel that this symmetric presentation offers a deeper insight in the structure of the logic than other styles of proof system, where the classical duality of negation is not as apparent in the rules.

### 74.1.4   Distinction between natural deduction and sequent calculus

Gentzen asserted a sharp distinction between his single-output natural deduction systems (NK and NJ) and his multiple-output sequent calculus systems (LK and LJ). He wrote that the intuitionistic natural deduction system NJ was somewhat ugly.[15] He said that the special role of the excluded middle in the classical natural deduction system NK is removed in the classical sequent calculus system LK.[16] He said that the sequent calculus LJ gave more symmetry than natural deduction NJ in the case of intuitionistic logic, as also in the case of classical logic (LK versus NK).[17] Then he said that in addition to these reasons, the sequent calculus with multiple succedent formulas is intended particularly for his principal theorem ("Hauptsatz").[18]

### 74.1.5  Origin of word "sequent"

The word "sequent" is taken from the word "Sequenz" in Gentzen's 1934 paper.[1] Kleene makes the following comment on the translation into English: "Gentzen says 'Sequenz', which we translate as 'sequent', because we have already used 'sequence' for any succession of objects, where the German is 'Folge'."[19]

## 74.2  The system LK

This section introduces the rules of the sequent calculus **LK** (which just stands for "klassische Prädikatenlogik"), as introduced by Gentzen in 1934.[20] A (formal) proof in this calculus is a sequence of sequents, where each of the sequents is derivable from sequents appearing earlier in the sequence by using one of the rules below.

### 74.2.1  Inference rules

The following notation will be used:

- $\vdash$ known as the turnstile, separates the *assumptions* on the left from the *propositions* on the right

- $A$ and $B$ denote formulae of first-order predicate logic (one may also restrict this to propositional logic),

- $\Gamma, \Delta, \Sigma$, and $\Pi$ are finite (possibly empty) sequences of formulae (in fact, the order of formulae do not matter; see subsection Structural Rules), called contexts,

  - when on the *left* of the $\vdash$, the sequence of formulas is considered *conjunctively* (all assumed to hold at the same time),

  - while on the *right* of the $\vdash$, the sequence of formulas is considered *disjunctively* (at least one of the formulas must hold for any assignment of variables),

- $t$ denotes an arbitrary term,

- $x$ and $y$ denote variables.

- a variable is said to occur free within a formula if it occurs outside the scope of quantifiers $\forall$ or $\exists$.

- $A[t/x]$ denotes the formula that is obtained by substituting the term $t$ for every free occurrence of the variable $x$ in formula $A$ with the restriction that the term $t$ must be free for the variable $x$ in $A$ (i.e., no occurrence of any variable in $t$ becomes bound in $A[t/x]$).

- $WL$ and $WR$ stand for *Weakening Left/Right*, $CL$ and $CR$ for *Contraction*, and $PL$ and $PR$ for *Permutation*.

*Restrictions: In the rules* $(\forall R)$ *and* $(\exists L)$, *the variable* $y$ *must not occur free within* $\Gamma$ *and* $\Delta$. *Alternatively, the variable* $y$ *must not appear anywhere in the respective lower sequents.*

### 74.2.2  An intuitive explanation

The above rules can be divided into two major groups: *logical* and *structural* ones. Each of the logical rules introduces a new logical formula either on the left or on the right of the turnstile $\vdash$. In contrast, the structural rules operate on the structure of the sequents, ignoring the exact shape of the formulae. The two exceptions to this general scheme are the axiom of identity (I) and the rule of (Cut).

Although stated in a formal way, the above rules allow for a very intuitive reading in terms of classical logic. Consider, for example, the rule $(\wedge L_1)$. It says that, whenever one can prove that $\Delta$ can be concluded from some sequence of formulae that contain $A$, then one can also conclude $\Delta$ from the (stronger) assumption, that $A \wedge B$ holds. Likewise, the rule $(\neg R)$

states that, if $\Gamma$ and A suffice to conclude $\Delta$ , then from $\Gamma$ alone one can either still conclude $\Delta$ or A must be false, i.e. $\neg A$ holds. All the rules can be interpreted in this way.

For an intuition about the quantifier rules, consider the rule $(\forall R)$ . Of course concluding that $\forall x A$ holds just from the fact that $A[y/x]$ is true is not in general possible. If, however, the variable y is not mentioned elsewhere (i.e. it can still be chosen freely, without influencing the other formulae), then one may assume, that $A[y/x]$ holds for any value of y. The other rules should then be pretty straightforward.

Instead of viewing the rules as descriptions for legal derivations in predicate logic, one may also consider them as instructions for the construction of a proof for a given statement. In this case the rules can be read bottom-up; for example, $(\wedge R)$ says that, to prove that $A \wedge B$ follows from the assumptions $\Gamma$ and $\Sigma$ , it suffices to prove that A can be concluded from $\Gamma$ and B can be concluded from $\Sigma$ , respectively. Note that, given some antecedent, it is not clear how this is to be split into $\Gamma$ and $\Sigma$ . However, there are only finitely many possibilities to be checked since the antecedent by assumption is finite. This also illustrates how proof theory can be viewed as operating on proofs in a combinatorial fashion: given proofs for both A and B, one can construct a proof for A∧B.

When looking for some proof, most of the rules offer more or less direct recipes of how to do this. The rule of cut is different: It states that, when a formula A can be concluded and this formula may also serve as a premise for concluding other statements, then the formula A can be "cut out" and the respective derivations are joined. When constructing a proof bottom-up, this creates the problem of guessing A (since it does not appear at all below). The cut-elimination theorem is thus crucial to the applications of sequent calculus in automated deduction: it states that all uses of the cut rule can be eliminated from a proof, implying that any provable sequent can be given a *cut-free* proof.

The second rule that is somewhat special is the axiom of identity (I). The intuitive reading of this is obvious: every formula proves itself. Like the cut rule, the axiom of identity is somewhat redundant: the completeness of atomic initial sequents states that the rule can be restricted to atomic formulas without any loss of provability.

Observe that all rules have mirror companions, except the ones for implication. This reflects the fact that the usual language of first-order logic does not include the "is not implied by" connective $\not\leftarrow$ that would be the De Morgan dual of implication. Adding such a connective with its natural rules would make the calculus completely left-right symmetric.

### 74.2.3   Example derivations

Here is the derivation of " $\vdash A \vee \neg A$ ", known as the *Law of excluded middle* (*tertium non datur* in Latin).

Next is the proof of a simple fact involving quantifiers. Note that the converse is not true, and its falsity can be seen when attempting to derive it bottom-up, because an existing free variable cannot be used in substitution in the rules $(\forall R)$ and $(\exists L)$ .

For something more interesting we shall prove $((A \rightarrow (B \vee C)) \rightarrow (((B \rightarrow \neg A) \wedge \neg C) \rightarrow \neg A))$ . It is straightforward to find the derivation, which exemplifies the usefulness of LK in automated proving.

These derivations also emphasize the strictly formal structure of the sequent calculus. For example, the logical rules as defined above always act on a formula immediately adjacent to the turnstile, such that the permutation rules are necessary. Note, however, that this is in part an artifact of the presentation, in the original style of Gentzen. A common simplification involves the use of multisets of formulas in the interpretation of the sequent, rather than sequences, eliminating the need for an explicit permutation rule. This corresponds to shifting commutativity of assumptions and derivations outside the sequent calculus, whereas LK embeds it within the system itself.

### 74.2.4   Structural rules

The structural rules deserve some additional discussion.

Weakening (W) allows the addition of arbitrary elements to a sequence. Intuitively, this is allowed in the antecedent because we can always restrict the scope of our proof (if all cars have wheels, then it's safe to say that all black cars have wheels); and in the succedent because we can always allow for alternative conclusions (if all cars have wheels, then it's safe to say that all cars have either wheels or wings).

Contraction (C) and Permutation (P) assure that neither the order (P) nor the multiplicity of occurrences (C) of elements of the sequences matters. Thus, one could instead of sequences also consider sets.

The extra effort of using sequences, however, is justified since part or all of the structural rules may be omitted. Doing so, one obtains the so-called substructural logics.

### 74.2.5 Properties of the system LK

This system of rules can be shown to be both sound and complete with respect to first-order logic, i.e. a statement $A$ follows semantically from a set of premises $\Gamma$ ($\Gamma \vDash A$) iff the sequent $\Gamma \vdash A$ can be derived by the above rules.[21]

In the sequent calculus, the rule of cut is admissible. This result is also referred to as Gentzen's *Hauptsatz* ("Main Theorem").[2][3]

## 74.3 Variants

The above rules can be modified in various ways:

### 74.3.1 Minor structural alternatives

There is some freedom of choice regarding the technical details of how sequents and structural rules are formalized. As long as every derivation in LK can be effectively transformed to a derivation using the new rules and vice versa, the modified rules may still be called LK.

First of all, as mentioned above, the sequents can be viewed to consist of sets or multisets. In this case, the rules for permuting and (when using sets) contracting formulae are obsolete.

The rule of weakening will become admissible, when the axiom (I) is changed, such that any sequent of the form $\Gamma, A \vdash A, \Delta$ can be concluded. This means that $A$ proves $A$ in any context. Any weakening that appears in a derivation can then be performed right at the start. This may be a convenient change when constructing proofs bottom-up.

Independent of these one may also change the way in which contexts are split within the rules: In the cases $(\wedge R), (\vee L)$, and $(\to L)$ the left context is somehow split into $\Gamma$ and $\Sigma$ when going upwards. Since contraction allows for the duplication of these, one may assume that the full context is used in both branches of the derivation. By doing this, one assures that no important premises are lost in the wrong branch. Using weakening, the irrelevant parts of the context can be eliminated later.

### 74.3.2 Absurdity

One can introduce $\perp$, the absurdity constant representing *false*, with the axiom:

$$\frac{}{\perp \vdash}$$

Or if, as described above, weakening is to be an admissible rule, then with the axiom:

$$\frac{}{\Gamma, \perp \vdash \Delta}$$

With $\perp$, negation can be subsumed as a special case of implication, via the definition $\neg A \iff A \to \perp$.

### 74.3.3  Substructural logics

Main article: Substructural logic

Alternatively, one may restrict or forbid the use of some of the structural rules. This yields a variety of substructural logic systems. They are generally weaker than LK (*i.e.*, they have fewer theorems), and thus not complete with respect to the standard semantics of first-order logic. However, they have other interesting properties that have led to applications in theoretical computer science and artificial intelligence.

### 74.3.4  Intuitionistic sequent calculus: System LJ

Surprisingly, some small changes in the rules of LK suffice to turn it into a proof system for intuitionistic logic.[22] To this end, one has to restrict to sequents with exactly one formula on the right-hand side, and modify the rules to maintain this invariant. For example, $(\vee L)$ is reformulated as follows (where C is an arbitrary formula):

$$\frac{\Gamma, A \vdash C \qquad \Sigma, B \vdash C}{\Gamma, \Sigma, A \vee B \vdash C} \ (\vee L)$$

The resulting system is called LJ. It is sound and complete with respect to intuitionistic logic and admits a similar cut-elimination proof. This can be used in proving disjunction and existence properties.

In fact, the only two rules in LK that need to be restricted to single-formula consequents are $(\to R)$ and $(\neg R)$ [23] (and the latter can be seen as a special case of the former, via $\bot$ as described above). When multi-formula consequents are interpreted as disjunctions, all of the other inference rules of LK are actually derivable in LJ, while the offending rule is

$$\frac{\Gamma, A \vdash B \vee C}{\Gamma \vdash (A \to B) \vee C}$$

This amounts to the propositional formula $(A \to (B \vee C)) \to ((A \to B) \vee C)$, a classical tautology that is not constructively valid.

## 74.4  See also

- Resolution (logic)

## 74.5  Notes

[1] Gentzen 1934, Gentzen 1935.

[2] Curry 1977, pp. 208–213, gives a 5-page proof of the elimination theorem. See also pages 188, 250.

[3] Kleene 2009, pp. 453, gives a very brief proof of the cut-elimination theorem.

[4] Curry 1977, pp. 189–244, calls Gentzen systems LC systems. Curry's emphasis is more on theory than on practical logic proofs.

[5] Kleene 2009, pp. 440–516. This book is much more concerned with the theoretical, metamathematical implications of Gentzen-style sequent calculus than applications to practical logic proofs.

[6] Kleene 2002, pp. 283–312, 331–361, defines Gentzen systems and proves various theorems within these systems, including Gödel's completeness theorem and Gentzen's theorem.

[7] Smullyan 1995, pp. 101–127, gives a brief theoretical presentation of Gentzen systems. He uses the tableau proof layout style.

[8] Curry 1977, pp. 184–244, compares natural deduction systems, denoted LA, and Gentzen systems, denoted LC. Curry's emphasis is more theoretical than practical.

[9] Suppes 1999, pp. 25–150, is an introductory presentation of practical natural deduction of this kind. This became the basis of System L.

[10] Lemmon 1965 is an elementary introduction to practical natural deduction based on the convenient abbreviated proof layout style System L based on Suppes 1999, pp. 25–150.

[11] Here, "whenever" is used as an informal abbreviation "for every assignment of values to the free variables in the judgment"

[12] Shankar, Natarajan; Owre, Sam; Rushby, John M.; Stringer-Calvert, David W. J. (2001-11-01). "PVS Prover Guide" (PDF). *User guide*. SRI International. Retrieved 2015-05-29.

[13] Buss 1998, p. 10

[14] For explanations of the disjunctive semantics for the right side of sequents, see Curry 1977, pp. 189–190, Kleene 2002, pp. 290, 297, Kleene 2009, p. 441, Hilbert & Bernays 1970, p. 385, Smullyan 1995, pp. 104–105 and Gentzen 1934, p. 180.

[15] Gentzen 1934, p. 188. "Der Kalkül *NJ* hat manche formale Unschönheiten."

[16] Gentzen 1934, p. 191. "In dem klassischen Kalkül *NK* nahm der Satz vom ausgeschlossenen Dritten eine Sonderstellung unter den Schlußweisen ein [...], indem er sich der Einführungs- und Beseitigungssystematik nicht einfügte. Bei dem im folgenden anzugebenden logistischen klassichen Kalkül *LK* wird diese Sonderstellung aufgehoben."

[17] Gentzen 1934, p. 191. "Die damit erreichte Symmetrie erweist sich als für die klassische Logik angemessener."

[18] Gentzen 1934, p. 191. "Hiermit haben wir einige Gesichtspunkte zur Begründung der Aufstellung der folgenden Kalküle angegeben. Im wesentlichen ist ihre Form jedoch durch die Rücksicht auf den nachher zu beweisenden 'Hauptsatz' bestimmt und kann daher vorläufig nicht näher begründet werden."

[19] Kleene 2002, p. 441.

[20] Gentzen 1934, pp. 190–193.

[21] Kleene 2002, p. 336, wrote in 1967 that "it was a major logical discovery by Gentzen 1934–5 that, when there is any (purely logical) proof of a proposition, there is a direct proof. The implications of this discovery are in theoretical logical investigations, rather than in building collections of proved formulas."

[22] Gentzen 1934, p. 194, wrote: "Der Unterschied zwischen *intuitionistischer* und *klassischer* Logik ist bei den Kalkülen *LJ* und *LK* äußerlich ganz anderer Art als bei *NJ* und *NK*. Dort bestand er in Weglassung bzw. Hinzunahme des Satzes vom ausgeschlossenen Dritten, während er hier durch die Sukzedensbedingung ausgedrückt wird." English translation: "The difference between *intuitionistic* and *classical* logic is in the case of the calculi *LJ* and *LK* of an extremely, totally different kind to the case of *NJ* and *NK*. In the latter case, it consisted of the removal or addition respectively of the excluded middle rule, whereas in the former case, it is expressed through the succedent conditions."

[23] Structural Proof Theory (CUP, 2001), Sara Negri and Jan van Plato

## 74.6 References

- Buss, Samuel R. (1998). "An introduction to proof theory". In Samuel R. Buss. *Handbook of proof theory*. Elsevier. pp. 1–78. ISBN 0-444-89840-9.

- Curry, Haskell Brooks (1977) [1963]. *Foundations of mathematical logic*. New York: Dover Publications Inc. ISBN 978-0-486-63462-3.

- Gentzen, Gerhard Karl Erich (1934). "Untersuchungen über das logische Schließen. I". *Mathematische Zeitschrift* **39** (2): 176–210. doi:10.1007/BF01201353.

- Gentzen, Gerhard Karl Erich (1935). "Untersuchungen über das logische Schließen. II". *Mathematische Zeitschrift* **39** (3): 405–431. doi:10.1007/bf01201363.

- Girard, Jean-Yves; Paul Taylor; Yves Lafont (1990) [1989]. *Proofs and Types*. Cambridge University Press (Cambridge Tracts in Theoretical Computer Science, 7). ISBN 0-521-37181-3.

- Hilbert, David; Bernays, Paul (1970) [1939]. *Grundlagen der Mathematik II* (Second ed.). Berlin, New York: Springer-Verlag. ISBN 978-3-642-86897-9.

- Kleene, Stephen Cole (2009) [1952]. *Introduction to metamathematics*. Ishi Press International. ISBN 978-0-923891-57-2.

- Kleene, Stephen Cole (2002) [1967]. *Mathematical logic*. Mineola, New York: Dover Publications. ISBN 978-0-486-42533-7.

- Lemmon, Edward John (1965). *Beginning logic*. Thomas Nelson. ISBN 0-17-712040-1.

- Smullyan, Raymond Merrill (1995) [1968]. *First-order logic*. New York: Dover Publications. ISBN 978-0-486-68370-6.

- Suppes, Patrick Colonel (1999) [1957]. *Introduction to logic*. Mineola, New York: Dover Publications. ISBN 978-0-486-40687-9.

## 74.7   External links

- Hazewinkel, Michiel, ed. (2001), "Sequent calculus", *Encyclopedia of Mathematics*, Springer, ISBN 978-1-55608-010-4

- A Brief Diversion: Sequent Calculus

- Interactive tutorial of the Sequent Calculus

# Chapter 75

# Setoid

In mathematics, a **setoid** (also called an **E-set**) is a set (or type) equipped with an equivalence relation.

Setoids are studied especially in proof theory and in type-theoretic foundations of mathematics. Often in mathematics, when one defines an equivalence relation on a set, one immediately forms the quotient set (turning equivalence into equality). In contrast, setoids may be used when a difference between identity and equivalence must be maintained, often with an interpretation of intensional equality (the equality on the original set) and extensional equality (the equivalence relation, or the equality on the quotient set).

## 75.1 Proof theory

In proof theory, particularly the proof theory of constructive mathematics based on the Curry–Howard correspondence, one often identifies a mathematical proposition with its set of proofs (if any). A given proposition may have many proofs, of course; according to the principle of proof irrelevance, normally only the truth of the proposition matters, not which proof was used. However, the Curry–Howard correspondence can turn proofs into algorithms, and differences between algorithms are often important. So proof theorists may prefer to identify a proposition with a *setoid* of proofs, considering proofs equivalent if they can be converted into one another through beta conversion or the like.

## 75.2 Type theory

In type-theoretic foundations of mathematics, setoids may be used in a type theory that lacks quotient types to model general mathematical sets. For example, in Per Martin-Löf's intuitionistic type theory, there is no type of real numbers, only a type of regular Cauchy sequences of rational numbers. To do real analysis in Martin-Löf's framework, therefore, one must work with a *setoid* of real numbers, the type of regular Cauchy sequences equipped with the usual notion of equivalence. Predicates and functions of real numbers need to be defined for regular Cauchy sequences and proven to be compatible with the equivalence relation. Typically (although it depends on the type theory used), the axiom of choice will hold for functions between types (intensional functions), but not for functions between setoids (extensional functions). The term "set" is variously used either as a synonym of "type" or as a synonym of "setoid".[1]

## 75.3 Constructive mathematics

In constructive mathematics, one often takes a setoid with an apartness relation instead of an equivalence relation, called a **constructive** setoid. One sometimes also considers a **partial** setoid using a partial equivalence relation or partial apartness. (see e.g. Barthe *et al.*, section 1)

## 75.4  See also

- Groupoid

## 75.5  Notes

[1] "Bishop's set theory" (PDF). p. 9.

## 75.6  References

- Hofmann, Martin (1995), "A simple model for quotient types", *Typed lambda calculi and applications (Edinburgh, 1995)*, Lecture Notes in Comput. Sci. **902**, Berlin: Springer, pp. 216–234, doi:10.1007/BFb0014055, MR 1477985.

- Barthe, Gilles; Capretta, Venanzio; Pons, Olivier (2003), "Setoids in type theory" (PDF), *Journal of Functional Programming* **13** (2): 261–293, doi:10.1017/S0956796802004501, MR 1985376.

- Setoid in *nLab*

## 75.7  External links

- Implementation of setoids in Coq

# Chapter 76

# Slow-growing hierarchy

In computability theory, computational complexity theory and proof theory, the **slow-growing hierarchy** is an ordinal-indexed family of slowly increasing functions $g\alpha$: $\mathbf{N} \to \mathbf{N}$ (where $\mathbf{N}$ is the set of natural numbers, $\{0, 1, ...\}$). It contrasts with the fast-growing hierarchy.

## 76.1 Definition

Let $\mu$ be a large countable ordinal such that a fundamental sequence is assigned to every limit ordinal less than $\mu$. The **slow-growing hierarchy** of functions $g\alpha$: $\mathbf{N} \to \mathbf{N}$, for $\alpha < \mu$, is then defined as follows:

- $g_0(n) = 0$
- $g_{k+1}(n) = g_k(n) + 1$
- $g_\alpha(n) = g_{\alpha[n]}(n)$ for limit ordinal $\alpha$.

Here $\alpha[n]$ denotes the $n^{\text{th}}$ element of the fundamental sequence assigned to the limit ordinal $\alpha$.

The article on the Fast-growing hierarchy describes a standardized choice for fundamental sequence for all $\alpha < \varepsilon_0$.

## 76.2 Relation to fast-growing hierarchy

The slow-growing hierarchy grows much more slowly than the fast-growing hierarchy. Even $g\varepsilon_0$ is only equivalent to $f_3$ and $g\alpha$ only attains the growth of $f\varepsilon_0$ (the first function that Peano arithmetic cannot prove total in the hierarchy) when $\alpha$ is the Bachmann–Howard ordinal.[1][2][3]

However, Girard proved that the slow-growing hierarchy eventually *catches up* with the fast-growing one.[1] Specifically, that there exists an ordinal $\alpha$ such that for all integers $n$

$$g\alpha(n) < f\alpha(n) < g\alpha(n + 1)$$

where $f\alpha$ are the functions in the fast-growing hierarchy. He further showed that the first $\alpha$ this holds for is the ordinal of the theory $ID_{<\omega}$ of arbitrary finite iterations of an inductive definition.[4] However for the assignment of fundamental sequences found in [2] the first match up occurs at the level $\varepsilon_0$.[5] For Buchholz style tree ordinals it could be shown that the first match up even occurs at $\omega^2$.

Extensions of the result proved[4] to considerably larger ordinals show that there are very few ordinals below the ordinal of transfinitely iterated $\Pi_1^1$-comprehension where the slow- and fast-growing hierarchy match up.[6]

The slow-growing hierarchy depends extremely sensitively on the choice of the underlying fundamental sequences.[5][7][8][8]

## 76.3   Relation to term rewriting

Cichon provided an interesting connection between the slow-growing hierarchy and derivation length for term rewriting.[2]

## 76.4   References

- Gallier, Jean H. (1991). "What's so special about Kruskal's theorem and the ordinal $\Gamma_0$? A survey of some results in proof theory". *Ann. Pure Appl. Logic* **53** (3): 199–260. doi:10.1016/0168-0072(91)90022-E. MR 1129778 PDF's: part 1 2 3. (In particular part 3, Section 12, pp. 59–64, "A Glimpse at Hierarchies of Fast and Slow Growing Functions".)

## 76.5   Notes

[1] Girard, Jean-Yves (1981). "$\Pi^1_2$-logic. I. Dilators". *Annals of Mathematical Logic* **21** (2): 75–219. doi:10.1016/0003-4843(81)90016-4. ISSN 0003-4843. MR 656793

[2] Cichon (1992). "Termination Proofs and Complexity Characterisations". In P. Aczel, H. Simmons, S. Wainer. *Proof Theory.* Cambridge University Press. pp. 173–193.

[3] Cichon, E. A.; Wainer, S. S. (1983). "The slow-growing and the Grzegorczyk hierarchies". *The Journal of Symbolic Logic* **48** (2): 399–408. doi:10.2307/2273557. ISSN 0022-4812. MR 704094

[4] Wainer, S. S. (1989). "Slow Growing Versus Fast Growing". *The Journal of Symbolic Logic* **54**(2): 608–614. doi:10.2307/2274 JSTOR 2274873.

[5] Weiermann, A (1997). "Sometimes slow growing is fast growing". *Annals of Pure and Applied Logic* **90**: 91. doi:10.1016/S0168-0072(97)00033-X.

[6] Weiermann, A. (1995). *Archives of Mathematical Logic* **34**: 313–330. Missing or empty |title= (help)

[7] Weiermann, A. (1999), "What makes a (pointwise) subrecursive hierarchy slow growing?" Cooper, S. Barry (ed.) et al., Sets and proofs. Invited papers from the Logic colloquium '97, European meeting of the Association for Symbolic Logic, Leeds, UK, July 6–13, 1997. Cambridge: Cambridge University Press. Lond. Math. Soc. Lect. Note Ser. 258; 403-423.

[8] Weiermann, A. (2001) $\Gamma_0$ may be minimal subrecursively inaccessible. Mathematical Logic Quarterly 47 (2001) 397-408.

# Chapter 77

# Soundness

In mathematical logic, a logical system has the **soundness** property if and only if its inference rules prove only formulas that are valid with respect to its semantics. In most cases, this comes down to its rules having the property of preserving *truth*, but this is not the case in general.

## 77.1   Of arguments

An argument is **sound** if and only if

1. The argument is valid, and 2. All of its premises are true.

For instance,

> All men are mortal.
>
> Socrates is a man.
>
> Therefore, Socrates is mortal.

The argument is valid (because the conclusion is true based on the premises, that is, that the conclusion follows the premises) and since the premises are in fact true, the argument is sound.

The following argument is valid but not sound:

> All organisms with wings can fly.
>
> Penguins have wings.
>
> Therefore, penguins can fly.

Since the first premise is actually false, the argument, though valid, is not sound.

## 77.2   Logical systems

Soundness is among the most fundamental properties of mathematical logic. The soundness property provides the initial reason for counting a logical system as desirable. The completeness property means that every validity (truth) is provable. Together they imply that all and only validities are provable.

Most proofs of soundness are trivial. For example, in an axiomatic system, proof of soundness amounts to verifying the validity of the axioms and that the rules of inference preserve validity (or the weaker property, truth). Most axiomatic

247

systems have only the rule of modus ponens (and sometimes substitution), so it requires only verifying the validity of the axioms and one rule of inference.

Soundness properties come in two main varieties: weak and strong soundness, of which the former is a restricted form of the latter.

### 77.2.1  Soundness

Soundness of a deductive system is the property that any sentence that is provable in that deductive system is also true on all interpretations or structures of the semantic theory for the language upon which that theory is based. In symbols, where $S$ is the deductive system, $L$ the language together with its semantic theory, and $P$ a sentence of $L$: if $\vdash_S P$, then also $\vDash_L P$.

### 77.2.2  Strong soundness

Strong soundness of a deductive system is the property that any sentence $P$ of the language upon which the deductive system is based that is derivable from a set $\Gamma$ of sentences of that language is also a logical consequence of that set, in the sense that any model that makes all members of $\Gamma$ true will also make $P$ true. In symbols where $\Gamma$ is a set of sentences of $L$: if $\Gamma \vdash_S P$, then also $\Gamma \vDash_L P$. Notice that in the statement of strong soundness, when $\Gamma$ is empty, we have the statement of weak soundness.

### 77.2.3  Arithmetic soundness

If $T$ is a theory whose objects of discourse can be interpreted as natural numbers, we say $T$ is *arithmetically sound* if all theorems of $T$ are actually true about the standard mathematical integers. For further information, see ω-consistent theory.

## 77.3  Relation to completeness

The converse of the soundness property is the semantic completeness property. A deductive system with a semantic theory is strongly complete if every sentence $P$ that is a semantic consequence of a set of sentences $\Gamma$ can be derived in the deduction system from that set. In symbols: whenever $\Gamma \vDash P$, then also $\Gamma \vdash P$. Completeness of first-order logic was first explicitly established by Gödel, though some of the main results were contained in earlier work of Skolem.

Informally, a soundness theorem for a deductive system expresses that all provable sentences are true. Completeness states that all true sentences are provable.

Gödel's first incompleteness theorem shows that for languages sufficient for doing a certain amount of arithmetic, there can be no effective deductive system that is complete with respect to the intended interpretation of the symbolism of that language. Thus, not all sound deductive systems are complete in this special sense of completeness, in which the class of models (up to isomorphism) is restricted to the intended one. The original completeness proof applies to *all* classical models, not some special proper subclass of intended ones.

## 77.4  See also

- Validity

- Completeness (logic)

## 77.5  References

- Hinman, P. (2005). *Fundamentals of Mathematical Logic*. A K Peters. ISBN 1-56881-262-0.

- Copi, Irving (1979), *Symbolic Logic* (5th ed.), Macmillan Publishing Co., ISBN 0-02-324880-7

- Boolos, Burgess, Jeffrey. *Computability and Logic*, 4th Ed, Cambridge, 2002.

## 77.6  External links

- Validity and Soundness in the *Internet Encyclopedia of Philosophy*.

# Chapter 78

# Soundness (interactive proof)

**Soundness** is a property of interactive proof systems that requires that no prover can make the verifier accept for a wrong statement $y \notin L$ except with some small probability. The upper bound of this probability is referred to as the soundness error of a proof system.

More formally, for every prover $(\tilde{\mathcal{P}})$, and every $y \notin L$:

$$\Pr[(\bot, (\text{accept})) \leftarrow (\tilde{\mathcal{P}})(y) \leftrightarrow (\mathcal{V})(y)] < 2^{-80}.$$

The above definition uses the somewhat arbitrary soundness error $2^{-80}$. As long as the soundness error is bounded by a polynomial fraction of the potential running time of the verifier (i.e. $\leq 1/\text{poly}(|y|)$ ), it is always possible to amplify soundness until the soundness error becomes negligible relative to the running time of the verifier. This is achieved by repeating the proof and accepting only if all proofs verify. After $\ell$ repetitions, a soundness error $\epsilon$ will be reduced to $\epsilon^{\ell}$ [1].

## 78.1   See also

- Interactive proof system

- Proof of knowledge

- Zero-knowledge proof

## 78.2   References

[1] Goldreich, Oded (2002), *Zero-Knowledge twenty years after its invention*, ECCC TR02-063.

# Chapter 79

# Structural proof theory

In mathematical logic, **structural proof theory** is the subdiscipline of proof theory that studies proof calculi that support a notion of analytic proof.

## 79.1   Analytic proof

Main article: analytic proof

The notion of analytic proof was introduced into proof theory by Gerhard Gentzen for the sequent calculus; the analytic proofs are those that are cut-free. His natural deduction calculus also supports a notion of analytic proof, as was shown by Dag Prawitz; the definition is slightly more complex — we say the analytic proofs are the normal forms, which are related to the notion of normal form in term rewriting.

## 79.2   Structures and connectives

The term *structure* in structural proof theory comes from a technical notion introduced in the sequent calculus: the sequent calculus represents the judgement made at any stage of an inference using special, extra-logical operators which we call structural operators: in $A_1, \ldots, A_m \vdash B_1, \ldots, B_n$ , the commas to the left of the turnstile are operators normally interpreted as conjunctions, those to the right as disjunctions, whilst the turnstile symbol itself is interpreted as an implication. However, it is important to note that there is a fundamental difference in behaviour between these operators and the logical connectives they are interpreted by in the sequent calculus: the structural operators are used in every rule of the calculus, and are not considered when asking whether the subformula property applies. Furthermore, the logical rules go one way only: logical structure is introduced by logical rules, and cannot be eliminated once created, while structural operators can be introduced and eliminated in the course of a derivation.

The idea of looking at the syntactic features of sequents as special, non-logical operators is not old, and was forced by innovations in proof theory: when the structural operators are as simple as in Getzen's original sequent calculus there is little need to analyse them, but proof calculi of deep inference such as display logic support structural operators as complex as the logical connectives, and demand sophisticated treatment.

## 79.3   Cut-elimination in the sequent calculus

Main article: Cut-elimination

251

## 79.4  Natural deduction and the formulae-as-types correspondence

Main article: Natural deduction


## 79.5  Logical duality and harmony

Main article: Logical harmony


## 79.6  Display logic

## 79.7  Calculus of structures

Main article: Calculus of structures


## 79.8  References

- Sara Negri; Jan Von Plato (2001). *Structural proof theory*. Cambridge University Press. ISBN 978-0-521-79307-0.

- Anne Sjerp Troelstra; Helmut Schwichtenberg (2000). *Basic proof theory* (2nd ed.). Cambridge University Press. ISBN 978-0-521-77911-1.

# Chapter 80

# Structural rule

For the type of rule used in linguistics, see Phrase structure rule.

In proof theory, a **structural rule** is an inference rule that does not refer to any logical connective, but instead operates on the judgment or sequents directly. Structural rules often mimic intended meta-theoretic properties of the logic. Logics that deny one or more of the structural rules are classified as substructural logics.

## 80.1   Common structural rules

Three common structural rules are:

- **Weakening**, where the hypotheses or conclusion of a sequent may be extended with additional members. In symbolic form weakening rules can be written as $\frac{\Gamma \vdash \Sigma}{\Gamma, A \vdash \Sigma}$ on the left of the turnstile, and $\frac{\Gamma \vdash \Sigma}{\Gamma \vdash A, \Sigma}$ on the right.

- **Contraction**, where two equal (or unifiable) members on the same side of a sequent may be replaced by a single member (or common instance). Symbolically: $\frac{\Gamma, A, A \vdash \Sigma}{\Gamma, A \vdash \Sigma}$ and $\frac{\Gamma \vdash A, A, \Sigma}{\Gamma \vdash A, \Sigma}$ . Also known as **factoring** in automated theorem proving systems using resolution.

- **Exchange**, where two members on the same side of a sequent may be swapped. Symbolically: $\frac{\Gamma_1, A, \Gamma_2, B, \Gamma_3 \vdash \Sigma}{\Gamma_1, B, \Gamma_2, A, \Gamma_3 \vdash \Sigma}$ and $\frac{\Gamma \vdash \Sigma_1, A, \Sigma_2, B, \Sigma_3}{\Gamma \vdash \Sigma_1, B, \Sigma_2, A, \Sigma_3}$ . (This is also known as the *permutation rule*.)

A logic without any of the above structural rules would interpret the sides of a sequent as pure sequences; with exchange, they are multisets; and with both contraction and exchange they are sets.

These are not the only possible structural rules. A famous structural rule is known as **cut**. Considerable effort is spent by proof theorists in showing that cut rules are superfluous in various logics. More precisely, what is shown is that cut is only (in a sense) a tool for abbreviating proofs, and does not add to the theorems that can be proved. The successful 'removal' of cut rules, known as *cut elimination*, is directly related to the philosophy of *computation as normalization* (see Curry–Howard correspondence); it often gives a good indication of the complexity of deciding a given logic.

## 80.2   See also

- Affine logic
- Linear logic
- Ordered logic

- Strict logic

# Chapter 81

# Takeuti's conjecture

In mathematics, **Takeuti's conjecture** is the conjecture of Gaisi Takeuti that a sequent formalisation of second-order logic has cut-elimination (Takeuti 1953). It was settled positively:

- By Tait, using a semantic technique for proving cut-elimination, based on work by Schütte (Tait 1966);

- Independently by Takahashi by a similar technique (Takahashi 1967);

- It is a corollary of Jean-Yves Girard's syntactic proof of strong normalization for System F.

Takeuti's conjecture is equivalent[1] to the consistency of second-order arithmetic and to the strong normalization of the Girard/Reynold's System F.

## 81.1   See also

- Hilbert's second problem

## 81.2   Notes

- ^ Equivalent in the sense that each of the statements can be derived from each other in the weak system PRA of arithmetic; consistency refers here to the truth of the Gödel sentence for second-order arithmetic. See consistency proof for more discussion.

## 81.3   References

- William W. Tait, 1966. A nonconstructive proof of Gentzen's Hauptsatz for second order predicate logic. In *Bulletin of the American Mathematical Society*, 72:980–983.

- Gaisi Takeuti, 1953. On a generalized logic calculus. In *Japanese Journal of Mathematics*, 23:39–96. An errata to this article was published in the same journal, 24:149–156, 1954.

- Moto-o Takahashi, 1967. A proof of cut-elimination in simple type theory. In *Japanese Mathematical Society*, 10:44–45.

# Chapter 82

# Tolerant sequence

In mathematical logic, a **tolerant sequence** is a sequence

$$T_1,...,T_n$$

of formal theories such that there are consistent extensions

$$S_1,...,S_n$$

of these theories with each $S_{i+1}$ interpretable in $S_i$. Tolerance naturally generalizes from sequences of theories to trees of theories. Weak interpretability can be shown to be a special, binary case of tolerance.

This concept, together with its dual concept of cotolerance, was introduced by Japaridze in 1992, who also proved that, for Peano arithmetic and any stronger theories with effective axiomatizations, tolerance is equivalent to $\Pi_1$-consistency.

## 82.1   See also

- Interpretability

- Cointerpretability

- Interpretability logic

## 82.2   References

- G.Japaridze, *The logic of linear tolerance*. Studia Logica 51 (1992), pp. 249–277.

- G.Japaridze, *A generalized notion of weak interpretability and the corresponding logic*. Annals of Pure and Applied Logic 61 (1993), pp. 113–160.

- G.Japaridze and D. de Jongh, *The logic of provability*. **Handbook of Proof Theory**. S.Buss, ed. Elsevier, 1998, pp. 476–546.

# Chapter 83

# Turnstile (symbol)

In mathematical logic and computer science the symbol ⊢ has taken the name **turnstile** because of its resemblance to a typical turnstile if viewed from above. It is also referred to as **tee** and is often read as "yields", "proves", "satisfies" or "entails". The symbol was first used by Gottlob Frege in his 1879 book on logic, *Begriffsschrift*.[1]

Martin-Löf analyzes the ⊢ symbol thus: "...[T]he combination of Frege's Urteilsstrich, judgement stroke [ | ], and Inhaltsstrich, content stroke [—], came to be called the assertion sign."[2] Frege's notation for a judgement of some content $A$

$$\vdash A$$

can be then be read

*I know  A is true"*.[3]

In the same vein, a conditional assertion

$$P \vdash Q$$

can be read as:

*From  P , I know that  Q*

In TeX, the turnstile symbol ⊢ is obtained from the command \vdash. In Unicode, the turnstile symbol (⊢) is called **right tack** and is at code point U+22A2.[4] On a typewriter, a turnstile can be composed from a vertical bar (|) and a dash (–). In LaTeX there is a turnstile package which issues this sign in many ways, and is capable of putting labels below or above it, in the correct places.[5]

## 83.1  Interpretations

The turnstile represents a binary relation. It has several different interpretations in different contexts:

- In metalogic, the study of formal languages; the turnstile represents syntactic consequence (or "derivability"). This is to say, that it shows that one string can be derived from another in a single step, according to the transformation rules (i.e. the syntax) of some given formal system.[6] As such, the expression

$P \vdash Q$

means that $Q$ is derivable from $P$ in the system.

Consistent with its use for derivability, a "$\vdash$" followed by an expression without anything preceding it denotes a theorem, which is to say that the expression can be derived from the rules using an empty set of axioms. As such, the expression

$\vdash Q$

means that $Q$ is a theorem in the system.

- In proof theory, the turnstile is used to denote "provability". For example, if $T$ is a formal theory and $S$ is a particular sentence in the language of the theory then

$T \vdash S$

means that $S$ is provable from $T$ .[7] This usage is demonstrated in the article on propositional calculus.

- In the typed lambda calculus, the turnstile is used to separate typing assumptions from the typing judgment.[8][9]

- In category theory, a reversed turnstile ($\dashv$), as in $F \dashv G$ , is used to indicate that the functor $F$ is left adjoint to the functor $G$ .

- In APL the symbol is called "right tack" and represents the ambivalent right identity function where both $X⊢Y$ and $⊢Y$ are $Y$. The reversed symbol "$\dashv$" is called "left tack" and represents the analogous left identity where $X⊣Y$ is $X$ and $⊣Y$ is $Y$.[10][11]

- In combinatorics, $\lambda \vdash n$ means that $\lambda$ is a partition of the integer $n$ . [12]

## 83.2 See also

- Double turnstile $\models$

- Sequent

- Sequent calculus

- List of logic symbols

- List of mathematical symbols

## 83.3 Notes

[1] Frege 1879

[2] Martin-Löf 1996, pp. 6,15

[3] Martin-Löf 1996, p. 15

[4] Unicode standard

[5] http://www.ctan.org/tex-archive/macros/latex/contrib/turnstile

[6] http://dingo.sbs.arizona.edu/~{}hammond/ling178-sp06/mathCh6.pdf

[7] Troelstra & Schwichtenberg 2000

[8] http://www.mscs.dal.ca/~{}selinger/papers/lambdanotes.pdf

[9] Schmidt 1994

[10] http://www.jsoftware.com/papers/APLDictionary.htm

[11] Iverson 1987

[12] p.287 of Stanley, Richard P.. Enumerative Combinatorics. 1st ed. Vol. 2. Cambridge: Cambridge University Press, 1999.

## 83.4 References

- Frege, Gottlob (1879). "Begriffsschrift: Eine der arithmetischen nachgebildete Formelsprache des reinen Denkens". Halle.

- Iverson, Kenneth (1987). "A Dictionary of APL".

- Martin-Löf, Per (1996). "On the meanings of the logical constants and the justifications of the logical laws" (PDF). *Nordic Journal of Philosophical Logic* **1** (1): 11–60. Lecture notes to a short course at Università degli Studi di Siena, April 1983.

- Schmidt, David (1994). "The Structure of Typed Programming Languages". MIT Press. ISBN 0-262-19349-3.

- Troelstra, A. S.; Schwichtenberg, H. (2000). "Basic Proof Theory" (second ed.). Cambridge University Press. ISBN 978-0-521-77911-1.

# Chapter 84

# Undecidable problem

In computability theory and computational complexity theory, an **undecidable problem** is a decision problem for which it is known to be impossible to construct a single algorithm that always leads to a correct yes-or-no answer.

A decision problem is any arbitrary yes-or-no question on an infinite set of inputs. Because of this, it is traditional to define the decision problem equivalently as the set of inputs for which the problem returns *yes*. These inputs can be natural numbers, but also other values of some other kind, such as strings of a formal language. Using some encoding, such as a Gödel numbering, the strings can be encoded as natural numbers. Thus, a decision problem informally phrased in terms of a formal language is also equivalent to a set of natural numbers. To keep the formal definition simple, it is phrased in terms of subsets of the natural numbers.

Formally, a decision problem is a subset of the natural numbers. The corresponding informal problem is that of deciding whether a given number is in the set. A decision problem $A$ is called decidable or effectively solvable if $A$ is a recursive set. A problem is called partially decidable, **semi-decidable**, solvable, or provable if $A$ is a recursively enumerable set. This means that there exists an algorithm that halts eventually when the answer is *yes* but may run for ever if the answer is *no*. Partially decidable problems and any other problems that are not decidable are called undecidable.

## 84.1   In computability theory

In computability theory, the halting problem is a decision problem which can be stated as follows:

> Given the description of an arbitrary program and a finite input, decide whether the program finishes running or will run forever.

Alan Turing proved in 1936 that a general algorithm running on a Turing machine that solves the halting problem for *all* possible program-input pairs necessarily cannot exist. Hence, the halting problem is *undecidable* for Turing machines.

## 84.2   Relationship with Gödel's incompleteness theorem

The concepts raised by Gödel's incompleteness theorems are very similar to those raised by the halting problem, and the proofs are quite similar. In fact, a weaker form of the First Incompleteness Theorem is an easy consequence of the undecidability of the halting problem. This weaker form differs from the standard statement of the incompleteness theorem by asserting that a complete, consistent and sound axiomatization of all statements about natural numbers is unachievable. The "sound" part is the weakening: it means that we require the axiomatic system in question to prove only *true* statements about natural numbers. It is important to observe that the statement of the standard form of Gödel's First Incompleteness Theorem is completely unconcerned with the question of truth, but only concerns the issue of whether it can be proven.

The weaker form of the theorem can be proved from the undecidability of the halting problem as follows. Assume that we have a consistent and complete axiomatization of all true first-order logic statements about natural numbers. Then we can build an algorithm that enumerates all these statements. This means that there is an algorithm $N(n)$ that, given a natural number $n$, computes a true first-order logic statement about natural numbers such that, for all the true statements, there is at least one $n$ such that $N(n)$ yields that statement. Now suppose we want to decide if the algorithm with representation $a$ halts on input $i$. We know that this statement can be expressed with a first-order logic statement, say $H(a, i)$. Since the axiomatization is complete it follows that either there is an $n$ such that $N(n) = H(a, i)$ or there is an $n'$ such that $N(n') = \neg H(a, i)$. So if we iterate over all $n$ until we either find $H(a, i)$ or its negation, we will always halt. This means that this gives us an algorithm to decide the halting problem. Since we know that there cannot be such an algorithm, it follows that the assumption that there is a consistent and complete axiomatization of all true first-order logic statements about natural numbers must be false.

## 84.3 Examples of undecidable problems

Main article: List of undecidable problems

Undecidable problems can be related to different topics, such as logic, abstract machines or topology. Note that since there are uncountably many undecidable problems, any list, even one of infinite length, is necessarily incomplete.

## 84.4 Examples of undecidable statements

There are two distinct senses of the word "undecidable" in contemporary use. The first of these is the sense used in relation to Gödel's theorems, that of a statement being neither provable nor refutable in a specified deductive system. The second sense is used in relation to computability theory and applies not to statements but to decision problems, which are countably infinite sets of questions each requiring a yes or no answer. Such a problem is said to be undecidable if there is no computable function that correctly answers every question in the problem set. The connection between these two is that if a decision problem is undecidable (in the recursion theoretical sense) then there is no consistent, effective formal system which proves for every question $A$ in the problem either "the answer to $A$ is yes" or "the answer to $A$ is no".

Because of the two meanings of the word undecidable, the term independent is sometimes used instead of undecidable for the "neither provable nor refutable" sense. The usage of "independent" is also ambiguous, however. It can mean just "not provable", leaving open whether an independent statement might be refuted.

Undecidability of a statement in a particular deductive system does not, in and of itself, address the question of whether the truth value of the statement is well-defined, or whether it can be determined by other means. Undecidability only implies that the particular deductive system being considered does not prove the truth or falsity of the statement. Whether there exist so-called "absolutely undecidable" statements, whose truth value can never be known or is ill-specified, is a controversial point among various philosophical schools.

One of the first problems suspected to be undecidable, in the second sense of the term, was the word problem for groups, first posed by Max Dehn in 1911, which asks if there is a finitely presented group for which no algorithm exists to determine whether two words are equivalent. This was shown to be the case in 1952.

The combined work of Gödel and Paul Cohen has given two concrete examples of undecidable statements (in the first sense of the term): The continuum hypothesis can neither be proved nor refuted in ZFC (the standard axiomatization of set theory), and the axiom of choice can neither be proved nor refuted in ZF (which is all the ZFC axioms *except* the axiom of choice). These results do not require the incompleteness theorem. Gödel proved in 1940 that neither of these statements could be disproved in ZF or ZFC set theory. In the 1960s, Cohen proved that neither is provable from ZF, and the continuum hypothesis cannot be proven from ZFC.

In 1970, Russian mathematician Yuri Matiyasevich showed that Hilbert's Tenth Problem, posed in 1900 as a challenge to the next century of mathematicians, cannot be solved. Hilbert's challenge sought an algorithm which finds all solutions of a Diophantine equation. A Diophantine equation is a more general case of Fermat's Last Theorem; we seek the integer roots of a polynomial in any number of variables with integer coefficients.  Since we have only one equation but $n$ variables, infinitely many solutions exist (and are easy to find) in the complex plane; however, the problem becomes impossible if solutions are constrained to integer values only.  Matiyasevich showed this problem to be unsolvable by mapping a Diophantine equation to a recursively enumerable set and invoking Gödel's Incompleteness Theorem.[1]

In 1936, Alan Turing proved that the halting problem—the question of whether or not a Turing machine halts on a given program—is undecidable, in the second sense of the term.  This result was later generalized by Rice's theorem.

In 1973, the Whitehead problem in group theory was shown to be undecidable, in the first sense of the term, in standard set theory.

In 1977, Paris and Harrington proved that the Paris-Harrington principle, a version of the Ramsey theorem, is undecidable in the axiomatization of arithmetic given by the Peano axioms but can be proven to be true in the larger system of second-order arithmetic.

Kruskal's tree theorem, which has applications in computer science, is also undecidable from the Peano axioms but provable in set theory.  In fact Kruskal's tree theorem (or its finite form) is undecidable in a much stronger system codifying the principles acceptable on basis of a philosophy of mathematics called predicativism.

Goodstein's theorem is a statement about the Ramsey theory of the natural numbers that Kirby and Paris showed is undecidable in Peano arithmetic.

Gregory Chaitin produced undecidable statements in algorithmic information theory and proved another incompleteness theorem in that setting.  Chaitin's theorem states that for any theory that can represent enough arithmetic, there is an upper bound $c$ such that no specific number can be proven in that theory to have Kolmogorov complexity greater than $c$.  While Gödel's theorem is related to the liar paradox, Chaitin's result is related to Berry's paradox.

In 2007, researchers Kurtz and Simon, building on earlier work by J.H. Conway in the 1970s, proved that a natural generalization of the Collatz problem is undecidable.[2]

## 84.5   See also

- Entscheidungsproblem

## 84.6   References

[1]  Matiyasevich, Yuri (1970). Диофантовость перечислимых множеств [Enumerable sets are Diophantine]. *Doklady Akademii Nauk SSSR* (in Russian) **191**: 279–282.

[2]  Kurtz, Stuart A.; Simon, Janos, "The Undecidability of the. Generalized Collatz Problem", in Proceedings of the 4th International Conference on Theory and Applications of Models of Computation, TAMC 2007, held in Shanghai, China in May 2007. ISBN 3-540-72503-2. doi:10.1007/978-3-540-72504-6_49

# Chapter 85

# Veblen function

In mathematics, the **Veblen functions** are a hierarchy of normal functions (continuous strictly increasing functions from ordinals to ordinals), introduced by Oswald Veblen in Veblen (1908). If $\varphi_0$ is any normal function, then for any non-zero ordinal $\alpha$, $\varphi\alpha$ is the function enumerating the common fixed points of $\varphi_\beta$ for $\beta<\alpha$. These functions are all normal.

## 85.1 The Veblen hierarchy

In the special case when $\varphi_0(\alpha)=\omega^\alpha$ this family of functions is known as the **Veblen hierarchy**. The function $\varphi_1$ is the same as the $\varepsilon$ function: $\varphi_1(\alpha)= \varepsilon\alpha$. If $\alpha < \beta$, then $\varphi_\alpha(\varphi_\beta(\gamma)) = \varphi_\beta(\gamma)$. From this and the fact that $\varphi_\beta$ is strictly increasing we get the ordering: $\varphi_\alpha(\beta) < \varphi_\gamma(\delta)$ if and only if either ( $\alpha = \gamma$ and $\beta < \delta$ ) or ( $\alpha < \gamma$ and $\beta < \varphi_\gamma(\delta)$ ) or ( $\alpha > \gamma$ and $\varphi_\alpha(\beta) < \delta$ ).

### 85.1.1 Fundamental sequences for the Veblen hierarchy

The fundamental sequence for an ordinal with cofinality $\omega$ is a distinguished strictly increasing $\omega$-sequence which has the ordinal as its limit. If one has fundamental sequences for $\alpha$ and all smaller limit ordinals, then one can create an explicit constructive bijection between $\omega$ and $\alpha$, (i.e. one not using the axiom of choice). Here we will describe fundamental sequences for the Veblen hierarchy of ordinals. The image of $n$ under the fundamental sequence for $\alpha$ will be indicated by $\alpha[n]$.

A variation of Cantor normal form used in connection with the Veblen hierarchy is — every nonzero ordinal number $\alpha$ can be uniquely written as $\alpha = \varphi_{\beta_1}(\gamma_1) + \varphi_{\beta_2}(\gamma_2) + \cdots + \varphi_{\beta_k}(\gamma_k)$, where $k>0$ is a natural number and each term after the first is less than or equal to the previous term, $\varphi_{\beta_m}(\gamma_m) \geq \varphi_{\beta_{m+1}}(\gamma_{m+1})$, and each $\gamma_m < \varphi_{\beta_m}(\gamma_m)$. If a fundamental sequence can be provided for the last term, then that term can be replaced by such a sequence to get $\alpha[n] = \varphi_{\beta_1}(\gamma_1) + \cdots + \varphi_{\beta_{k-1}}(\gamma_{k-1}) + (\varphi_{\beta_k}(\gamma_k)[n])$.

For any $\beta$, if $\gamma$ is a limit with $\gamma < \varphi_\beta(\gamma)$, then let $\varphi_\beta(\gamma)[n] = \varphi_\beta(\gamma[n])$.

No such sequence can be provided for $\varphi_0(0) = \omega^0 = 1$ because it does not have cofinality $\omega$.

For $\varphi_0(\gamma + 1) = \omega^{\gamma+1} = \omega^\gamma \cdot \omega$, we choose $\varphi_0(\gamma + 1)[n] = \varphi_0(\gamma) \cdot n = \omega^\gamma \cdot n$.

For $\varphi_{\beta+1}(0)$, we use $\varphi_{\beta+1}(0)[0] = 0$ and $\varphi_{\beta+1}(0)[n + 1] = \varphi_\beta(\varphi_{\beta+1}(0)[n])$, i.e. 0, $\varphi_\beta(0)$, $\varphi_\beta(\varphi_\beta(0))$, etc..

For $\varphi_{\beta+1}(\gamma + 1)$, we use $\varphi_{\beta+1}(\gamma + 1)[0] = \varphi_{\beta+1}(\gamma) + 1$ and $\varphi_{\beta+1}(\gamma + 1)[n + 1] = \varphi_\beta(\varphi_{\beta+1}(\gamma + 1)[n])$.

Now suppose that $\beta$ is a limit:

If $\beta < \varphi_\beta(0)$, then let $\varphi_\beta(0)[n] = \varphi_{\beta[n]}(0)$.

For $\varphi_\beta(\gamma + 1)$, use $\varphi_\beta(\gamma + 1)[n] = \varphi_{\beta[n]}(\varphi_\beta(\gamma) + 1)$.

Otherwise, the ordinal cannot be described in terms of smaller ordinals using $\varphi$ and this scheme does not apply to it.

### 85.1.2   The $\Gamma$ function

The function $\Gamma$ enumerates the ordinals $\alpha$ such that $\varphi\alpha(0) = \alpha$. $\Gamma_0$ is the Feferman–Schütte ordinal, i.e. it is the smallest $\alpha$ such that $\varphi\alpha(0) = \alpha$.

For $\Gamma_0$, a fundamental sequence could be chosen to be $\Gamma_0[0] = 0$ and $\Gamma_0[n+1] = \varphi_{\Gamma_0[n]}(0)$.

For $\Gamma_{\beta+1}$, let $\Gamma_{\beta+1}[0] = \Gamma_\beta + 1$ and $\Gamma_{\beta+1}[n+1] = \varphi_{\Gamma_{\beta+1}[n]}(0)$.

For $\Gamma_\beta$ where $\beta < \Gamma_\beta$ is a limit, let $\Gamma_\beta[n] = \Gamma_{\beta[n]}$.

## 85.2   Generalizations

### 85.2.1   Finitely many variables

In this section it is more convenient to think of $\varphi\alpha(\beta)$ as a function $\varphi(\alpha,\beta)$ of two variables. Veblen showed how to generalize the definition to produce a function $\varphi(\alpha n, \alpha_{n-1}, \ldots, \alpha_0)$ of several variables, namely:

- $\varphi(\alpha) = \omega^\alpha$ for a single variable,

- $\varphi(0, \alpha_{n-1}, \ldots, \alpha_0) = \varphi(\alpha_{n-1}, \ldots, \alpha_0)$, and

- for $\alpha > 0$, $\gamma \mapsto \varphi(\alpha n, \ldots, \alpha_{i+1}, \alpha, 0, \ldots, 0, \gamma)$ is the function enumerating the common fixed points of the functions $\xi \mapsto \varphi(\alpha n, \ldots, \alpha_{i+1}, \beta, \xi, 0, \ldots, 0)$   for all $\beta < \alpha$.

For example, $\varphi(1,0,\gamma)$ is the $\gamma$-th fixed point of the functions $\xi \mapsto \varphi(\xi,0)$, namely $\Gamma_\gamma$; then $\varphi(1,1,\gamma)$ enumerates the fixed points of that function, i.e., of the $\xi \mapsto \Gamma\xi$ function; and $\varphi(2,0,\gamma)$ enumerates the fixed points of all the $\xi \mapsto \varphi(1,\xi,0)$. Each instance of the generalized Veblen functions is continuous in the *last nonzero* variable (i.e., if one variable is made to vary and all later variables are kept constantly equal to zero).

The ordinal $\varphi(1,0,0,0)$ is sometimes known as the Ackermann ordinal. The limit of the $\varphi(1,0,\ldots,0)$ where the number of zeroes ranges over $\omega$, is sometimes known as the "small" Veblen ordinal.

### 85.2.2   Transfinitely many variables

More generally, Veblen showed that $\varphi$ can be defined even for a transfinite sequence of ordinals $\alpha_\beta$, provided that all but a finite number of them are zero. Notice that if such a sequence of ordinals is chosen from those less than an uncountable regular cardinal $\kappa$, then the sequence may be encoded as a single ordinal less than $\kappa^\kappa$. So one is defining a function $\varphi$ from $\kappa^\kappa$ into $\kappa$.

The definition can be given as follows: let $\alpha$ be a transfinite sequence of ordinals (i.e., an ordinal function with finite support) *which ends in zero* (i.e., such that $\alpha_0 = 0$), and let $\alpha[0 \mapsto \gamma]$ denote the same function where the final 0 has been replaced by $\gamma$. Then $\gamma \mapsto \varphi(\alpha[0 \mapsto \gamma])$ is defined as the function enumerating the common fixed points of all functions $\xi \mapsto \varphi(\beta)$ where $\beta$ ranges over all sequences which are obtained by decreasing the smallest-indexed nonzero value of $\alpha$ and replacing some smaller-indexed value with the indeterminate $\xi$ (i.e., $\beta = \alpha[\iota_0 \mapsto \zeta, \iota \mapsto \xi]$ meaning that for the smallest index $\iota_0$ such that $\alpha\iota_0$ is nonzero the latter has been replaced by some value $\zeta < \alpha\iota_0$ and that for some smaller index $\iota < \iota_0$, the value $\alpha\iota = 0$ has been replaced with $\xi$).

For example, if $\alpha = (\omega \mapsto 1)$ denotes the transfinite sequence with value 1 at $\omega$ and 0 everywhere else, then $\varphi(\omega \mapsto 1)$ is the smallest fixed point of all the functions $\xi \mapsto \varphi(\xi, 0, \ldots, 0)$ with finitely many final zeroes (it is also the limit of the $\varphi(1,0,\ldots,0)$ with finitely many zeroes, the small Veblen ordinal).

The smallest ordinal $\alpha$ such that $\alpha$ is greater than $\varphi$ applied to any function with support in $\alpha$ (i.e., which cannot be reached "from below" using the Veblen function of transfinitely many variables) is sometimes known as the "large" Veblen ordinal.

# 85.3 References

- Hilbert Levitz, *Transfinite Ordinals and Their Notations: For The Uninitiated*, expository article (8 pages, in PostScript)

- Pohlers, Wolfram (1989), *Proof theory*, Lecture Notes in Mathematics **1407**, Berlin: Springer-Verlag, ISBN 3-540-51842-8, MR 1026933

- Schütte, Kurt (1977), *Proof theory*, Grundlehren der Mathematischen Wissenschaften **225**, Berlin-New York: Springer-Verlag, pp. xii+299, ISBN 3-540-07911-4, MR 0505313

- Takeuti, Gaisi (1987), *Proof theory*, Studies in Logic and the Foundations of Mathematics **81** (Second ed.), Amsterdam: North-Holland Publishing Co., ISBN 0-444-87943-9, MR 0882549

- Smorynski, C. (1982), "The varieties of arboreal experience", *Math. Intelligencer* **4**(4): 182–189, doi:10.1007/BF0 contains an informal description of the Veblen hierarchy.

- Veblen, Oswald (1908), "Continuous Increasing Functions of Finite and Transfinite Ordinals", *Transactions of the American Mathematical Society* **9** (3): 280–292, doi:10.2307/1988605, JSTOR 1988605

- Miller, Larry W. (1976), "Normal Functions and Constructive Ordinal Notations", *The Journal of Symbolic Logic* **41** (2): 439–459, doi:10.2307/2272243, JSTOR 2272243

# Chapter 86

# VIPER microprocessor

**VIPER** is a 32-bit microprocessor design created by Royal Signals and Radar Establishment in the 1980s, intended for use in safety-critical systems such as avionics.[1] It was the first commercial microprocessor design to be formally proven correct, although there was some controversy surrounding this claim and the definition of proof.

## 86.1 References

[1] Churchley, Andrew (1991-11-30). *Microprocessor Based Protection Systems*. Springer. p. 64. ISBN 9781851666119. Retrieved 23 July 2012.

## 86.2 External links

- http://oai.dtic.mil/oai/oai?verb=getRecord&metadataPrefix=html&identifier=ADA194561

- http://www.nature.com/nature/journal/v352/n6335/abs/352467a0.html

- http://monoskop.org/images/4/49/MacKenzie_Donald_Knowing_Machines_Essays_on_Technical_Change.pdf

# Chapter 87

# Weak interpretability

In mathematical logic, **weak interpretability** is a notion of translation of logical theories, introduced together with interpretability by Alfred Tarski in 1953.

Assume T and S are formal theories. Slightly simplified, T is said to be **weakly interpretable** in S if, and only if, the language of T can be translated into the language of S in such a way that the translation of every theorem of T is consistent with S. Of course, there are some natural conditions on admissible translations here, such as the necessity for a translation to preserve the logical structure of formulas.

A generalization of weak interpretability, tolerance, was introduced by Giorgi Japaridze in 1992.

## 87.1  See also

- Interpretability logic.

## 87.2  References

- Tarski, Alfred (1953), *Undecidable theories*, Studies in Logic and the Foundations of Mathematics, Amsterdam: North-Holland Publishing Company, MR 0058532. Written in collaboration with Andrzej Mostowski and Raphael M. Robinson.

- Dzhaparidze, Giorgie (1993), "A generalized notion of weak interpretability and the corresponding modal logic", *Annals of Pure and Applied Logic* **61** (1-2): 113–160, doi:10.1016/0168-0072(93)90201-N, MR 1218658.

- Dzhaparidze, Giorgie(1992), "The logic of linear tolerance",*Studia Logica***51**(2): 249–277,doi:10.1007/BF0037 MR 1185914

- Japaridze, Giorgi; de Jongh, Dick (1998), "The logic of provability", in Buss, Samuel R., *Handbook of Proof Theory*, Stud. Logic Found. Math. **137**, Amsterdam: North-Holland, pp. 475–546, doi:10.1016/S0049-237X(98)80022-0, MR 1640331

## 87.3   Text and image sources, contributors, and licenses

### 87.3.1   Text

- **Proof theory** *Source:* https://en.wikipedia.org/wiki/Proof_theory?oldid=676991676 *Contributors:* Mav, Bryan Derksen, The Anome, Toby Bartels, Youandme, Michael Hardy, Llywrch, Dominus, Rotem Dan, Charles Matthews, Dysprosia, Markhurd, Hyacinth, Jni, Giftlite, Markus Krötzsch, Jorend, Kntg, Leibniz, Pj.de.bruin, Luqui, Number 0, Brian0918, Chalst, Nortexoid, Msh210, Krappie, Wtmitchell, Gene Nygaard, Oleg Alexandrov, Jtauber, Porcher, Mathbot, Comiscuous, Tillmo, Chobot, YurikBot, Hairy Dude, Tong~enwiki, Arthur Rubin, Sardanaphalus, SmackBot, Cabe6403, Jahiegel, Vina-iwbot~enwiki, Byelf2007, Lambiam, Dbtfz, Kuru, Rizome~enwiki, Dicklyon, JRSpriggs, CBM, Gregbard, Nick Number, Escarbot, LaForge, David Eppstein, Yonaa, J.delanoy, Policron, Alan U. Kennington, JohnBlackburne, Hqb, Qxz, Magmi, VanishedUserABC, Radagast3, OKBot, Kumioko, Anchor Link Bot, CBM2, ClueBot, Mannypabla, Thisthat12345, Beach drifter, Good Olfactory, Addbot, Matěj Grabovský, Dima125, Yobot, Ptbotgourou, JRB-Europe, MattTait, Citation bot, Tbvdm, Sa'y, Foobarnix, Gamewizard71, Dbmikus, WildBot, EmausBot, ZéroBot, Donner60, ClueBot NG, Rezabot, Helpful Pixie Bot, Brad7777, Rsmbf, Begadkepat, Kkval0, JHU1959, The Editor of All Things Wikipedia and Anonymous: 48

- **Epsilon calculus** *Source:* https://en.wikipedia.org/wiki/Epsilon_calculus?oldid=635469584 *Contributors:* Michael Hardy, AugPi, Altenmann, Tobias Bergemann, Oleg Alexandrov, Lambiam, Gregbard, Nick Number, Hotfeba, Negi(afk), Chricho, Jochen Burghardt, Aubreybardo, Monkbot and Anonymous: 4

- **Analytic proof** *Source:* https://en.wikipedia.org/wiki/Analytic_proof?oldid=627143857 *Contributors:* Charles Matthews, Iwehrman, Chalst, Mairi, Oleg Alexandrov, CmdrObot, CBM, Gregbard, Omnipaedista, Proof Theorist, Bgeron, Widr, Rhaycock, Seppi333 and Anonymous: 3

- **Bachmann–Howard ordinal** *Source:* https://en.wikipedia.org/wiki/Bachmann%E2%80%93Howard_ordinal?oldid=680486245 *Contributors:* AxelBoldt, Michael Hardy, Dominus, Gro-Tsen, Mike Rosoft, Sligocki, Gene Nygaard, Rjwilmsi, R.e.b., JRSpriggs, Headbomb, Unzerlegbarkeit, Yobot, Citation bot, Citation bot 1, Alf.laylah.wa.laylah and Anonymous: 1

- **Bounded quantifier** *Source:* https://en.wikipedia.org/wiki/Bounded_quantifier?oldid=607157910 *Contributors:* EmilJ, Ruud Koot, Hairy Dude, JRSpriggs, CBM, Cydebot, Philosophy.dude, Unzerlegbarkeit, Yobot, Pcap, Citation bot, Chharvey, Tijfo098, Helpful Pixie Bot and Anonymous: 1

- **Büchi arithmetic** *Source:* https://en.wikipedia.org/wiki/B%C3%BCchi_arithmetic?oldid=622648400 *Contributors:* Bgwhite, Hebrides, Arthur MILCHIOR and Deltahedron

- **Church–Kleene ordinal** *Source:* https://en.wikipedia.org/wiki/Church%E2%80%93Kleene_ordinal?oldid=551346192 *Contributors:* Michael Hardy, Gro-Tsen, Rjwilmsi, R.e.b., JRSpriggs, Headbomb, Yobot, Citation bot, VladimirReshetnikov, Citation bot 1 and Anonymous: 1

- **Cirquent calculus** *Source:* https://en.wikipedia.org/wiki/Cirquent_calculus?oldid=676096767 *Contributors:* Michael Hardy, FrescoBot, Ano and Arley82

- **Completeness (logic)** *Source:* https://en.wikipedia.org/wiki/Completeness_(logic)?oldid=679536734 *Contributors:* Michael Hardy, Hyacinth, Kri, Lambiam, Gregbard, Jochen Burghardt, Brirush, There is a T101 in your kitchen, SiddMahen and Anonymous: 3

- **Completeness of atomic initial sequents** *Source:* https://en.wikipedia.org/wiki/Completeness_of_atomic_initial_sequents?oldid=456261617 *Contributors:* RDBury, CBM, Noamz and Anonymous: 1

- **Conservative extension** *Source:* https://en.wikipedia.org/wiki/Conservative_extension?oldid=633223419 *Contributors:* Docu, Charles Matthews, Luqui, Zenohockey, EmilJ, Mairi, Tim Smith, Mathbot, Tillmo, PM Poon, Figaro, MalafayaBot, Turms, Geh, JRSpriggs, CBM, Sdorrance, Gregbard, Pit-trout, Neithan Agarwaen, Alksentrs, Hans Adler, Addbot, Yobot, ZéroBot, Knutknutus, Yann.hourdel and Anonymous: 9

- **Conservativity theorem** *Source:* https://en.wikipedia.org/wiki/Conservativity_theorem?oldid=632731011 *Contributors:* Charles Matthews, Mairi, Oleg Alexandrov, Meloman, Banes, RDBury, BeteNoir, Mets501, JRSpriggs, CBM, WhatamIdoing, Lightbot, Mini-floh and Anonymous: 1

- **Consistency** *Source:* https://en.wikipedia.org/wiki/Consistency?oldid=669765886 *Contributors:* Michael Hardy, Dominus, Kku, Tim Retout, Charles Matthews, Hyacinth, Chealer, Chancemill, Robinh, Tobias Bergemann, Filemon, Giftlite, Beland, Luqui, Chalst, Dreish, Oleg Alexandrov, Ylem, Isnow, Jobnikon, Amorrow, NatusRoma, Jevon, Rbonvall, Intgr, Robbyslaughter, Tillmo, Gslin, SmackBot, Radagast83, Jon Awbrey, Tilin, PhiJ, Lambiam, Wvbailey, Levineps, Iridescent, Zero sharp, DavidBaelde, JRSpriggs, CRGreathouse, CBM, Myasuda, Gregbard, DumbBOT, Boemanneke, Thijs!bot, Dqd, JAnDbot, Magioladitis, MetsBot, Vinograd19, Natsirtguy, Regicollis, Kumioko (renamed), ClueBot, The Thing That Should Not Be, Qwfp, Bleeve, Feministo, Addbot, Maschelos, Tide rolls, LuK3, Luckas-bot, Yobot, Vanished user rt41as76lk, Ningauble, AnomieBOT, Akhran, Galoubet, VladimirReshetnikov, Noamz, Whassan, Antares5245, FrescoBot, RedBot, Masti-Bot, Gopher p, Morton Shumway, EmausBot, John of Reading, Shshahryari, Bomazi, Tijfo098, Wcherowi, JimsMaher, Masssly, Helpful Pixie Bot, Lowercase sigmabot, Justincheng12345-bot, Tapped-out, BurkeFT and Anonymous: 40

- **Curry–Howard correspondence** *Source:* https://en.wikipedia.org/wiki/Curry%E2%80%93Howard_correspondence?oldid=682721130 *Contributors:* Taral, Toby Bartels, Genneth, Edward, Michael Hardy, Dominus, Chinju, AugPi, Charles Matthews, Doradus, Phil Boswell, Anthony, Iwehrman, Jleedev, Tobias Bergemann, Ancheta Wis, Giftlite, DefLog~enwiki, SethTisue, Leibniz, Ben Standeven, Chalst, Nickj, Cmdrjameson, Karlheg, Txa, Anthony Appleyard, Jamiemichelle, Oleg Alexandrov, Linas, Ruud Koot, Dionyziz, Marudubshinki, Rjwilmsi, Mathbot, Jrtayloriv, Wavelength, Hairy Dude, Michael Slone, Eaefremov, Nnxion, SmackBot, Eskimbot, Mhss, Cybercobra, Jon Awbrey, Igrant, Physis, Iridescent, JRSpriggs, Chris55, CRGreathouse, ShelfSkewed, Gregbard, Goldenowl, Seunghun, Oerjan, Rowandavies, RebelRobot, Four Dog Night, David Eppstein, JaGa, Gwern, N4nojohn, SparsityProblem, McM.bot, SieBot, Laocoön11, Classicalecon, MilesAgain, Hugo Herbelin, Tre2, Addbot, LaaknorBot, Legobot, Yobot, Ptbotgourou, Pcap, EricP, Citation bot, Xqbot, Noamz, FrescoBot, Goheeca, OriumX, RedBot, Francis Lima, Burritoburritoburrito, Mattghg, RjwilmsiBot, EmausBot, John of Reading, ZéroBot, Elaz85, ClueBot NG, Helpful Pixie Bot, CitationCleanerBot, BattyBot, Jochen Burghardt, Lambda Fairy, Andyhowlett, Monkbot and Anonymous: 65

- **Cut-elimination theorem** *Source:* https://en.wikipedia.org/wiki/Cut-elimination_theorem?oldid=612476377 *Contributors:* Michael Hardy, David.Monniaux, Iwehrman, Giftlite, Sam Hocevar, Gauge, Chalst, Vipul, Sligocki, Triddle, GregorB, Liulk, Trovatore, Pacogo7, RDBury, Neil Leslie, Mhss, John, DavidBaelde, JRSpriggs, E-boy, Gregbard, Hannes Eder, Johnadam789, Alan U. Kennington, Rootwhisk, L0mars01, Addbot, DOI bot, Lightbot, Omnipaedista, Citation bot 1, MPeterHenry, BertSeghers, Tijfo098, Janburse and Anonymous: 16

Transcontinental, SashatoBot, Dan Gluck, Joseph Solis in Australia, Hilverd, Zero sharp, JRSpriggs, CBM, Myasuda, Gregbard, Spewin, Thijs!bot, Headbomb, Jbaranao, Morphriz, Meeples, TXiKiBoT, EuTuga, Da Joe, IsleLaMotte, Martarius, Hans Adler, El bot de la dieta, Thingg, Hugo Herbelin, Marc van Leeuwen, MrOllie, Lightbot, Legobot, OrgasGirl, Omnipaedista, BrideOfKripkenstein, Citation bot 1, WikitanvirBot, Jaydiem, Brad7777, ChrisGualtieri, Jochen Burghardt, Epicgenius, Salspaugh, 2.71828182845904523austen, Whiterray and Anonymous: 33

- **Gödel's incompleteness theorems** *Source:* https://en.wikipedia.org/wiki/G%C3%B6del'{}s_incompleteness_theorems?oldid=685961148 *Contributors:* AxelBoldt, Joao, Chenyu, LC~enwiki, Lee Daniel Crocker, Mav, Bryan Derksen, The Anome, Tarquin, Jan Hidders, Andre Engels, Danny, MadSurgeon, SimonP, Camembert, Genneth, Olivier, Michael Hardy, Tim Starling, Kwertii, Dominus, Lousyd, MartinHarper, Wapcaplet, Chinju, TakuyaMurata, GTBacchus, Eric119, CesarB, HarmonicSphere, ThirdParty, Ootachi, Suisui, Den fjättrade ankan~enwiki, Mark Foskey, Александър, BuzzB, AugPi, Tim Retout, Rotem Dan, Evercat, Madir, Gamma~enwiki, Rzach, Charles Matthews, Timwi, Dcoetzee, Reddi, Dysprosia, Wikid, Doradus, Markhurd, Lfwlfw, Hyacinth, Bevo, Joseaperez, .mau., Gakrivas, Pakaran, Ldo, David.Monniaux, Dmytro, Owen, Aleph4, Altenmann, Gandalf61, MathMartin, Bethenco, Rasmus Faber, Bkell, Mjscud, Aetheling, Ruakh, Tobias Bergemann, Jimpaz, Ancheta Wis, Tosha, Matt Gies, Giftlite, Lethe, Lupin, Fropuff, Mellum, Waltpohl, Geoffroy~enwiki, Wikiwikifast, Andris, Bovlb, Sundar, Prosfilaes, Siroxo, Khalid hassani, C17GMaster, Neilc, Utcursch, Andycjp, Sigfpe, Toytoy, Gdr, SarekOfVulcan, Gdm, Lightst, Antandrus, Beland, Amoss, APH, DragonflySixtyseven, Mike Storm, Elroch, Sam Hocevar, Asbestos, Karl Dickman, Gazpacho, D6, Sysy, 4pq1injbok, Guanabot, FT2, Cacycle, Smyth, Francis Davey, Maksym Ye., Paul August, Chalst, Crisófilax, EmilJ, Lance Williams, DG~enwiki, AshtonBenson, Jumbuck, Keenan Pepper, Hu, Wtmitchell, Simplebrain, Cal 1234, GabrielF, Gene Nygaard, MIT Trekkie, TXlogic, Oleg Alexandrov, Revived, Billhpike, OwenX, Rodrigo Rocha~enwiki, Pol098, Ruud Koot, Apokrif, Frungi, Waldir, Marudubshinki, MACherian, Mandarax, Graham87, BD2412, Qwertyus, Rjwilmsi, Tim!, Kinu, WoodenTaco, RCSB, Pasky, R.e.b., Magidin, FlaBot, VKokielov, Docbug, Mathbot, Jrtayloriv, Celendin, Sperxios, Xelloss~enwiki, Eric.dane~enwiki, Chobot, YurikBot, Wavelength, Hairy Dude, Dmharvey, Arado, Rcaetano, Me and, Ksnortum, JabberWok, KSmrq, Bergsten, SpuriousQ, Thoreaulylazy, Vibritannia, Ytcracker, Aatu, Trovatore, Długosz, JoeBruno, Dogcow, Nick, Anetode, Philosofool, Guruparan18, Hakeem.gadi, Liyang, Arthur Rubin, Curpsbot-unicodify, GrinBot~enwiki, Finell, SolarMcPanel, SmackBot, RDBury, Selfworm, BeteNoir, InverseHypercube, Melchoir, SaxTeacher, Pokipsy76, Mandelum, Biedermann, Brick Thrower, Srnec, The Rhymesmith, Chris the speller, MagnusW, Irving Anellis, RoyArne, MalafayaBot, Kostmo, Jdthood, Charles Moss, Grover cleveland, AndySimpson, Trifon Triantafillidis, John wesley, Kid A~enwiki, LoveMonkey, Rhkramer, Byelf2007, Zchenyu, Lambiam, Wvbailey, Evildictaitor, Illythr, Mets501, ASKingquestions, Sgutkind, Nbhatla, Quaeler, Dan Gluck, Jason.grossman, Dreftymac, Joseph Solis in Australia, Zero sharp, Wfructose, Matthew Kornya, JRSpriggs, CRGreathouse, Geremia, Diegueins, CBM, Nicolaennio, Gregbard, Nilfanion, Sopoforic, Cydebot, Pce3@ij.net, Steel, Peterdjones, TicketMan, Mon4, Blaisorblade, Michael C Price, DumbBOT, Robertinventor, Morgaladh, Clickheretologin, Thijs!bot, Headbomb, Tamalet, Towopedia, Turkeyphant, Kborer, Stan the fisher, Joegoodbud, Gioto, Blue Tie, Ad88110, NBeale, Husond, Avaya1, Gavia immer, JamesBWatson, Renosecond, Tedickey, Baccyak4H, K95, David Eppstein, Exiledone, Pavel Jelínek, Infovarius, Franp9am, Eliko, Abecedare, Fredeaker, Warut, CeilingCrash, DadaNeem, Policron, Mad7777, DavidCBryant, Merzul, Quux0r, Germanium, VolkovBot, DDSaeger, AlnoktaBOT, TXiKiBoT, Sacramentis, Lou2261, CaptinJohn, Broadbot, David in DC, Lifeisfun0007, Latulla, Popopp, Gbawden, Davebuehler, YohanN7, SieBot, Simplifier, Iamthedeus, Gerakibot, Noaqiyeum, Likebox, Flyer22 Reborn, Scouto2, OKBot, Valeria.depaiva, D14C050, IsleLaMotte, Anchor Link Bot, S2000magician, Tesi1700, CBM2, ReluctantPhilosopher, Beeblebrox, ClueBot, Admiral Norton, DFRussia, Pi zero, Juustensson, Razimantv, Niceguyedc, Libett, Ajoykt, Ademh, Byates5637, Bender2k14, Sun Creator, Coinmanj, AmirOnWiki, Tnxman307, Snacks, Hans Adler, Muro Bot, Darkicebot, Palnot, Gerhardvalentin, TravisAF, Xamce, Addbot, Yousou, Harttwood47, RPHv, Mpholmes, EjsBot, Vishnava, Download, LaaknorBot, דניאל ב., Legobot, Luckas-bot, Yobot, Ht686rg90, TaBOT-zerem, Nallimbot, Third Merlin, FeydHuxtable, AnomieBOT, 9258fahsflkh917fas, Materialscientist, Citation bot, Twiceuponatime, Markworthen, LilHelpa, Obersachsebot, Xqbot, Psyoptix, False vacuum, VladimirReshetnikov, Shaun.mk, Kristjan.Jonasson, Andrewjameskirk, Mark Renier, Mfwitten, Citation bot 1, Þjóðólfr, Sptzimas, Tkuvho, Aunin, MarcelB612, MPeterHenry, Standardfact, H.ehsaan, RjwilmsiBot, Fictionalist, Kozation, EmausBot, Nameguy101, Bijuro, AvicAWB, Negyek, Dominique.devriese, Herk1955, Llightex, CasualUser1729, ClueBot NG, Kevin Gorman, Helpful Pixie Bot, Wbm1058, Lowercase sigmabot, BG19bot, Slippingspy, Modelpractice, Cyberpower678, Minsbot, Marktoiii0, Kiewbra, Cyberbot II, Deltahedron, Fernandodelucia, Harri jensen, LFOlsnes-Lea, Paul1andrews, EvergreenFir, Dinadineke, Eng.alireda, 22merlin, Zeus000000, Latinosopher and Anonymous: 370

- **Hardy hierarchy** *Source:* https://en.wikipedia.org/wiki/Hardy_hierarchy?oldid=447604733 *Contributors:* Michael Hardy, Sligocki, R.e.s., CBM, Cydebot, Headbomb and Logan

- **Herbrand's theorem** *Source:* https://en.wikipedia.org/wiki/Herbrand'{}s_theorem?oldid=686689335 *Contributors:* Charles Matthews, Gandalf61, Giftlite, Paul August, Ntmatter, Giraffedata, Versageek, Linas, R.e.b., NavarroJ, YurikBot, Pacogo7, SmackBot, RDBury, Cronholm144, Mets501, JRSpriggs, CmdrObot, CBM, Gregbard, Julian Mendez, Hamaryns, Maximiliano.Guerra, Taemyr, IsleLaMotte, Ideal gas equation, Hans Adler, Ceilican, 1ForTheMoney, MystBot, Addbot, Luckas-bot, Yobot, Playmobilonhishorse, Thiscmd, BG19bot and Anonymous: 15

- **Hilbert system** *Source:* https://en.wikipedia.org/wiki/Hilbert_system?oldid=677643445 *Contributors:* Michael Hardy, Bogdanb, Rich Farmbrough, ZeroOne, Chalst, Flamingspinach, Tizio, Jayme, Spacepotato, Mahahahaneapneap, SmackBot, Mhss, SchfiftyThree, Henning Makholm, Wvbailey, STyx, Physis, Makyen, Vagary, JRSpriggs, CBM, Gregbard, Marco.caminati, Hqb, Outmind~enwiki, Outs, Anchor Link Bot, PixelBot, Hans Adler, Chaosdruid, Hugo Herbelin, Addbot, Yobot, AnomieBOT, ArthurBot, Tayloj, PeterSchueller, Dendropithecus, Omnipaedista, Citation bot 1, ClueBot NG, Helpful Pixie Bot, Samaritan01, Monkbot, Firedrake93 and Anonymous: 21

- **Hilbert's program** *Source:* https://en.wikipedia.org/wiki/Hilbert'{}s_program?oldid=669418894 *Contributors:* Michael Hardy, Charles Matthews,Jay, Siroxo, Nomeata, Icairns, Discospinster, Rich Farmbrough, Pjacobi, Chalst, Mairi, Duesentrieb, Grick, Zaraki ~enwiki, Oleg Alexan-drov, Nuno Tavares, Isnow, Waldir, Rjwilmsi, R.e.b., Mathrick, Bluebot, JCSantos, Hibbleton, Angela26, Lambiam, Geh, Mets501, Pej-man47, JRSpriggs, Galex, CBM, Gregbard, M a s, JAnDbot, Infrangible, Robertgreer, Zweidinge, Likebox, Jdaloner, DesolateReality, Clas-sicalecon, DumZiBoT, Marc van Leeuwen, Addbot, Vasiľ, Luckas-bot, KamikazeBot, AnomieBOT, FuzTheCat, Noamz, CES1596, Tkuvho,Mokhtari34, TheKing44 and Anonymous: 24

- **Independence (mathematical logic)** *Source:* https://en.wikipedia.org/wiki/Independence_(mathematical_logic)?oldid=641772009 *Contributors:* Charles Matthews, Hyacinth, Thue, Gandalf61, LX, Barnaby dawson, ESkog, Oleg Alexandrov, Roboto de Ajvol, YurikBot, Trovatore, CRGreathouse, CBM, Gregbard, JAnDbot, Pavel Jelínek, DesolateReality, IsleLaMotte, Addbot, Theking17825, Yobot, Pcap, Erik9bot, EffeX2, LucienBOT, Ebony Jackson, RA0808, Masssly, Daysrr, Nathanielfirst and Anonymous: 12

- **Interpretability** *Source:* https://en.wikipedia.org/wiki/Interpretability?oldid=612104705 *Contributors:* Ahoerstemeier, Dysprosia, Vanished user 1234567890, Kntg, EmilJ, PWilkinson, Oleg Alexandrov, MrShamrock, CBM, Gregbard, Singularity, Matthew Yeager, DesolateReality, Hans Adler, Qwfp, Lightbot, Yobot, BrideOfKripkenstein and Anonymous: 1

- **Japaridze's polymodal logic** *Source:* https://en.wikipedia.org/wiki/Japaridze'{ }s_polymodal_logic?oldid=680711428 *Contributors:* Michael Hardy, Ruud Koot, David Eppstein, Robvanvee, BG19bot, Miszatomic and ProvLog

- **Judgment (mathematical logic)** *Source:* https://en.wikipedia.org/wiki/Judgment_(mathematical_logic)?oldid=600394251 *Contributors:* Hardy, BD2412, Physis, Gregbard, Deadbeef, GoingBatty, JPaestpreornJeolhlna, PlaidPolarity and Anonymous: 2

- **Lambda-mu calculus** *Source:* https://en.wikipedia.org/wiki/Lambda-mu_calculus?oldid=570879470 *Contributors:* Michael Hardy, Rjwilmsi, SmackBot, Imz, E-boy, CBM, Radjenef, Mate Juhasz, Omicron18, JaGa, Benjamin Barenblat, Hugo Herbelin, Pcap and Anonymous: 5

- **Large countable ordinal** *Source:* https://en.wikipedia.org/wiki/Large_countable_ordinal?oldid=677151249 *Contributors:* AxelBoldt, Toby Bartels, Michael Hardy, Dominus, Charles Matthews, Aleph4, Lethe, Gro-Tsen, Luqui, Ben Standeven, AshtonBenson, Sligocki, Rjwilmsi, R.e.b., Hairy Dude, Trovatore, Crasshopper, SmackBot, Michael Kinyon, Zero sharp, JRSpriggs, CBM, Gregbard, Headbomb, JustAGal, Magioladitis, Albmont, Robin S, Peterwshor, Miaoku, MystBot, Addbot, Yobot, VladimirReshetnikov, FrescoBot, Trappist the monk, Chharvey, Helpful Pixie Bot and Anonymous: 18

- **LowerUnits** *Source:* https://en.wikipedia.org/wiki/LowerUnits?oldid=603221241 *Contributors:* Gregbard, Dthomsen8, Ad Orientem and Ezequiel234

- **Mathematical fallacy** *Source:* https://en.wikipedia.org/wiki/Mathematical_fallacy?oldid=686207328 *Contributors:* AxelBoldt, Bryan Derksen, Zundark, Arvindn, Michael Hardy, Dominus, Eric119, Minesweeper, Ijon, LittleDan, UserGoogol, Schneelocke, Charles Matthews, Timwi, Dcoetzee, Dysprosia, Jitse Niesen, Wik, Wiwaxia, Fredrik, Altenmann, Gandalf61, Merovingian, Henrygb, Tobias Bergemann, Giftlite, Wolfkeeper, Paul Pogonyshev, Guanaco, Matt Crypto, Mdob, Chowbok, Bongbang, Starx, Peter Kwok, Gdabski, Gazpacho, TheJames, Paul August, ESkog, Lankiveil, Spoon!, Nandhp, JRM, Wood Thrush, I9Q79oL78KiL0QTFHgyc, Martinultima, Tsirel, JYolkowski, Anders Kaseorg, SurrealWarrior, Splat, Mikeo, Drbreznjev, Axeman89, Feezo, StradivariusTV, Apokrif, Waldir, BD2412, Jshadias, Josh Parris, Eyu100, R.e.b., Tomtheman5, Tedd, Alexb@cut-the-knot.com, Mathbot, Celestianpower, King of Hearts, Sbrools, DVdm, X42bn6, RussBot, IanManka, Gaius Cornelius, Pnrj, Cheeser1, Simxp, Syko, Brentt, KnightRider~enwiki, SmackBot, Incnis Mrsi, Melchoir, Rokfaith, Fulldecent, Bluebot, Thumperward, SchfiftyThree, Tavianator, RyanEberhart, Calbaer, Fuhghettaboutit, Pwjb, Turms, Louisng114, Henning Makholm, Ged UK, Byelf2007, Zchenyu, Lambiam, Polihale, Omnedon, Illythr, Cstella23, Kpengboy, Mets501, Hyperwiz, Dr.K., Iridescent, AlsatianRain, JRSpriggs, George100, Whyareall, CRGreathouse, CBM, JPadron, Cydebot, Reywas92, WillowW, MC10, Steel, Crossmr, Carifio24, Kacie Jane, Odie5533, Krzysiu Jarzyna, Englishmerd, Yurell, Kilva, Pallas44, Jojan, AntiVandalBot, Seaphoto, Joe Schmedley, MarvinCZ, Dylan Lake, Husond, Oxinabox, MER-C, Boleslaw, Drhlajos, Some Guy123, Timanderso, Albmont, Email4mobile, Catgut, MetsBot, Error792, Cpl Syx, Dravick, Patstuart, Connor Behan, Ztobor, MartinBot, Ariel., Xoran99, Arjun01, Comperr, Anaxial, J.delanoy, AstroHurricane001, Uncle Dick, Laurusnobilis, Paidgenius, GEWilker, Soccersabo, Useight, TWiStErRob, RJASE1, Kimandy, Science4sail, Indubitably, Nousernamesleft, Wannger27, Anonymous Dissident, Amahdy, Kmhkmh, Geoffreyfishing, Secretss, Dmcq, Sue Rangell, Misha Mullov-Abbado, Oboeboy, Paradoctor, Phe-bot, Keilana, Happysailor, Dragnmn, Taemyr, 0rrAvenger, Kudret abi, 𝌆𝌆𝌆𝌆~enwiki, Kortaggio, Tuntable, ClueBot, EoGuy, Mild Bill Hiccup, Xenon54, Oxnard27, Doloco, Fletcher17, Lartoven, Ykhwong, H.Marxen, Djk3, XLinkBot, Marc van Leeuwen, Fastily, Pichpich, Tongrongtian, Gwandoya, Gerhardvalentin, Charles Sturm, AlexFekken, Luca Antonelli, Nickolai kazimir, Addbot, Joe-Moron2000, 067012732s, CanadianLinuxUser, Fluffernutter, Favonian, Barak Sh, 3qwertbbb7, Calculuslover, Tide rolls, NKapustin, Yobot, TaBOT-zerem, Timeroot, Spenalzo, AnomieBOT, Pkukiss, Jim1138, AdjustShift, Terminatore, Georgepowell2008, Xqbot, TechBot, The Evil IP address, Point-set topologist, POTUS270, Jetpackboy14, Pottersson, Pinethicket, Number Googol, Patwotrik, Tcnuk, Jujutacular, Barras, Double sharp, Trappist the monk, Niketmalik, Phatency, Le Docteur, Thewriter006, Tbhotch, Sideways713, Whisky drinker, Martianpackets, Mr. Anon515, EmausBot, John of Reading, Slawekb, Anoop.dixith, Derekleungtszhei, Quondum, L Kensington, JonRichfield, ClueBot NG, Rtucker913, Helpful Pixie Bot, Hguy, BG19bot, Hawkwindeb, Mocky3497, Undersum, BattyBot, Avengingbandit, Rupert'sscribe, Saung Tadashi, Lugia2453, Mmitchell10, Yehianumb, Gtklocker, Bilorv, That kiwi guy, Mario Castelán Castro, Stishuk.hf, PiotrGrochowski000, Gov vj, Abhishekx7 and Anonymous: 288

- **Metalanguage** *Source:* https://en.wikipedia.org/wiki/Metalanguage?oldid=684802420 *Contributors:* Andres, Charles Matthews, Dcoetzee, Nickg, RedWolf, Benc, Wile E. Heresiarch, BenFrantzDale, Monedula, Neilc, Knutux, Ja malcolm, Almit39, Sam Hocevar, Lacrimosus, Discospinster, Kb, Visualerror, Aaronbrick, Ntmatter, John Vandenberg, Pyrrhos, Sam Korn, Guaca, Typobox, Hawky, Graham87, Qwertyus, KYPark, Venullian, Mathbot, Wars, Lmatt, Pricey3000, Chobot, YurikBot, Wavelength, Tomisti, Nzzl, Curpsbot-unicodify, Trickstar, SmackBot, FocalPoint, Nazgjunk, Lewstherin, Bn, JoseREMY, 16@r, Dicklyon, Graham Hurley, George100, InvisibleK, CBM, Gregbard, Peterdjones, Infinito, Mentifisto, VictorAnyakin, JAnDbot, Tedickey, Tonyfaull, LookingGlass, Joshua Davis, Maurice Carbonaro, VolkovBot, Shinju, VanishedUserABC, Carn29, Brainfsck, Ivan Štambuk, PanagosTheOther, ClueBot, Tyurp, Ordinaterr, DragonBot, Aitias, Anticipation of a New Lover's Arrival, The, Addbot, Jarble, Luckas-bot, Yobot, Empro2, GrouchoBot, FrescoBot, GoodSpeller, Brightkingdom, Lars Washington, Dude1818, FoxBot, Morton Shumway, EmausBot, ZéroBot, ClueBot NG, Snotbot, Steamerandy, Myconix, Sweeter49 and Anonymous: 68

- **Natural deduction** *Source:* https://en.wikipedia.org/wiki/Natural_deduction?oldid=665869249 *Contributors:* The Cunctator, Edward, Michael Hardy, Chinju, EdH, Charles Matthews, Markhurd, Hyacinth, Phil Boswell, Robbot, Benwing, Cholling, Jleedev, Ancheta Wis, Ido50, Marc Mongenet, Kaustuv, Alikhtarov, Guanabot, Rspeer, Francis Davey, Bender235, Glenn Willen, Chalst, Cmdrjameson, Jeltz, Oleg Alexandrov, Joriki, Ruud Koot, Waldir, BD2412, Oterhaar~enwiki, Rjwilmsi, Koavf, Margosbot~enwiki, William Lovas, Hairy Dude, Clemente~enwiki, MSully4321, Welshbyte, That Guy, From That Show!, SmackBot, Imz, Aij, Mhss, Cybercobra, Dbtfz, Physis, Comicist, Iridescent, Gregbard, Blaisorblade, TheDean, Dgies, Rowandavies, Mate Juhasz, David Eppstein, Vesa Linja-aho, Alan U. Kennington, Aleph42, Ontoraul, VanishedUserABC, Givegains, Valeria.depaiva, CBM2, Watchduck, Hugo Herbelin, Liviusbarbatus~enwiki, Addbot, DOI bot, LinkFA-Bot, Lightbot, Luckas-bot, Yobot, ArthurBot, Ansa211, FrescoBot, Abrahanfer, Cleves, Curb Chain, Chiguri, Mpiedrav, Jomey and Anonymous: 60

- **Ω-consistent theory** *Source:* https://en.wikipedia.org/wiki/%CE%A9-consistent_theory?oldid=685818686 *Contributors:* Michael Hardy, GT-Bacchus, Jitse Niesen, Sam, Chalst, EmilJ, Gene Nygaard, NekoDaemon, Bgwhite, Hairy Dude, Trovatore, Lambiam, Wvbailey, Mets501, Zero sharp, JRSpriggs, Vaughan Pratt, CBM, Gregbard, Xantharius, Masaki K, Dispenser, Likebox, Zero over zero, PixelBot, Addbot, Yobot, Citation bot, Tkuvho, Btilm, Stj6, AManWithNoPlan, Jack Greenmaven, BG19bot, BinaryFriend, Dexbot and Anonymous: 5

- **Ordinal analysis** *Source:* https://en.wikipedia.org/wiki/Ordinal_analysis?oldid=648058414 *Contributors:* AxelBoldt, Dominus, Greenrd, Tobias Bergemann, Gro-Tsen, Ben Standeven, Sligocki, R.e.b., JRSpriggs, CBM, Headbomb, A3nm, Synthebot, Hugo Herbelin, Unzerlegbarkeit, Yobot, Citation bot, VladimirReshetnikov, Citation bot 1, Trappist the monk and Anonymous: 14

- **Ordinal notation** *Source:* https://en.wikipedia.org/wiki/Ordinal_notation?oldid=685367063 *Contributors:* Michael Hardy, Tobias Bergemann, Ben Standeven, Peter M Gerdes, Sligocki, Dolfrog, Rjwilmsi, R.e.b., R.e.s., JRSpriggs, JustAGal, Robin S, PaulTanenbaum, Firestonetireguy, Yobot, Citation bot, VladimirReshetnikov, BrideOfKripkenstein, Citation bot 1 and Anonymous: 4

- **Paraconsistent mathematics** *Source:* https://en.wikipedia.org/wiki/Paraconsistent_mathematics?oldid=602434485 *Contributors:* JohnOwens, Charles Matthews, Henrygb, Eduardoporcher, Mcsee, Porcher, Joelr31, SmackBot, CBM, Gregbard, David Eppstein, Philg88, EmbraceParadox, Aubreybardo and Anonymous: 3

- **Peano-Russell notation** *Source:* https://en.wikipedia.org/wiki/Peano-Russell_notation?oldid=644431770 *Contributors:* CBM, Gregbard, CBryant, Hotfeba, Xiaonanln, Jan Schreiber, FrescoBot, 18percent gray and Anonymous: 3

- **Presburger arithmetic** *Source:* https://en.wikipedia.org/wiki/Presburger_arithmetic?oldid=686290075 *Contributors:* Damian Yerrick, AxelBoldt, LC~enwiki, 0, Taw, Andre Engels, XJaM, Vkuncak, Michael Hardy, Chinju, Zeno Gantner, Paddu, Ajk, Tim Retout, Wik, Anon-Moos, David.Monniaux, Ruakh, Giftlite, Gdr, Mike Rosoft, Smimram, Guanabot, Pavel Vozenilek, Petrus~enwiki, Ben Standeven, Pmetzger, Spayrard, EmilJ, HasharBot~enwiki, Zenosparadox, RJFJR, Oleg Alexandrov, Graham87, BD2412, Rjwilmsi, R.e.b., Trovatore, Jpbowen, JCSantos, Janm67, Clements, Lambiam, Nagle, Mets501, Zero sharp, ILikeThings, CRGreathouse, CBM, Myasuda, Gregbard, Erxnmedia, Thenub314, Nyq, DWIII, Rogator, Sapphic, VVVBot, C. lorenz, Addbot, DOI bot, AnnaFrance, Derekoppen, Matěj Grabovský, Miym, Confront, Citation bot 1, OriumX, Erwinrcat, EmausBot, ZéroBot, Quondum, Helpful Pixie Bot, BG19bot, BattyBot, Jochen Burghardt, There is a T101 in your kitchen and Anonymous: 44

- **Primitive recursive functional** *Source:* https://en.wikipedia.org/wiki/Primitive_recursive_functional?oldid=607160223 *Contributors:* CBM, Cydebot, Yobot and Anonymous: 2

- **Proof (truth)** *Source:* https://en.wikipedia.org/wiki/Proof_(truth)?oldid=661799360 *Contributors:* Toby Bartels, Michael Hardy, Gandalf61, Kwamikagami, BlastOButter42, Woohookitty, BD2412, Rjwilmsi, Mayumashu, RussBot, SmackBot, Byelf2007, Vaughan Pratt, CRGreathouse, CBM, Gregbard, DumbBOT, Bongwarrior, Theodore.norvell, Chiswick Chap, Technopat, MustbeAmoocow, VanishedUserABC, Radagast3, Moonriddengirl, Dodger67, Excirial, Addbot, Favonian, Alfie66, Citation bot, False vacuum, Sławomir Biały, Citation bot 1, 9E2, Reaper Eternal, EmausBot, Zacchro, JSquish, Tijfo098, ClueBot NG, Masssly, Widr, Helpful Pixie Bot, Lowercase sigmabot, Ssonday002, Dolphin33438, Hihahe, Superdudereturns, HamboGlider, Jshaps1, Pickles123 and Anonymous: 53

- **Proof calculus** *Source:* https://en.wikipedia.org/wiki/Proof_calculus?oldid=675793646 *Contributors:* Silverfish, Ganymead, Boojum, Chalst, Chrajohn, GregorB, BD2412, Qwertyus, Jpbowen, Mhss, Lambiam, MartinBot, Katharineamy, The Tetrast, SilvonenBot, Addbot, Download, Lightbot, Yobot, Tijfo098, Arley82 and Anonymous: 7

- **Proof compression** *Source:* https://en.wikipedia.org/wiki/Proof_compression?oldid=683617033 *Contributors:* Michael Hardy, Mr. Stradivarius, Ceilican, LilHelpa, BG19bot, ChrisGualtieri, Mark viking, Ezequiel234 and Anonymous: 1

- **Proof mining** *Source:* https://en.wikipedia.org/wiki/Proof_mining?oldid=442037556 *Contributors:* Michael Hardy, Jayme, CRGreathouse, CBM, Classicalecon, Pauloboliva, Unzerlegbarkeit, Yobot, Willy xD and Anonymous: 1

- **Proof net** *Source:* https://en.wikipedia.org/wiki/Proof_net?oldid=615489541 *Contributors:* Michael Hardy, Silverfish, Charles Matthews, Kaustuv, Chalst, Oleg Alexandrov, WoodenTaco, Ott2, SmackBot, Physis, CBM, Gregbard, Magioladitis, A3nm, Selinger, Safulop, Addbot, 9258fahsflkh917fas, Greatfermat, Frietjes, ChrisGualtieri, Soujak and Anonymous: 3

- **Proof procedure** *Source:* https://en.wikipedia.org/wiki/Proof_procedure?oldid=635471315 *Contributors:* Silverfish, Charles Matthews, Jleedev, Chalst, Nortexoid, Noogz, Tizio, Jpbowen, SmackBot, Wossi, Lambiam, CBM, Gregbard, VanishedUserABC, Radagast3, Mild Bill Hiccup, Gamewizard71, Proof Theorist and Brirush

- **Proof-theoretic semantics** *Source:* https://en.wikipedia.org/wiki/Proof-theoretic_semantics?oldid=650356165 *Contributors:* Edward, Michael Hardy, Chalst, Velvetsmog, Porcher, Trovatore, Mhss, Gregbard, Cydebot, John254, Nick Number, Hjoole, Unara, The Wiki ghost, Cerabot~enwiki, Leftarrow and Anonymous: 4

- **Provability logic** *Source:* https://en.wikipedia.org/wiki/Provability_logic?oldid=680563966 *Contributors:* Edward, Charles Matthews, Kntg, Chalst, EmilJ, Nortexoid, PWilkinson, Ruud Koot, SLi, Trovatore, Black Falcon, Nahaj, Chris the speller, OneSixOne, CBM, Gregbard, David Eppstein, VanishedUserABC, DumZiBoT, Addbot, D'ohBot, ZéroBot, Mogism, Brirush, ProvLog and Anonymous: 4

- **$\Psi_0(\Omega\omega)$** *Source:* https://en.wikipedia.org/wiki/%CE%A8%E2%82%80(%CE%A9%CF%89)?oldid=627092207 *Contributors:* Michael Hardy, Sligocki, Ketiltrout, R.e.b., SmackBot, JRSpriggs, WOSlinker, CBM2, Trappist the monk, Weux082690 and Anonymous: 3

- **Pure type system** *Source:* https://en.wikipedia.org/wiki/Pure_type_system?oldid=676796313 *Contributors:* Michael Hardy, Greenrd, Phil Boswell, Kaustuv, Rich Farmbrough, Bender235, Ben Standeven, NotAbel, SmackBot, Cybercobra, Lambiam, Dougher, Cobi, Enoksrd, Functor salad, Dekart, Addbot, Yobot, Pcap, AnomieBOT, Omnipaedista, Citation bot 1, MarcelB612, RjwilmsiBot, Gf uip, Clayrat, Helpful Pixie Bot, Monkbot and Anonymous: 7

- **Realizability** *Source:* https://en.wikipedia.org/wiki/Realizability?oldid=676252867 *Contributors:* Edward, Chinju, Greenrd, Fram, Mhss, CBM, Gregbard, Cydebot, David Eppstein, Hotfeba, SchreiberBike, Hugo Herbelin, Addbot, DOI bot, Yobot, FrescoBot, Citation bot 1, Gamewizard71, Tagib, CitationCleanerBot, Monkbot and Anonymous: 9

- **Redundant proof** *Source:* https://en.wikipedia.org/wiki/Redundant_proof?oldid=607430635 *Contributors:* Michael Hardy, Bearcat, Boomur, D.Lazard, Ad Orientem, Mark viking and Ezequiel234

- **Resolution inference** *Source:* https://en.wikipedia.org/wiki/Resolution_inference?oldid=598608676 *Contributors:* Michael Hardy, Gregbard, Ceilican, BG19bot, Ezequiel234 and Anonymous: 1

- **Resolution proof compression by splitting** *Source:* https?????wikipedia.??????Resolution_proof_compression_??_splitting?oldid=68461 *Contributors:* Michael Hardy, Yobot, Maximo.marcos, Fshtea and Jodosma

- **Resolution proof reduction via local context rewriting** *Source:* https://en.wikipedia.org/wiki/Resolution_proof_reduction_via_local_context_rewriting?oldid=632209691 *Contributors:* Michael Hardy, Rjwilmsi, Joel7687, Chris the speller, Arjayay, Yobot, BG19bot, Maximo.marcos, Mark viking and Anonymous: 1

- **Reverse mathematics** *Source:* https://en.wikipedia.org/wiki/Reverse_mathematics?oldid=672097662 *Contributors:* Zundark, Michael Hardy, Bcrowell, Angela, Charles Matthews, Dcoetzee, Dmytro, Tobias Bergemann, Nick8325, Gene Ward Smith, Gro-Tsen, Boojum, Marcos, Stevenzenith, Fenice, Pt, EmilJ, Tsirel, Alai, Oleg Alexandrov, Daira Hopwood, Isnow, Rjwilmsi, R.e.b., FlaBot, Algebraist, YurikBot, KSmrq, Wknight94, Stet01, Grover cleveland, Ugur Basak Bot~enwiki, JRSpriggs, CRGreathouse, CmdrObot, CBM, Cydebot, Omicron18, Althai, Gwern, Hagman, AlleborgoBot, CBM2, Addbot, Yobot, Mon oncle, AnomieBOT, Citation bot, Antendren, Citation bot 1, Petecrawford, Trappist the monk, ZéroBot, Staszek Lem, ClueBot NG, Brad7777 and Anonymous: 19

- **Self-verifying theories** *Source:* https://en.wikipedia.org/wiki/Self-verifying_theories?oldid=606600583 *Contributors:* Charles Matthews, TravelingDude, Anville, Karnan, Metahacker, Chalst, Mairi, Cohesion, Oleg Alexandrov, Porcher, Trovatore, BranStark, CBM, Gregbard, Acroterion, Joeoettinger, Thehotelambush, Emk, Addbot, Unzerlegbarkeit and Anonymous: 4

- **Sequent** *Source:* https://en.wikipedia.org/wiki/Sequent?oldid=667026525 *Contributors:* Zundark, The Anome, Dysprosia, Rholton, Iwehrman, Tobias Bergemann, Snobot, Ancheta Wis, Psb777, Markus Krötzsch, Discospinster, Paul August, EmilJ, Ntmatter, Diego Moya, Oleg Alexandrov, Waldir, Marudubshinki, Qwertyus, Rjwilmsi, Salix alba, John Baez, Mathbot, YurikBot, Light current, Otto ter Haar, Mhss, Clconway, Dbtfz, Physis, Zero sharp, Entropyfails, CBM, Gregbard, Julian Mendez, Egriffin, Policron, Alan U. Kennington, Alejandrocaro35, Addbot, Yobot, AnomieBOT, Jellystones, Noamz, FrescoBot, 777sms, LoMaPh and Anonymous: 22

- **Sequent calculus** *Source:* https://en.wikipedia.org/wiki/Sequent_calculus?oldid=686658960 *Contributors:* Zundark, The Anome, DrBob, Michael Hardy, Erik Zachte, Chinju, Angela, Doradus, Sanxiyn, Gandalf61, Snobot, Ancheta Wis, Giftlite, Markus Krötzsch, Dedalus (usurped), Leibniz, Spayrard, Chalst, Thüringer, Oleg Alexandrov, Joriki, Linas, Benhocking, GregorB, Waldir, Rjwilmsi, Tizio, Salix alba, Mkehrt, Mathbot, Polux2001, Jpbowen, Tony1, Pacogo7, Canley, Mhss, Clconway, Allan McInnes, Lregnier, Henning Makholm, Shushruth, Lambiam, Dbtfz, Physis, Lim Wei Quan, Skelta, E-boy, Sky-surfer, Julian Mendez, Hamaryns, Magnus Bakken, Policron, Alan U. Kennington, JohnBlackburne, Aleph42, Jamelan, IsleLaMotte, Sun Creator, Addbot, DOI bot, Lightbot, Yobot, Goodmorningworld, Citation bot, GrouchoBot, Dendropithecus, Noamz, Undsoweiter, FrescoBot, Citation bot 1, Albertzeyer, Erwinrcat, LoStrangolatore, Chharvey, MathNlogic, Tijfo098, Janburse, BG19bot, Jetbeard, Brad7777, Jochen Burghardt, Naereen, There is a T101 in your kitchen, Monkbot and Anonymous: 43

- **Setoid** *Source:* https://en.wikipedia.org/wiki/Setoid?oldid=667056247 *Contributors:* Toby Bartels, Michael Hardy, Charles Matthews, Greenrd, Smimram, Viriditas, Oleg Alexandrov, Linas, BD2412, MarSch, Salix alba, Hairy Dude, SmackBot, Cybercobra, STyx, Michael Kinyon, CRGreathouse, Sagaciousuk, JackSparrow Ninja, David Eppstein, MartinBot, Classicalecon, Hans Adler, ChrisGualtieri, Catsrfurrytheory and Anonymous: 4

- **Slow-growing hierarchy** *Source:* https://en.wikipedia.org/wiki/Slow-growing_hierarchy?oldid=622488324 *Contributors:* The Anome, Ixfd64, Ben Standeven, Sligocki, Chris the speller, Cydebot, Headbomb, David Eppstein, Cheesefondue, H.Marxen, Addbot, !Silent, Citation bot, Xqbot, RjwilmsiBot, John of Reading and Anonymous: 5

- **Soundness** *Source:* https://en.wikipedia.org/wiki/Soundness?oldid=683108003 *Contributors:* The Anome, Ixfd64, Eric119, AugPi, Rossami, Hyacinth, Mpost89, Ancheta Wis, Markus Krötzsch, Kpalion, Andycjp, Chalst, Art LaPella, EmilJ, Saturnight, Nortexoid, Jumbuck, Raboof, Rh~enwiki, Omphaloscope, Oleg Alexandrov, MattGiuca, Justin Custer, Koavf, Mathbot, Kri, Vonkje, YurikBot, Hairy Dude, SmackBot, Gilliam, Skizzik, Bluebot, NYKevin, Mikezhao, Cybercobra, Richard001, Luxgratia, Lambiam, Bjankuloski06en~enwiki, Jenadeleh, Esurnir, CBM, Simeon, Gregbard, Thijs!bot, Nick Number, Thenub314, Arno Matthias, The dark lord trombonator, Deleet, IllaZilla, Tomaxer, Ohiostandard, SieBot, Sullen skies, Kumioko (renamed), DesolateReality, Tomas e, Dhulme, Alexbot, Hans Adler, Addbot, Tide rolls, Citation bot, MauritsBot, Je ne détiens pas la vérité universelle, NoldorinElf, LucienBOT, Amirhoseinaliakbarian, GregKaye, EmausBot, ZéroBot, Josve05a, JonRichfield, Helpful Pixie Bot, Epicgenius, Mohamed-Ahmed-FG, JHU1959, Kiwifist and Anonymous: 43

- **Soundness (interactive proof)** *Source:* https://en.wikipedia.org/wiki/Soundness_(interactive_proof)?oldid=588653087 *Contributors:* Michael Hardy, Malcolma, Gadget850, SmackBot, Markulf, David Eppstein, J824h and Anonymous: 2

- **Structural proof theory** *Source:* https://en.wikipedia.org/wiki/Structural_proof_theory?oldid=492152266 *Contributors:* Edward, Charles Matthews, Joy, Seggy, Rich Farmbrough, Chalst, AllyUnion, Linas, Waldir, Hairy Dude, SmackBot, Huperniketes, Zero sharp, Erik9bot, Vincent Aravantinos, WildBot, Tijfo098 and Helpful Pixie Bot

- **Structural rule** *Source:* https://en.wikipedia.org/wiki/Structural_rule?oldid=668593634 *Contributors:* Charles Matthews, Hyacinth, Ruakh, Kaustuv, STHayden, RDBury, Mhss, Fplay, Byelf2007, Soumyasch, Gregbard, Pi zero, Hans Adler, Addbot, Milksea, Noamz, Mattg82, Erik9bot, RobinK, TomT0m, Mark viking, W. P. Uzer and Anonymous: 4

- **Takeuti's conjecture** *Source:* https://en.wikipedia.org/wiki/Takeuti'{ }s_conjecture?oldid=627015510 *Contributors:* Michael Hardy, Takuya-Murata, Charles Matthews, Chalst, Mairi, CambridgeBayWeather, ArglebargleIV, Gregbard, David Eppstein, Cobi, Tradereddy, AlptaBot, Anne Bauval, Omnipaedista, FrescoBot, Proof Theorist and Anonymous: 3

- **Tolerant sequence** *Source:* https://en.wikipedia.org/wiki/Tolerant_sequence?oldid=612105965 *Contributors:* Charles Matthews, Dysprosia, Andrewman327, Kntg, PWilkinson, Oleg Alexandrov, John Broughton, Hans Adler and Anonymous: 2

- **Turnstile (symbol)** *Source:* https://en.wikipedia.org/wiki/Turnstile_(symbol)?oldid=681118086 *Contributors:* Hyacinth, Ancheta Wis, Urhixidur, Porges, Night Gyr, Oleg Alexandrov, Davidkazuhiro, Apokrif, Flamingspinach, Waldir, Koavf, William Lovas, Pburka, Hakeem.gadi, SimonMorgan, SmackBot, Unschool, CBM, Gregbard, Cydebot, Egriffin, Arthur Buchsbaum, Saibod, Methossant, Eeky, Plastikspork, Alejandrocaro35, Yobot, Phil Last, SporkBot, BG19bot, Leonren, JPaestpreornJeolhlna and Anonymous: 11

- **Undecidable problem** *Source:* https://en.wikipedia.org/wiki/Undecidable_problem?oldid=680528340 *Contributors:* Michael Hardy, Ixfd64, Aleph4, Psychonaut, Giftlite, Mike Rosoft, TedPavlic, Longhair, John Quiggin, Sligocki, Woohookitty, Shreevatsa, Oliphaunt, UsaSatsui, Dtrebbien, Trovatore, Googl, Bibliomaniac15, SmackBot, Swatjester, Stephen B Streater, Gregbard, Ultimus, Erxnmedia, Albany NY, Pádraig Coogan, Ratfox, VolkovBot, Phil Bridger, Vanished user kijsdion3i4jf, Stevenrasnick, Jordan Gray, Identityandconsulting, Addbot, Mahtab mk, Diego Queiroz, LucienBOT, SchreyP, ZéroBot, D.Lazard, ClueBot NG, Helpful Pixie Bot, Architectual, ChrisGualtieri, Jochen Burghardt, Froglich, Ginsuloft, Monkbot, PErdos, Loraof and Anonymous: 19

- **Veblen function** *Source:*  https://en.wikipedia.org/wiki/Veblen_function?oldid=683500150 *Contributors:* Tobias Bergemann, Giftlite, Gro-Tsen, Sligocki, Rjwilmsi, R.e.b., Dicklyon, Zero sharp, JRSpriggs, Headbomb, Synthebot, Cheesefondue, Addbot, Citation bot, ZéroBot, ClueBot NG, SuperJedi224 and Anonymous: 6
- **VIPER microprocessor** *Source:* https://en.wikipedia.org/wiki/VIPER_microprocessor?oldid=630629915 *Contributors:* Xezbeth, Whpq, Jimvin, Hebrides, Andy Dingley, MarkMLl, Fadesga, HughesJohn and Enfcer
- **Weak interpretability** *Source:*  https://en.wikipedia.org/wiki/Weak_interpretability?oldid=612106359 *Contributors:* Ahoerstemeier, Dysprosia, Antandrus, Kntg, PWilkinson, Oleg Alexandrov, GAYNIGGER ON WHEELS, SmackBot, CBM, ClydeC, David Eppstein, Princess Tiswas, Hans Adler, Lightbot and Anonymous: 1

## 87.3.2   Images

- **File:Ambox_important.svg** *Source:* https://upload.wikimedia.org/wikipedia/commons/b/b4/Ambox_important.svg *License:* Public domain *Contributors:* Own work, based off of Image:Ambox scales.svg *Original artist:* Dsmurat (talk · contribs)
- **File:CardContin.svg** *Source:* https://upload.wikimedia.org/wikipedia/commons/7/75/CardContin.svg *License:* Public domain *Contributors:* en:Image:CardContin.png *Original artist:* en:User:Trovatore, recreated by User:Stannered
- **File:Commodore-64-Computer.png** *Source:* https://upload.wikimedia.org/wikipedia/commons/3/34/Commodore-64-Computer.png *License:* Public domain *Contributors:* Own work *Original artist:* Evan-Amos
- **File:Coq_plus_comm_screenshot.jpg** *Source:* https://upload.wikimedia.org/wikipedia/commons/8/8b/Coq_plus_comm_screenshot.jpg *License:* CC-BY-SA-3.0 *Contributors:* snapshot of LGPL software CoqIDE ran in Gnome *Original artist:* Hugo Herbelin
- **File:Deduction_architecture.png** *Source:* https://upload.wikimedia.org/wikipedia/commons/4/47/Deduction_architecture.png *License:* GFDL *Contributors:* Own work *Original artist:* Physis
- **File:Edit-clear.svg** *Source:* https://upload.wikimedia.org/wikipedia/en/f/f2/Edit-clear.svg *License:* Public domain *Contributors:* The *Tango! Desktop Project. Original artist:*
  The people from the Tango! project. And according to the meta-data in the file, specifically: "Andreas Nilsson, and Jakub Steiner (although minimally)."
- **File:Emoji_u1f510.svg** *Source:* https://upload.wikimedia.org/wikipedia/commons/3/35/Emoji_u1f510.svg *License:* Apache License 2.0 *Contributors:* https://code.google.com/p/noto/ *Original artist:* Google
- **File:Fallacy_of_the_isosceles_triangle2.svg** *Source:* https://upload.wikimedia.org/wikipedia/commons/a/ab/Fallacy_of_the_isosceles svg *License:* CC BY-SA 4.0 *Contributors:* Own work *Original artist:* Turms
- **File:First_order_natural_deduction.png** *Source:* https://upload.wikimedia.org/wikipedia/commons/e/e0/First_order_natural_deduction.png *License:* CC-BY-SA-3.0 *Contributors:* ? *Original artist:* ?
- **File:Logic_portal.svg** *Source:* https://upload.wikimedia.org/wikipedia/commons/7/7c/Logic_portal.svg *License:* CC BY-SA 3.0 *Contributors:* Own work *Original artist:* Watchduck (a.k.a. Tilman Piesk)
- **File:Question_book-new.svg** *Source:* https://upload.wikimedia.org/wikipedia/en/9/99/Question_book-new.svg *License:* Cc-by-sa-3.0 *Contributors:*
  Created from scratch in Adobe Illustrator. Based on Image:Question book.png created by User:Equazcion *Original artist:* Tkgd2007
- **File:Scale_of_justice_2.svg** *Source:* https://upload.wikimedia.org/wikipedia/commons/0/0e/Scale_of_justice_2.svg *License:* Public domain *Contributors:* Own work *Original artist:* DTR
- **File:Text_document_with_red_question_mark.svg** *Source:* https://upload.wikimedia.org/wikipedia/commons/a/a4/Text_document_with_red_question_mark.svg *License:* Public domain *Contributors:* Created by bdesham with Inkscape; based upon Text-x-generic.svg from the Tango project. *Original artist:* Benjamin D. Esham (bdesham)
- **File:Venn1001.svg** *Source:* https://upload.wikimedia.org/wikipedia/commons/4/47/Venn1001.svg *License:* Public domain *Contributors:* ? *Original artist:* ?
- **File:Wiki_letter_w_cropped.svg** *Source:* https://upload.wikimedia.org/wikipedia/commons/1/1c/Wiki_letter_w_cropped.svg *License:* CC-BY-SA-3.0 *Contributors:*
- Wiki_letter_w.svg *Original artist:* Wiki_letter_w.svg: Jarkko Piiroinen
- **File:Wikibooks-logo-en-noslogan.svg** *Source:* https://upload.wikimedia.org/wikipedia/commons/d/df/Wikibooks-logo-en-noslogan.svg *License:* CC BY-SA 3.0 *Contributors:* Own work *Original artist:* User:Bastique, User:Ramac et al.
- **File:Wikiquote-logo.svg** *Source:* https://upload.wikimedia.org/wikipedia/commons/f/fa/Wikiquote-logo.svg *License:* Public domain *Contributors:* ? *Original artist:* ?
- **File:Wiktionary-logo-en.svg** *Source:* https://upload.wikimedia.org/wikipedia/commons/f/f8/Wiktionary-logo-en.svg *License:* Public domain *Contributors:* Vector version of Image:Wiktionary-logo-en.png. *Original artist:* Vectorized by Fvasconcellos (talk · contribs), based on original logo tossed together by Brion Vibber

## 87.3.3   Content license

- Creative Commons Attribution-Share Alike 3.0

www.ingramcontent.com/pod-product-compliance
Lightning Source LLC
Chambersburg PA
CBHW080800180526
45168CB00006B/2274